CLASSIFICATION AND REGRESSION TREES

Lovingly dedicated to our children
Jessica, Rebecca, Kymm;
Melanie;
Elyse, Adam, Rachel, Stephen;
Daniel and Kevin

ERSEN ARSEVEN
(ACI)

CLASSIFICATION AND REGRESSION TREES

Leo Breiman
University of California, Berkeley

Jerome H. Friedman
Stanford University

Richard A. Olshen
Stanford University

Charles J. Stone
University of California, Berkeley

CHAPMAN & HALL/CRC

Boca Raton London New York Washington, D.C.

Library of Congress Cataloging-in-Publication Data

Main entry under title:
Classification and regression trees.
(The Wadsworth statistics/probability series)
Bibliography: p.
Includes Index.
ISBN 0-412-04841-8
1. Discriminant analysis. 2. Regression analysis.
3. Trees (Graph theory) I. Breiman, Leo. II. Title:
Regression trees. II. Series.
QA278.65.C54 1984
519.5'36—dc20 83-19708
CIP

Direct all inquiries to CRC Press LLC, 2000 N.W. Corporate Blvd., Boca Raton, Florida 33431.

First CRC Press reprint 1998

International Standard Book Number 0-412-04841-8
Library of Congress Card Number 83-19708
Printed in the United States of America 4 5 6 7 8 9 0
Printed on acid-free paper

CONTENTS

PREFACE

The tree methodology discussed in this book is a child of the computer age. Unlike many other statistical procedures which were moved from pencil and paper to calculators and then to computers, this use of trees was unthinkable before computers.

Binary trees give an interesting and often illuminating way of looking at data in classification or regression problems. They should not be used to the exclusion of other methods. We do not claim that they are always better. They do add a flexible nonparametric tool to the data analyst's arsenal.

Both practical and theoretical sides have been developed in our study of tree methods. The book reflects these two sides. The first eight chapters are largely expository and cover the use of trees as a data analysis method. These were written by Leo Breiman with the exception of Chapter 6 by Richard Olshen. Jerome Friedman developed the software and ran the examples.

Chapters 9 through 12 place trees in a more mathematical context and prove some of their fundamental properties. The first three of these chapters were written by Charles Stone and the last was jointly written by Stone and Olshen.

Trees, as well as many other powerful data analytic tools (factor analysis, nonmetric scaling, and so forth) were originated by social scientists motivated by the need to cope with actual problems and data. Use of trees in regression dates back to the AID (Automatic Interaction Detection) program developed at the Institute for Social Research, University of Michigan, by Morgan and Sonquist in the early 1960s. The ancestor classification program is THAID, developed at the institute in the early 1970s by Morgan and Messenger. The research and developments described in this book are aimed at strengthening and extending these original methods.

Our work on trees began in 1973 when Breiman and Friedman, independently of each other, "reinvented the wheel" and began to use tree methods in classification. Later, they joined forces and were joined in turn by Stone, who contributed significantly to the methodological development. Olshen was an early user of tree methods in medical applications and contributed to their theoretical development.

Our blossoming fascination with trees and the number of ideas passing back and forth and being incorporated by Friedman into CART (Classification and Regression Trees) soon gave birth to the idea of a book on the subject. In 1980 conception occurred. While the pregnancy has been rather prolonged, we hope that the baby appears acceptably healthy to the members of our statistical community.

The layout of the book is

Chapters 1 to 5	Tree structured methodology in classification
Chapters 6, 7	Examples of trees used in classification
Chapters 8	Use of trees in regression
Chapters 9 to 12	Theoretical framework for tree structured methods.

Readers are encouraged to contact Richard Olshen regarding the availability of CART software.

ACKNOWLEDGMENTS

Three other people were instrumental in our research: William Meisel, who early on saw the potential in tree structured methods and encouraged their development; Laurence Rafsky, who participated in some of the early exchanges of ideas; and Louis Gordon, who collaborated with Richard Olshen in theoretical work. Many helpful comments were supplied by Peter Bickel, William Eddy, John Hartigan, and Paul Tukey, who all reviewed an early version of the manuscript.

Part of the research, especially that of Breiman and Friedman, was supported by the Office of Naval Research (Contract No. N00014-82-K-0054), and we appreciate our warm relations with Edward Wegman and Douglas De Priest of that agency. Stone's work was supported partly by the Office of Naval Research on the same contract and partly by the National Science Foundation (Grant No. MCS 80-02732). Olshen's work was supported by the National Science Foundation (Grant No. MCS 79-06228) and the National Institutes of Health (Grant No. CA-26666).

We were fortunate in having the services of typists Ruth Suzuki, Rosaland Englander, Joan Pappas, and Elaine Morici, who displayed the old-fashioned virtues of patience, tolerance, and competence.

We are also grateful to our editor, John Kimmel of Wadsworth, for his abiding faith that eventually a worthy book would emerge, and to the production editor, Andrea Cava, for her diligence and skillful supervision.

1 BACKGROUND

At the University of California, San Diego Medical Center, when a heart attack patient is admitted, 19 variables are measured during the first 24 hours. These include blood pressure, age, and 17 other ordered and binary variables summarizing the medical symptoms considered as important indicators of the patient's condition.

The goal of a recent medical study (see Chapter 6) was the development of a method to identify high risk patients (those who will not survive at least 30 days) on the basis of the initial 24-hour data.

Figure 1.1 is a picture of the tree structured classification rule that was produced in the study. The letter *F* means not high risk; *G* means high risk.

This rule classifies incoming patients as *F* or *G* depending on the yes-no answers to at most three questions. Its simplicity raises the suspicion that standard statistical classification methods may give classification rules that are more accurate. When these were tried, the rules produced were considerably more intricate, but less accurate.

The methodology used to construct tree structured rules is the major story of this monograph.

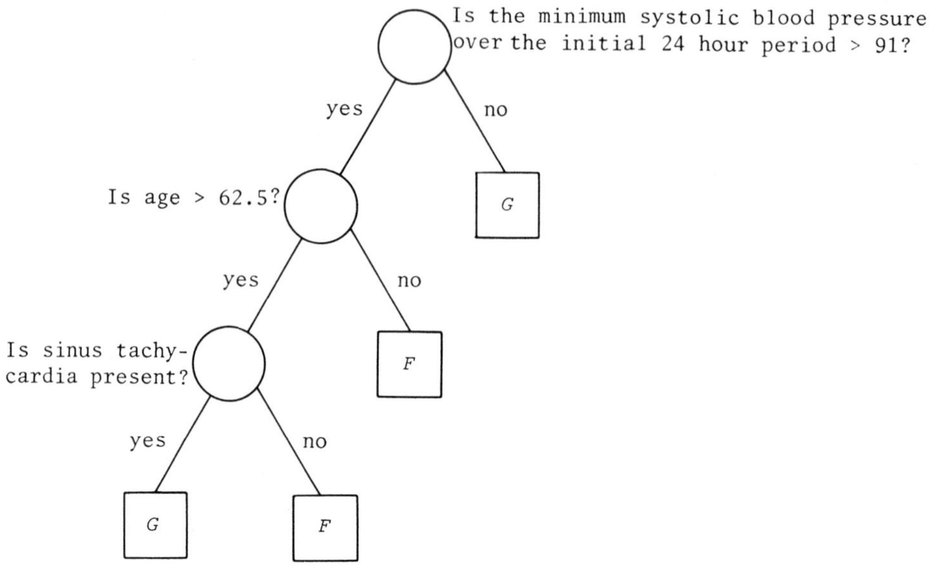

FIGURE 1.1

1.1 CLASSIFIERS AS PARTITIONS

The general classification problem is similar to the medical diagnosis problem sketched above. Measurements are made on some case or object. Based on these measurements, we then want to predict which class the case is in.

For instance, days in the Los Angeles basin are classified according to the ozone levels:

Class 1: nonalert (low ozone)
Class 2: first-stage alert (moderate ozone)
Class 3: second-stage alert (high ozone)

During the current day, measurements are made on many meteorological variables, such as temperature, humidity, upper atmospheric conditions, and on the current levels of a number of airborne pollutants. The purpose of a project funded by the California Air Resources Board (Zeldin and Cassmassi, 1978) was to explore methods for using the current-day measurements to predict the classification of the following day.

An EPA project had this goal: The exact analysis of a complex chemical compound into its atomic constituents is slow and costly. Measuring its mass spectra can be done quickly and at relatively low cost. Can the measured mass spectra be used to accurately predict whether, for example, the compound is in

class 1 (contains one or more chlorine atoms), or

class 2 (contains no chlorine)?

(See Chapter 7 for more discussion.)

In these problems, the goal is the same. Given a set of measurements on a case or object, find a systematic way of predicting what class it is in. In any problem, a *classifier* or a *classification rule* is a systematic way of predicting what class a case is in.

To give a more precise formulation, arrange the set of measurements on a case in a preassigned order; i.e., take the measurements to be x_1, x_2, ..., where, say, x_1 is age, x_2 is blood pressure, etc. Define the measurements $(x_1, x_2, \ldots)$ made on a case as the *measurement vector* $\mathbf{x}$ corresponding to the case. Take the *measurement space* X to be defined as containing all possible measurement vectors.

For example, in the heart attack study, X is a 19-dimensional space such that the first coordinate x_1 (age) ranges, say, over all integer values from 0 to 200; the second coordinate, blood pressure, might be defined as continuously ranging from 50 to 150. There can be a number of different definitions of X. What is important is that any definition of X have the property that the measurement vector $\mathbf{x}$ corresponding to any case we may wish to classify be a point in the space X.

Suppose that the cases or objects fall into J classes. Number the classes 1, 2, ..., J and let C be the set of classes; that is, $C = \{1, \ldots, J\}$.

A systematic way of predicting class membership is a rule that assigns a class membership in C to every measurement vector $\mathbf{x}$ in X. That is, given any $\mathbf{x} \in X$, the rule assigns one of the classes $\{1, \ldots, J\}$ to $\mathbf{x}$.

DEFINITION 1.1. *A classifier or classification rule is a function $d(\mathbf{x})$ defined on X so that for every $\mathbf{x}$, $d(\mathbf{x})$ is equal to one of the numbers* $1, 2, \ldots, J$.

Another way of looking at a classifier is to define A_j as the subset of X on which $d(\mathbf{x}) = j$; that is,

$$A_j = \{\mathbf{x};\ d(\mathbf{x}) = j\}.$$

The sets $A_1, \ldots, A_j$ are disjoint and $X = \bigcup_j A_j$. Thus, the A_j form a partition of X. This gives the equivalent

DEFINITION 1.2. *A classifier is a partition of X into J disjoint subsets $A_1, \ldots, A_j$, $X = \bigcup_j A_j$ such that for every $\mathbf{x} \in A_j$ the predicted class is j.*

1.2 USE OF DATA IN CONSTRUCTING CLASSIFIERS

Classifiers are not constructed whimsically. They are based on past experience. Doctors know, for example, that elderly heart attack patients with low blood pressure are generally high risk. Los Angelenos know that one hot, high pollution day is likely to be followed by another.

In systematic classifier construction, past experience is summarized by a *learning sample*. This consists of the measurement data on N cases observed in the past together with their actual classification.

In the medical diagnostic project the learning sample consisted of the records of 215 heart attack patients admitted to the hospital, all of whom survived the initial 24-hour period. The records contained the outcome of the initial 19 measurements together with an identification of those patients that did not survive at least 30 days.

The learning sample for the ozone classification project contained 6 years (1972-1977) of daily measurements on over 400 meteorological variables and hourly air pollution measurements at 30 locations in the Los Angeles basin.

The data for the chlorine project consisted of the mass spectra of about 30,000 compounds having known molecular structure. For each compound the mass spectra can be expressed as a measurement vector of dimension equal to the molecular weight. The set of 30,000 measurement vectors was of variable dimensionality, ranging from about 50 to over 1000.

We assume throughout the remainder of this monograph that the construction of a classifier is based on a learning sample, where

DEFINITION 1.3. *A learning sample consists of data* $(\mathbf{x}_1, j_1), \ldots, (\mathbf{x}_N, j_N)$ *on N cases where* $\mathbf{x}_n \in X$ *and* $j_n \in \{1, \ldots, J\}$, $n = 1, \ldots, N$. *The learning sample is denoted by* $\mathcal{L}$; i.e.,

$$\mathcal{L} = \{(\mathbf{x}_1, j_1), \ldots, (\mathbf{x}_N, j_N)\}.$$

We distinguish two general types of variables that can appear in the measurement vector.

DEFINITION 1.4. *A variable is called ordered or numerical if its measured values are real numbers. A variable is categorical if it takes values in a finite set not having any natural ordering.*

A categorical variable, for instance, could take values in the set {red, blue, green}. In the medical data, blood pressure and age are ordered variables.

Finally, define

DEFINITION 1.5. *If all measurement vectors* $\mathbf{x}_n$ *are of fixed dimensionality, we say that the data have standard structure.*

In the medical and ozone projects, a fixed set of variables is measured on each case (or day); the data have standard structure. The mass spectra data have nonstandard structure.

1.3 THE PURPOSES OF CLASSIFICATION ANALYSIS

Depending on the problem, the basic purpose of a classification study can be either to produce an accurate classifier or to uncover the predictive structure of the problem. If we are aiming at the latter, then we are trying to get an understanding of what variables or interactions of variables drive the phenomenon—that is, to give simple characterizations of the conditions (in terms of the measurement variables $\mathbf{x} \in \mathcal{X}$) that determine when an object is in one class rather than another. These two are not exclusive. Most often, in our experience, the goals will be both accurate prediction and understanding. Sometimes one or the other will have greater emphasis.

In the mass spectra project, the emphasis was on prediction. The purpose was to develop an efficient and accurate on-line algorithm that would accept as input the mass spectrum of an unknown compound and classify the compound as either chlorine containing or not.

The ozone project shared goals. The work toward understanding which meteorological variables and interactions between them were associated with alert-level days was an integral part of the development of a classifier.

The tree structured classification rule of Figure 1.1 gives some interesting insights into the medical diagnostic problem. All cases with blood pressure less than or equal to 91 are predicted high risks. For cases with blood pressure greater than 91, the classification depends only on age and whether sinus tachycardia is present. For the purpose of distinguishing between high and low risk cases, once age is recorded, only two variables need to be measured.

An important criterion for a good classification procedure is that it not only produce accurate classifiers (within the limits of the data) but that it also *provide insight and understanding into the predictive structure of the data.*

Many of the presently available statistical techniques were designed for small data sets having standard structure with all variables of the same type; the underlying assumption was that the phenomenon is homogeneous. That is, that the same relationship between variables held over all of the measurement space. This led to models where only a few parameters were necessary to trace the effects of the various factors involved.

With large data sets involving many variables, more structure can be discerned and a variety of different approaches tried. But largeness by itself does not necessarily imply a richness of structure.

What makes a data set interesting is not only its size but also its complexity, where complexity can include such considerations as:

High dimensionality
A mixture of data types
Nonstandard data structure

and, perhaps most challenging, nonhomogeneity; that is, different relationships hold between variables in different parts of the measurement space.

Along with complex data sets comes "the curse of dimensionality" (a phrase due to Bellman, 1961). The difficulty is that the higher the dimensionality, the sparser and more spread apart are the data points. Ten points on the unit interval are not distant neighbors. But 10 points on a 10-dimensional unit rectangle are like oases in the desert.

For instance, with 100 points, constructing a 10-cell histogram on the unit interval is a reasonable procedure. In M dimensions, a histogram that uses 10 intervals in each dimension produces 10^M cells. For even moderate M, a very large data set would be needed to get a sensible histogram.

Another way of looking at the "curse of dimensionality" is the number of parameters needed to specify distributions in M dimensions:

Normal: $O(M^2)$
Binary: $O(2^M)$

Unless one makes the very strong assumption that the variables are independent, the number of parameters usually needed to specify an M-dimensional distribution goes up much faster than $O(M)$. To put this another way, *the complexity of a data set increases rapidly with increasing dimensionality.*

With accelerating computer usage, complex, high dimensional data bases, with variable dimensionality or mixed data types, nonhomogeneities, etc., are no longer odd rarities.

In response to the increasing dimensionality of data sets, the most widely used multivariate procedures all contain some sort of dimensionality reduction process. Stepwise variable selection and variable subset selection in regression and discriminant analysis are examples.

Although the drawbacks in some of the present multivariate reduction tools are well known, they are a response to a clear need. To analyze and understand complex data sets, methods are needed which in some sense select salient features of the data, discard the background noise, and feed back to the analyst understandable summaries of the information.

1.4 ESTIMATING ACCURACY

Given a classifier, that is, given a function $d(\mathbf{x})$ defined on X taking values in C, we denote by $R^*(d)$ its "*true misclassifica-*

tion rate." The question raised in this section is: What is truth and how can it be estimated?

One way to see how accurate a classifier is (that is, to estimate $R^*(d)$) is to test the classifier on subsequent cases whose correct classification has been observed. For instance, in the ozone project, the classifier was developed using the data from the years 1972-1975. Then its accuracy was estimated by using the 1976-1977 data. That is, $R^*(d)$ was estimated as the proportion of days in 1976-1977 that were misclassified when $d(\mathbf{x})$ was used on the previous day data.

In one part of the mass spectra project, the 30,000 spectra were randomly divided into one set of 20,000 and another of 10,000. The 20,000 were used to construct the classifier. The other 10,000 were then run through the classifier and the proportion misclassified used as an estimate of $R^*(d)$.

The value of $R^*(d)$ can be conceptualized in this way: Using $\mathcal{L}$, construct d. Now, draw another very large (virtually infinite) set of cases from the same population as $\mathcal{L}$ was drawn from. Observe the correct classification for each of these cases, and also find the predicted classification using $d(\mathbf{x})$. The proportion misclassified by d is the value of $R^*(d)$.

To make the preceding concept precise, a probability model is needed. Define the space $X \times C$ as a set of all couples $(\mathbf{x}, j)$ where $\mathbf{x} \in X$ and j is a class label, $j \in C$. Let $P(A, j)$ be a probability on $X \times C$, $A \subset X$, $j \in C$ (niceties such as Borel measurability will be ignored). The interpretation of $P(A, j)$ is that a case drawn at random from the relevant population has probability $P(A, j)$ that its measurement vector $\mathbf{x}$ is in A and its class is j. Assume that the learning sample $\mathcal{L}$ consists of N cases $(\mathbf{x}_1, j_1)$, $\ldots$, $(\mathbf{x}_N, j_N)$ independently drawn at random from the distribution $P(A, j)$. Construct $d(\mathbf{x})$ using $\mathcal{L}$. Then define $R^*(d)$ as the probability that d will misclassify a new sample drawn from the same distribution as $\mathcal{L}$.

DEFINITION 1.6 *Take* $(\mathbf{X}, Y)$, $\mathbf{X} \in \mathcal{X}$, $Y \in \mathcal{C}$, *to be a new sample from the probability distribution* $P(A, j)$; *i.e.,*

(i) $P(\mathbf{X} \in A, Y = j) = P(A, j)$,

(ii) $(\mathbf{X}, Y)$ *is independent of* $\mathcal{L}$.

Then define

$$R^*(d) = P\big(d(\mathbf{X}) \neq Y\big).$$

In evaluating the probability $P\big(d(\mathbf{X}) \neq Y\big)$, the set $\mathcal{L}$ is considered fixed. A more precise notation is $P\big(d(\mathbf{X}) \neq Y|\mathcal{L}\big)$, the probability of misclassifying the new sample given the learning sample $\mathcal{L}$.

This model must be applied cautiously. Successive pairs of days in the ozone data are certainly not independent. Its usefulness is that it gives a beginning conceptual framework for the definition of "truth."

How can $R^*(d)$ be estimated? There is no difficulty in the examples of simulated data given in this monograph. The data in $\mathcal{L}$ are sampled independently from a desired distribution using a pseudorandom number generator. After $d(\mathbf{x})$ is constructed, 5000 additional cases are drawn from the same distribution independently of $\mathcal{L}$ and classified by d. The proportion misclassified among those 5000 is the estimate of $R^*(d)$.

In actual problems, only the data in $\mathcal{L}$ are available with little prospect of getting an additional large sample of classified cases. Then $\mathcal{L}$ must be used both to construct $d(\mathbf{x})$ and to estimate $R^*(d)$. We refer to such estimates of $R^*(d)$ as *internal estimates*. A summary and large bibliography concerning such estimates is in Toussaint (1974).

Three types of internal estimates will be of interest to us. The first, least accurate, and most commonly used is the *resubstitution* estimate.

After the classifier d is constructed, the cases in $\mathcal{L}$ are run through the classifier. The proportion of cases misclassified is the resubstitution estimate. To put this in equation form:

DEFINITION 1.7. *Define the indicator function* $\chi(\cdot)$ *to be 1 if the statement inside the parentheses is true, otherwise zero.*

The resubstitution estimate, denoted $R(d)$, is

$$R(d) = \frac{1}{N} \sum_{n=1}^{N} \chi(d(\mathbf{x}_n) \neq j_n). \tag{1.8}$$

The problem with the resubstitution estimate is that it is computed using the same data used to construct d, instead of an independent sample. All classification procedures, either directly or indirectly, attempt to minimize $R(d)$. Using the subsequent value of $R(d)$ as an estimate of $R^*(d)$ can give an overly optimistic picture of the accuracy of d.

As an exaggerated example, take $d(\mathbf{x})$ to be defined by a partition $A_1, \ldots, A_j$ such that A_j contains all measurement vectors $\mathbf{x}_n$ in $\mathcal{L}$ with $j_n = j$ and the vectors $\mathbf{x} \in X$ not equal to some $\mathbf{x}_n$ are assigned in an arbitrary random fashion to one or the other of the A_j. Then $R(d) = 0$, but it is hard to believe that $R^*(d)$ is anywhere near zero.

The second method is *test sample* estimation. Here the cases in $\mathcal{L}$ are divided into two sets $\mathcal{L}_1$ and $\mathcal{L}_2$. Only the cases in $\mathcal{L}_1$ are used to construct d. Then the cases in $\mathcal{L}_2$ are used to estimate $R^*(d)$. If N_2 is the number of cases in $\mathcal{L}_2$, then the test sample estimate, $R^{ts}(d)$, is given by

$$R^{ts}(d) = \frac{1}{N_2} \sum_{(\mathbf{x}_n, j_n) \in \mathcal{L}_2} \chi(d(\mathbf{x}_n) \neq j_n). \tag{1.9}$$

In this method, care needs to be taken so that the cases in $\mathcal{L}_2$ can be considered as independent of the cases in $\mathcal{L}_1$ and drawn from the same distribution. The most common procedure used to help ensure these properties is to draw $\mathcal{L}_2$ at random from $\mathcal{L}$. Frequently, $\mathcal{L}_2$ is taken as 1/3 of the cases in $\mathcal{L}$, but we do not know of any theoretical justification for this 2/3, 1/3 split.

The test sample approach has the drawback that it reduces effective sample size. In a 2/3, 1/3 split, only 2/3 of the data are used to construct d, and only 1/3 to estimate $R^*(d)$. If the sample size is large, as in the mass spectra problem, this is a minor difficulty, and test sample estimation is honest and efficient.

For smaller sample sizes, another method, called *V-fold cross-validation*, is preferred (see the review by M. Stone, 1977). The cases in $\mathcal{L}$ are randomly divided into V subsets of as nearly equal size as possible. Denote these subsets by $\mathcal{L}_1, \ldots, \mathcal{L}_V$. Assume that the procedure for constructing a classifier can be applied to any learning sample. For every v, $v = 1, \ldots, V$, apply the procedure using as learning sample $\mathcal{L} - \mathcal{L}_v$, i.e., the cases in $\mathcal{L}$ not in $\mathcal{L}_v$, and let $d^{(v)}(\mathbf{x})$ be the resulting classifier. Since none of the cases in $\mathcal{L}_v$ has been used in the construction of $d^{(v)}$, a test sample estimate for $R^*(d^{(v)})$ is

$$R^{ts}(d^{(v)}) = \frac{1}{N_v} \sum_{(\mathbf{x}_n, j_n) \in \mathcal{L}_v} \chi(d^{(v)}(\mathbf{x}_n) \neq j_n), \tag{1.10}$$

where $N_v \simeq N/V$ is the number of cases in $\mathcal{L}_v$. Now using the same procedure again, construct the classifier d using all of $\mathcal{L}$.

For V large, each of the V classifiers is constructed using a learning sample of size $N(1 - 1/V)$ nearly as large as $\mathcal{L}$. The basic assumption of cross-validation is that the procedure is "stable." That is, that the classifiers $d^{(v)}$, $v = 1, \ldots, V$, each constructed using almost all of $\mathcal{L}$, have *misclassification rates* $R^*(d^{(v)})$ *nearly equal to* $R^*(d)$. Guided by this heuristic, define the V-fold cross-validation estimate $R^{CV}(d)$ as

$$R^{CV}(d) = \frac{1}{V} \sum_{v=1}^{V} R^{ts}(d^{(v)}). \tag{1.11}$$

N-fold cross-validation is the "leave-one-out" estimate. For each n, $n = 1, \ldots, N$, the nth case is set aside and the classifier constructed using the other $N - 1$ cases. Then the nth case is used as a single-case test sample and $R^*(d)$ estimated by (1.11).

Cross-validation is parsimonious with data. Every case in $\mathcal{L}$ is used to construct d, and every case is used exactly once in a test sample. In tree structured classifiers tenfold cross-validation has been used, and the resulting estimators have been satisfactorily close to $R^*(d)$ on simulated data.

The *bootstrap* method can also be used to estimate $R^*(d)$, but may not work well when applied to tree structured classifiers (see Section 11.7).

1.5 THE BAYES RULE AND CURRENT CLASSIFICATION PROCEDURES

The major guide that has been used in the construction of classifiers is the concept of the Bayes rule. If the data are drawn from a probability distribution $P(A, j)$, then the form of the most accurate rule can be given in terms of $P(A, j)$. This rule is called the *Bayes rule* and is denoted by $d_B(\mathbf{x})$.

To be more precise, suppose that $(\mathbf{X}, Y)$, $\mathbf{X} \in \mathcal{X}$, $Y \in \mathcal{C}$, is a random sample from the probability distribution $P(A, j)$ on $\mathcal{X} \times \mathcal{C}$; i.e., $P(\mathbf{X} \in A, Y = j) = P(A, j)$.

DEFINITION 1.12. *$d_B(\mathbf{x})$ is a Bayes rule if for any other classifier $d(\mathbf{x})$,*

$$P(d_B(\mathbf{X}) \neq Y) \leq P(d(\mathbf{X}) \neq Y).$$

Then the Bayes misclassification rate is

$$R_B = P(d_B(\mathbf{X}) \neq Y).$$

To illustrate how $d_B(\mathbf{x})$ can be derived from $P(A, j)$, we give its form in an important special case.

DEFINITION 1.13. *Define the prior class probabilities $\pi(j)$, $j = 1, \ldots, J$, as*

$$\pi(j) = P(Y = j)$$

and the probability distribution of the jth class measurement vectors by

$$P(A|j) = P(A, j)/\pi(j).$$

ASSUMPTION 1.14. *$\mathcal{X}$ is M-dimensional euclidean space and for every j, $j = 1, \ldots, J$, $P(A|j)$ has the probability density $f_j(\mathbf{x})$; i.e., for sets $A \subset \mathcal{X}$,*

$$P(A|j) = \int_A f_j(\mathbf{x})d\mathbf{x}.$$

Then,

THEOREM 1.15. *Under Assumption 1.14 the Bayes rule is defined by*

$$d_B(\mathbf{x}) = j \text{ on } A_j = \{\mathbf{x};\ f_j(\mathbf{x})\pi(j) = \max_i f_i(\mathbf{x})\pi(i)\} \tag{1.16}$$

and the Bayes misclassification rate is

$$R_B = 1 - \int \max_j[f_j(\mathbf{x})\pi(j)]d\mathbf{x}. \tag{1.17}$$

Although d_B is called the Bayes rule, it is also recognizable as a maximum likelihood rule: Classify $\mathbf{x}$ as that j for which $f_j(\mathbf{x})\pi(j)$ is maximum. As a minor point, note that (1.16) does not uniquely define $d_B(\mathbf{x})$ on points $\mathbf{x}$ such that $\max_j f_j(\mathbf{x})\pi(j)$ is achieved by two or more different j's. In this situation, define $d_B(\mathbf{x})$ arbitrarily to be any one of the maximizing j's.

The proof of Theorem 1.15 is simple. For any classifier d, under Assumption 1.14,

$$\begin{aligned} P(d(\mathbf{X}) = Y) &= \textstyle\sum_{j=1}^J P(d(\mathbf{X}) = j|Y = j)\pi(j) \\ &= \textstyle\sum_{j=1}^J \int_{\{d(\mathbf{x})=j\}} f_j(\mathbf{x})\pi(j)d\mathbf{x} \\ &= \int[\textstyle\sum_{j=1}^J \chi(d(\mathbf{x}) = j)f_j(\mathbf{x})\pi(j)]d\mathbf{x}. \end{aligned}$$

For a fixed value of $\mathbf{x}$

$$\textstyle\sum_{j=1}^J \chi(d(\mathbf{x}) = j)f_j(\mathbf{x})\pi(j) \le \max_j [f_j(\mathbf{x})\pi(j)],$$

and equality is achieved if $d(\mathbf{x})$ equals that j for which $f_j(\mathbf{x})\pi(j)$ is a maximum. Therefore, the rule d_B given in (1.16) has the property that for any other classifier d,

$$P(d(\mathbf{X}) = Y) \leq P(d_B(\mathbf{X}) = Y) = \int \max_j [f_j(\mathbf{x})\pi(j)]d\mathbf{x}.$$

This shows that d_B is a Bayes rule and establishes (1.17) as the correct equation for the Bayes misclassification rate.

In the simulated examples we use later on, the data are generated from a known probability distribution. For these examples, d_B was derived and then the values of R_B computed. Since R_B is the minimum misclassification rate attainable, knowing R_B and comparing it with the accuracy of the tree structured classifiers give some idea of how effective they are.

In practice, neither the $\pi(j)$ nor the $f_j(\mathbf{x})$ are known. The $\pi(j)$ can either be estimated as the proportion of class j cases in $\mathcal{L}$ or their values supplied through other knowledge about the problem. The thorny issue is getting at the $f_j(\mathbf{x})$. The three most commonly used classification procedures

Discriminant analysis
Kernel density estimation
Kth nearest neighbor

attempt, in different ways, to approximate the Bayes rule by using the learning sample $\mathcal{L}$ to get estimates of $f_j(\mathbf{x})$.

Discriminant analysis assumes that all $f_j(\mathbf{x})$ are multivariate normal densities with common covariance matrix Γ and different means vectors $\{\mu_j\}$. Estimating Γ and the μ_j in the usual way gives estimates $\hat{f}_j(\mathbf{x})$ of the $f_j(\mathbf{x})$. These are substituted into the Bayes optimal rule to give the classification partition

$$A_j = \{\mathbf{x};\ \hat{f}_j(\mathbf{x})\pi(j) = \max_i \hat{f}_i(\mathbf{x})\pi(i)\}.$$

A stepwise version of linear discrimination is the most widely used method. It is usually applied without regard to lack of normality. It is not set up to handle categorical variables, and these are dealt with by the artifice of coding them into dummy variables. Our reaction to seeing the results of many runs on different data sets of this program is one of surprise that it does as well as it does. It provides insight into the structure of the data through the use of the discrimination coordinates (see Gnanadesikan, 1977, for a good discussion). However, the form of classifier for the J class problem (which depends on the maximum of J linear combinations) is difficult to interpret.

Density estimation and kth nearest neighbor methods are more recent arrivals generated, in part, by the observation that not all data sets contained classes that were normally distributed with common covariance matrices.

The density method uses a nonparametric estimate of each of the densities $f_j(\mathbf{x})$, most commonly done using a kernel type of estimate (see Hand, 1982). Briefly, a metric $\|\mathbf{x}\|$ on X is defined and a kernel function $K(\|\mathbf{x}\|) \geq 0$ selected which has a peak at $\|\mathbf{x}\| = 0$ and goes to zero as $\|\mathbf{x}\|$ becomes large, satisfying

$$\int K(\|\mathbf{x}\|)d\mathbf{x} = 1.$$

Then $f_j(\mathbf{x})$ is estimated by

$$\hat{f}_j(\mathbf{x}) = \frac{1}{N_j} \sum K(\|\mathbf{x} - \mathbf{x}_n\|),$$

where N_j is the number of cases in the jth class and the sum is over the N_j measurement vectors $\mathbf{x}_n$ corresponding to cases in the jth class.

The kth nearest neighbor rule, due to Fix and Hodges (1951), has this simple form: Define a metric $\|\mathbf{x}\|$ on X and fix an integer $k > 0$. At any point $\mathbf{x}$, find the k nearest neighbors to $\mathbf{x}$ in $\mathcal{L}$. Classify $\mathbf{x}$ as class j if more of the k nearest neighbors are in class j than in any other class. (This is equivalent to using density estimates for f_j based on the number of class j points among the k nearest neighbors.)

The kernal density estimation and kth nearest neighbor methods make minimal assumptions about the form of the underlying distribution. But there are serious limitations common to both methods.

1. They are sensitive to the choice of the metric $\|\mathbf{x}\|$, and there is usually no intrinsically preferred definition.
2. There is no natural or simple way to handle categorical variables and missing data.
3. They are computationally expensive as classifiers; $\mathcal{L}$ must be stored, the interpoint distances and $d(\mathbf{x})$ recomputed for each new point $\mathbf{x}$.
4. Most serious, they give very little usable information regarding the structure of the data.

Surveys of the literature on these and other methods of classification are given in Kanal (1974) and Hand (1981).

The use of classification trees did not come about as an abstract exercise. Problems arose that could not be handled in an easy or natural way by any of the methods discussed above. The next chapter begins with a description of one of these problems.

2

INTRODUCTION TO TREE CLASSIFICATION

2.1 THE SHIP CLASSIFICATION PROBLEM

The ship classification project (Hooper and Lucero, 1976) involved recognition of six ship classes through their radar range profiles. The data were gathered by an airplane flying in large circles around ships of six distinct structural types. The electronics in the airborne radar gave the intensity of the radar return as a function of distance (or range) at 2-foot intervals from the airplane to the objects reflecting the radar pulses.

Over each small time period, then, the airplane took a profile of the ship, where the profile consisted of the intensity of the radar returns from various parts of the ship versus the distance of these parts from the airplane.

The intensity of the radar returns from the ocean was small. When the profile was smoothed, it was not difficult to detect the ranges where the ship returns began and ended. The data were normalized so that one end of the smoothed profile corresponded to $x = 0$ and the other end to $x = 1$. The resulting radar range profiles were then continuous curves on the interval $0 \leq x \leq 1$, going to zero at the endpoints and otherwise positive (see Figure 2.1).

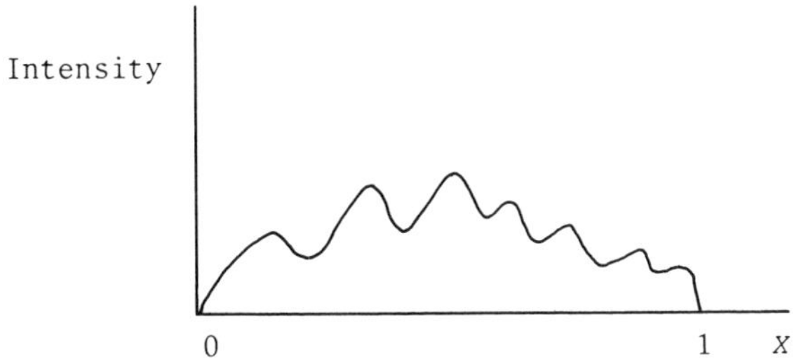

FIGURE 2.1 Typical range profile.

The peaks on each profile correspond to major structural elements on the ship that act as reflectors.

Unfortunately, the shape of the profile changes with the angle θ between the centerline of the ship and the airplane (see Figure 2.2).

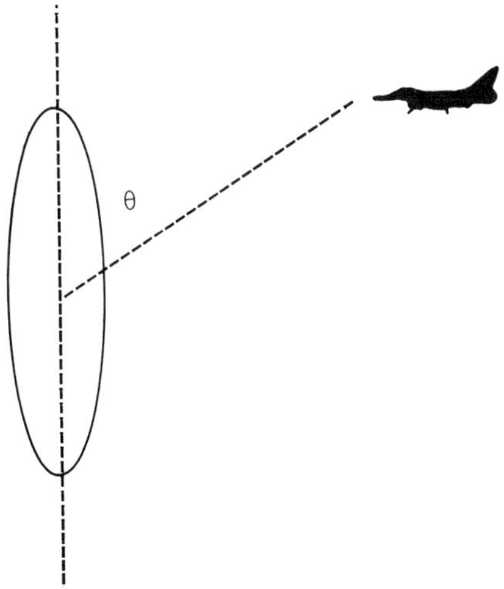

FIGURE 2.2

At the broadside angles (θ = 90°, θ = 270°), the points on the ship closest and farthest from the airplane may differ only by a few dozen feet. The profile contains very little information and may have only one peak. At bow and stern ($\theta = 0°$, $\theta = 180°$) the profile has the most detail.

The data consisted of a number of profiles for each of the different ship classes taken at angles about 20° apart around the compass. The goal was to construct a classifier which could take as input a profile at an unknown angle from one of the six classes and produce reliable predictions of class membership.

After some initial inspection, it was noticed that while the profiles changed with angle, the positions of the peaks stayed relatively constant. That is, in the profiles belonging to a given ship class, as long as a peak did not disappear, its x-coordinate stayed about the same (if bow and stern were appropriately labeled).

One of the initial difficulties in the project was reduction of dimensionality. Much of the information in any given profile was redundant. Profile heights corresponding to neighboring x values were highly correlated. In view of the initial look at the profiles, the decision was made to extract from each profile the vector of locations of the local maxima. Thus, to each profile was associated a vector of the form $(x_1, x_2, \ldots)$, where x_1 was the position of the first local maximum, etc.

This brought a new difficulty. The data had variable dimensionality, ranging from a low of 1 to a high of 15. None of the available classification methods seemed appropriate to this data structure. The most satisfactory solution was the tree structured approach outlined in the following sections.

2.2 TREE STRUCTURED CLASSIFIERS

Tree structured classifiers, or, more correctly, *binary* tree structured classifiers, are constructed by repeated splits of subsets of X into two descendant subsets, beginning with X itself. This process, for a hypothetical six-class tree, is pictured in Figure 2.3. In Figure 2.3, X_2 and X_3 are disjoint, with

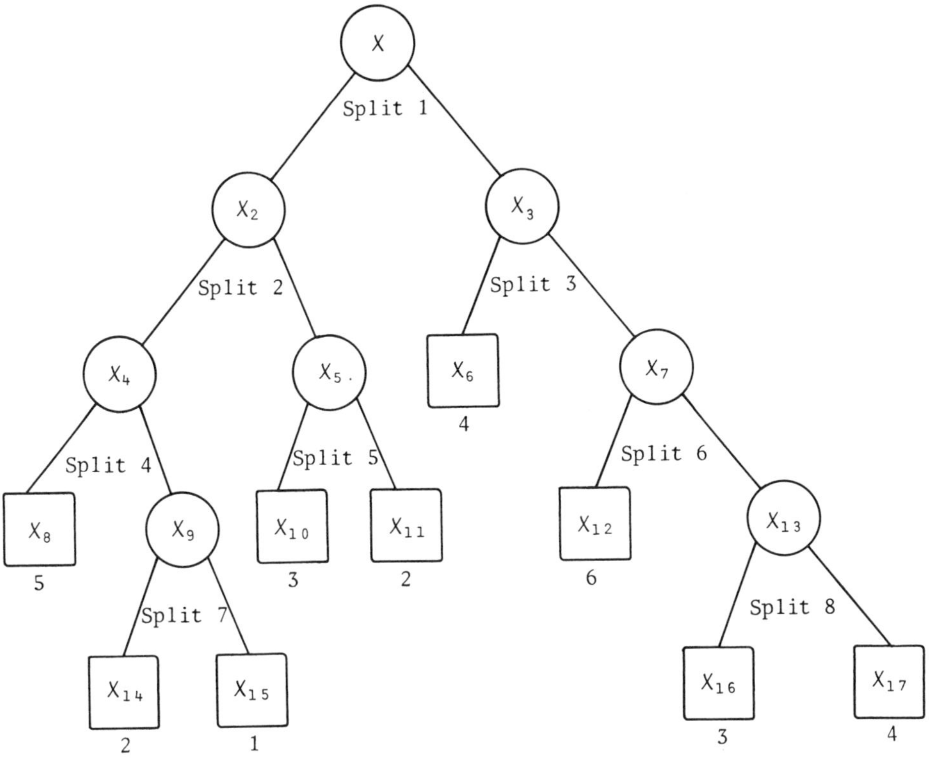

FIGURE 2.3

$X = X_2 \cup X_3$. Similarly, X_4 and X_5 are disjoint with $X_4 \cup X_5 = X_2$, and $X_6 \cup X_7 = X_3$. Those subsets which are not split, in this case X_6, X_8, X_{10}, X_{11}, X_{12}, X_{14}, X_{15}, X_{16}, and X_{17}, are called *terminal* subsets. This is indicated, here and in subsequent figures, by a rectangular box; nonterminal subsets are indicated by circles.

The terminal subsets form a partition of X. Each terminal subset is designated by a class label. There may be two or more terminal subsets with the same class label. The partition corresponding to the classifier is gotten by putting together all terminal subsets corresponding to the same class. Thus,

$$A_1 = X_{15} \qquad A_2 = X_{11} \cup X_{14}$$

$$A_3 = X_{10} \cup X_{16} \qquad A_4 = X_6 \cup X_{17}$$

$$A_5 = X_8 \qquad A_6 = X_{12}.$$

The splits are formed by conditions on the coordinates of $\mathbf{x} = (x_1, x_2, \ldots)$. For example, split 1 of X into X_2 and X_3 could be of the form

$$X_2 = \{\mathbf{x};\ x_4 \leq 7\}, \quad X_3 = \{\mathbf{x};\ x_4 > 7\}. \tag{2.1}$$

Split 3 of X_3 into X_6 and X_7 could be of the form

$$X_6 = \{\mathbf{x} \in X_3;\ x_3 + x_5 \leq -2\}$$
$$X_7 = \{\mathbf{x} \in X_3;\ x_3 + x_5 > -2\}.$$

The tree classifier predicts a class for the measurement vector $\mathbf{x}$ in this way: From the definition of the first split, it is determined whether $\mathbf{x}$ goes into X_2 or X_3. For example, if (2.1) is used, $\mathbf{x}$ goes into X_2 if $x_4 \leq 7$, and into X_3 if $x_4 > 7$. If $\mathbf{x}$ goes into X_3, then from the definition of split 3, it is determined whether $\mathbf{x}$ goes into X_6 or X_7.

When $\mathbf{x}$ finally moves into a terminal subset, its predicted class is given by the class label attached to that terminal subset.

At this point we change terminology to that of tree theory. From now on,

A node t = a subset of X

and

The root node $t_1 = X$.

Terminal subsets become *terminal nodes*, and nonterminal subsets are *nonterminal nodes*. Figure 2.3 becomes relabeled as shown in Figure 2.4.

The entire construction of a tree, then, revolves around three elements:

1. *The selection of the splits*
2. *The decisions when to declare a node terminal or to continue splitting it*
3. *The assignment of each terminal node to a class*

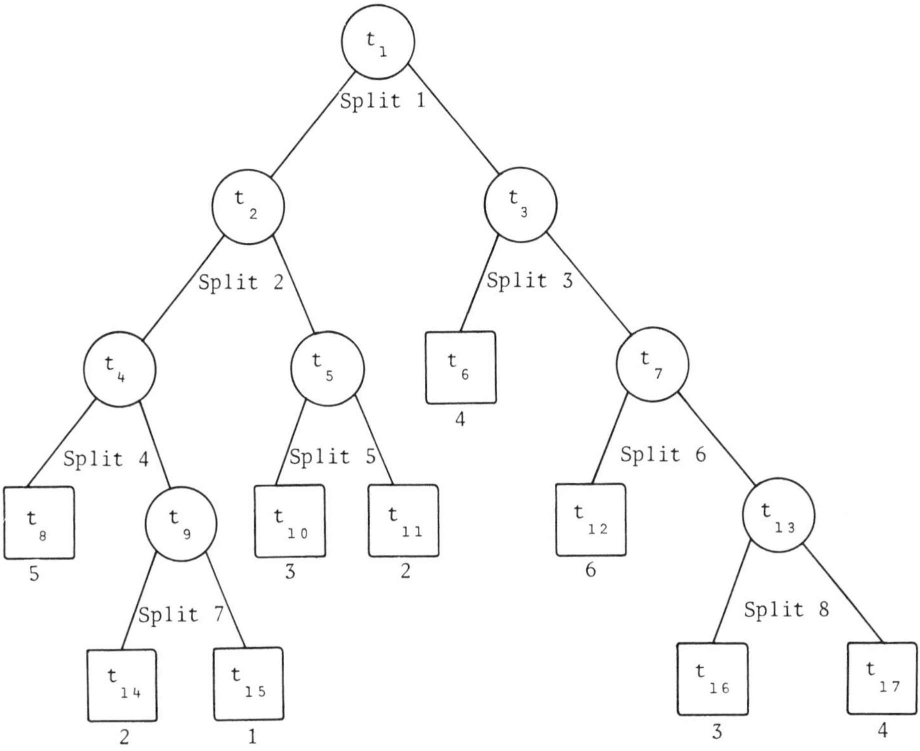

FIGURE 2.4

The crux of the problem is how to use the data $\mathcal{L}$ to determine the splits, the terminal nodes, and their assignments. It turns out that the class assignment problem is simple. The whole story is in finding good splits and in knowing when to stop splitting.

2.3 CONSTRUCTION OF THE TREE CLASSIFIER

The first problem in tree construction is how to use $\mathcal{L}$ to determine the binary splits of X into smaller and smaller pieces. The fundamental idea is to select each split of a subset so that the data in each of the descendant subsets are "purer" than the data in the parent subset.

For instance, in the six-class ship problem, denote by p_1, ..., p_6 the proportions of class 1, ..., 6 profiles in any node. For the root node t_1, $(p_1, \ldots, p_6) = \left(\frac{1}{6}, \frac{1}{6}, \ldots, \frac{1}{6}\right)$. A good split of t_1 would be one that separates the profiles in $\mathcal{L}$ so that all profiles in classes 1, 2, 3 go to the left node and the profiles in 4, 5, 6 go to the right node (Figure 2.5).

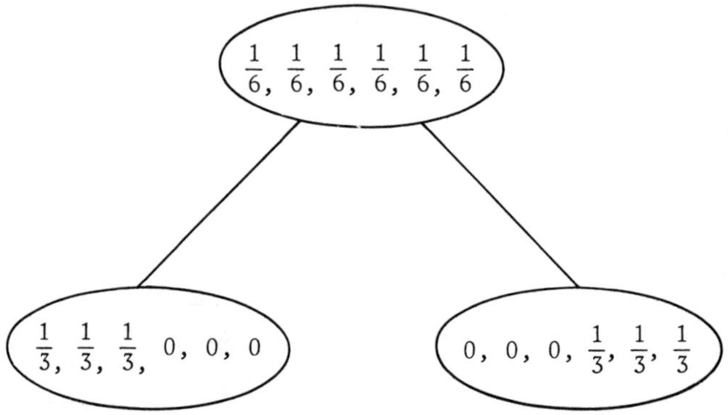

FIGURE 2.5

Once a good split of t_1 is found, then a search is made for good splits of each of the two descendant nodes t_2, t_3.

This idea of finding splits of nodes so as to give "purer" descendant nodes was implemented in this way:

1. Define the node proportions $p(j|t)$, $j = 1, \ldots, 6$, to be the proportion of the cases $\mathbf{x}_n \in t$ belonging to class j, so that

   $$p(1|t) + \cdots + p(6|t) = 1.$$

2. Define a measure $i(t)$ of the impurity of t as a nonnegative function ϕ of the $p(1|t), \ldots, p(6|t)$ such that

   $$\phi\left(\frac{1}{6}, \frac{1}{6}, \ldots, \frac{1}{6}\right) = \text{maximum},$$

   $$\phi(1, 0, 0, 0, 0, 0) = 0,\ \phi(0, 1, 0, 0, 0, 0) = 0,\ \ldots,$$
   $$\phi(0, 0, 0, 0, 0, 1) = 0$$

That is, the node impurity is largest when all classes are equally mixed together in it, and smallest when the node contains only one class.

For any node t, suppose that there is a candidate split s of the node which divides it into t_L and t_R such that a proportion p_L of the cases in t go into t_L and a proportion p_R go into t_R (Figure 2.6).

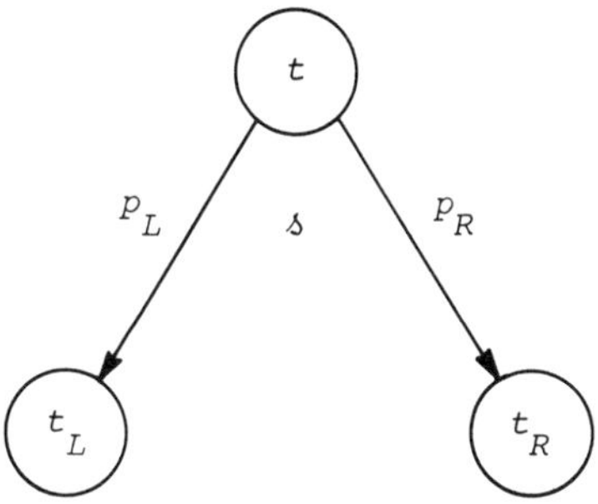

FIGURE 2.6

Then the goodness of the split is defined to be the decrease in impurity

$$\Delta i(s, t) = i(t) - p_L i(t_L) - p_R i(t_R).$$

The final step is:

3. Define a candidate set S of binary splits s at each node. Generally, it is simpler to conceive of the set S of splits as being generated by a set of questions $\mathcal{Q}$, where each question in $\mathcal{Q}$ is of the form

   Is $\mathbf{x} \in A$?, $A \subset X$.

   Then the associated split s sends all $\mathbf{x}_n$ in t that answer "yes" to t_L and all $\mathbf{x}_n$ in t that answer "no" to t_R.

   In the ship project the node impurity was defined as

$$i(t) = - \sum_{1}^{6} p(j|t) \log p(j|t).$$

There is no convincing justification for this specific form of $i(t)$. It was selected simply because it was a familiar function having the properties required by step 2. In later work, other definitions of $i(t)$ became preferable.

The set of questions $\mathcal{Q}$ were of the form:

> Do you have a local maximum in the interval $[a, b]$, where $a \leq b$ and a, b range from 0 to 1 in steps of .01?

Thus, a typical question in $\mathcal{Q}$ was: Do you have a local max in [.31, .53]? Put slightly differently, since the reduced data vectors consisted of the local max positions, the question was: Do you have any coordinate in the range [.31, .53]?

This gave a total of

$$\frac{100 \cdot 101}{2} \simeq 5000$$

questions, so that S contained a similar number of candidate splits.

The tree was grown in the following way: At the root node t_1, a search was made through all 5000 candidate splits to find that split s^* which gave the largest decrease in impurity; i.e.,

$$\Delta i(s^*, t_1) = \max_{s \in S} \Delta i(s, t_1).$$

Then t_1 was split into t_2 and t_3 using the split s^* and the same search procedure for the best $s \in S$ repeated on both t_2 and t_3 separately.

To terminate the tree growing, a heuristic rule was designed. When a node t was reached such that no significant decrease in impurity was possible, then t was not split and became a terminal node.

The class character of a terminal node was determined by the plurality rule. Specifically, if

$$p(j_0|t) = \max_j p(j|t),$$

then t was designated as a class j_0 terminal node.

The tree was used as a classifier in the obvious way. If a profile belonging to an unknown class was dropped into the tree and ended up in a class j terminal node, it was classified as j.

The procedure just described, given the limitations of the data, was considered a relatively successful solution to the problem. The first few nodes of the actual tree are pictured in Figure 2.7.

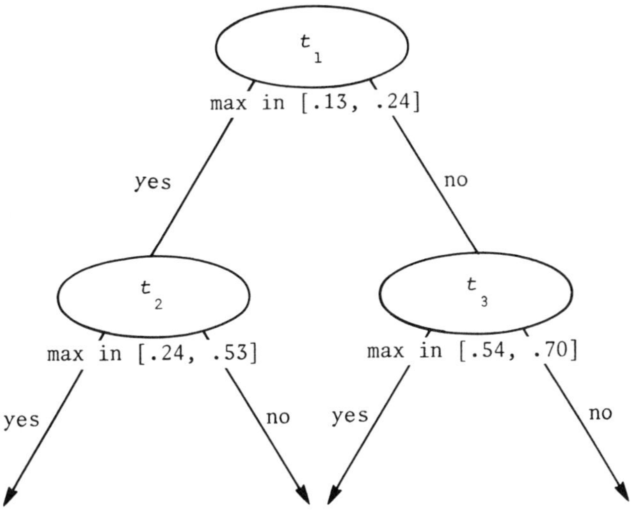

FIGURE 2.7

2.4 INITIAL TREE GROWING METHODOLOGY

Before discussing later methodological developments, we formulate more completely the initial methods used for constructing tree classifiers.

In the learning sample $\mathcal{L}$ for a J class problem, let N_j be the number of cases in class j. Often the prior probabilities $\{\pi(j)\}$ are taken to be the proportions $\{N_j/N\}$. But the learning sample proportions may not reflect the proportions expected in future cases. In the mass spectra problem, only one-tenth of the compounds in the data base contained chlorine. But since the unknowns the classifier would face were those already suspected of containing chlorine, the priors $\left(\frac{1}{2}, \frac{1}{2}\right)$ were used.

At any rate, the set of priors $\{\pi(j)\}$ are either estimated from the data as $\{N_j/N\}$ or supplied by the analyst.

In a node t, let $N(t)$ be the total number of cases in $\mathcal{L}$ with $\mathbf{x}_n \in t$, and $N_j(t)$ the number of class j cases in t. The proportion of the class j cases in $\mathcal{L}$ falling into t is $N_j(t)/N_j$. For a given set of priors, $\pi(j)$ is interpreted as the probability that a class j case will be presented to the tree. Therefore, we take

$$p(j, t) = \pi(j)N_j(t)/N_j \tag{2.2}$$

as the resubstitution estimate for the probability that a case will both be in class j and fall into node t.

The resubstitution estimate $p(t)$ of the probability that any case falls into node t is defined by

$$p(t) = \sum_j p(j, t). \tag{2.3}$$

The resubstitution estimate of the probability that a case is in class j given that it falls into node t is given by

$$p(j|t) = p(j, t)/p(t) \tag{2.4}$$

and satisfies

$$\sum_j p(j|t) = 1.$$

When $\{\pi(j)\} = \{N_j/N\}$, then $p(j|t) = N_j(t)/N(t)$, so the $\{p(j|t)\}$ are the relative proportions of class j cases in node t.

Note that throughout this book lower case p will denote an estimated probability and upper case P a theoretical probability.

The four elements needed in the initial tree growing procedure were

1. A set $\mathcal{Q}$ of binary questions of the form $\{\text{Is } \mathbf{x} \in A?\}$, $A \subset X$
2. A goodness of split criterion $\Phi(s, t)$ that can be evaluated for any split s of any node t

3. A stop-splitting rule
4. A rule for assigning every terminal node to a class

The set $\mathcal{Q}$ of binary questions generates a set S of splits s of every node t. Those cases in t answering "yes" go to the left descendant node t_L and those answering "no" to the right descendant t_R. In fact, if the question is $\{\text{Is } \mathbf{x} \in A?\}$, then $t_L = t \cap A$ and $t_R = t \cap A^c$, where A^c is the complement of A in X. At each intermediate node t, the split selected is that split s^* which maximizes $\Phi(s, t)$.

2.4.1 The Standard Set of Questions

If the data have standard structure, the class $\mathcal{Q}$ of questions can be standardized. Assume that the measurement vectors have the form

$$\mathbf{x} = (x_1, \ldots, x_M),$$

where M is the fixed dimensionality and the variables $x_1, \ldots, x_M$ can be a mixture of ordered and categorical types. The *standardized* set of questions $\mathcal{Q}$ is defined as follows:

1. Each split depends on the value of only a *single* variable.
2. For each ordered variable x_m, $\mathcal{Q}$ includes all questions of the form

   $\{\text{Is } x_m \leq c?\}$

   for all c ranging over $(-\infty, \infty)$.
3. If x_m is categorical, taking values, say, in $\{b_1, b_2, \ldots, b_L\}$, then $\mathcal{Q}$ includes all questions of the form

   $\{\text{Is } x_m \in S?\}$

   as S ranges over all subsets of $\{b_1, b_2, \ldots, b_L\}$.

The splits in items 2 and 3 for all M variables constitute the standardized set.

So, for example, if $M = 4$, x_1, x_2, x_3 are ordered, and $x_4 \in \{b_1, b_2, b_3\}$, then $\mathcal{Q}$ includes all questions of the form

Is $x_1 \leq 3.2$?
Is $x_3 \leq -6.8$?
Is $x_4 \in \{b_1, b_3\}$?

and so on. There are not an infinite number of distinct splits of the data. For example, if x_1 is ordered, then the data points in $\mathcal{L}$ contain at most N distinct values $x_{1,1}, x_{1,2}, \ldots, x_{1,N}$ of x_1. There are at most N different splits generated by the set of questions {Is $x_1 \leq c$?}. These are given by {Is $x_1 \leq c_n$?}, $n = 1, \ldots, N' \leq N$, where the c_n are taken halfway between consecutive distinct data values of x_1.

For a categorical variable x_m, since $\{x_m \in S\}$ and $\{x_m \notin S\}$ generate the same split with t_L and t_R reversed, if x_m takes on L distinct values, then $2^{L-1} - 1$ splits are defined on the values of x_m.

At each node the tree algorithm searches through the variables one by one, beginning with x_1 and continuing up to x_M. For each variable it finds the best split. Then it compares the M best single variable splits and selects the best of the best.

The computer program CART (Classification and Regression Trees) incorporates this standardized set of splits. Since most of the problems ordinarily encountered have a standard data structure, it has become a flexible and widely used tool.

When fixed-dimensional data have only ordered variables, another way of looking at the tree structured procedure is as a recursive partitioning of the data space into rectangles.

Consider a two-class tree using data consisting of two ordered variables x_1, x_2 with $0 \leq x_i \leq 1$, $i = 1, 2$. Suppose that the tree diagram looks like Figure 2.8.

An equivalent way of looking at this tree is that it divides the unit square as shown in Figure 2.9.

From this geometric viewpoint, the tree procedure recursively partitions X into rectangles such that the populations within each rectangle become more and more class homogeneous.

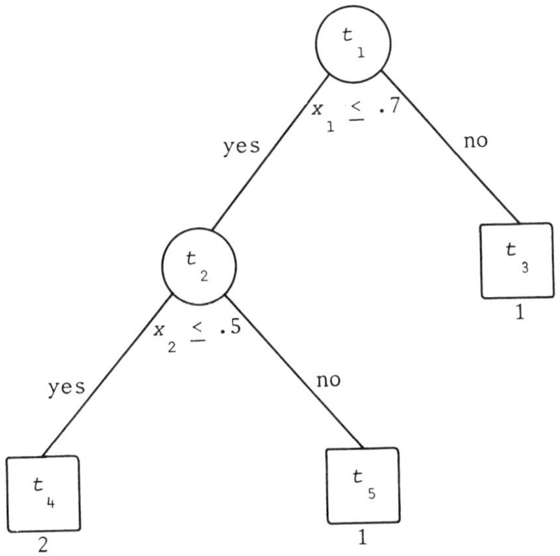

FIGURE 2.8

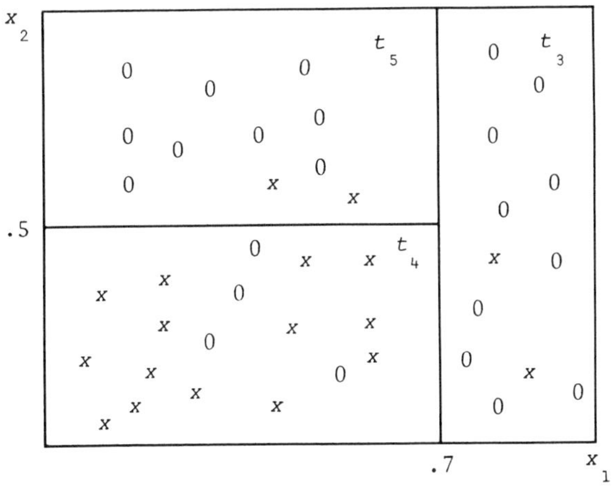

0 = class 1, x = class 2

FIGURE 2.9

2.4.2 *The Splitting and Stop-Splitting Rule*

The goodness of split criterion was originally derived from an impurity function.

DEFINITION 2.5. *An impurity function is a function* ϕ *defined on the set of all J-tuples of numbers* $(p_1, \ldots, p_J)$ *satisfying* $p_j \geq 0$, $j = 1, \ldots, J$, $\Sigma_j p_j = 1$ *with the properties*

(i) ϕ *is a maximum only at the point* $\left(\frac{1}{J}, \frac{1}{J}, \ldots, \frac{1}{J}\right)$,

(ii) ϕ *achieves its minimum only at the points* $(1, 0, \ldots, 0)$, $(0, 1, 0, \ldots, 0)$, $\ldots$, $(0, 0, \ldots, 0, 1)$,

(iii) ϕ *is a symmetric function of* $p_1, \ldots, p_J$.

DEFINITION 2.6. *Given an impurity function* ϕ, *define the impurity measure* $i(t)$ *of any node* t *as*

$$i(t) = \phi(p(1|t), p(2|t), \ldots, p(J|t)).$$

If a split s of a node t sends a proportion p_R of the data cases in t to t_R and the proportion p_L to t_L, define the decrease in impurity to be

$$\Delta i(s, t) = i(t) - p_R i(t_R) - p_L i(t_L).$$

Then take the goodness of split $\Phi(s, t)$ to be $\Delta i(s, t)$.

Suppose we have done some splitting and arrived at a current set of terminal nodes. The set of splits used, together with the order in which they were used, determines what we call a *binary tree* T. Denote the current set of terminal nodes by $\tilde{T}$; set $I(t) = i(t)p(t)$, and define the *tree impurity* $I(T)$ by

$$I(T) = \sum_{t \in \tilde{T}} I(t) = \sum_{t \in \tilde{T}} i(t)p(t).$$

It is easy to see that selecting the splits that maximize $\Delta i(s, t)$ is equivalent to selecting those splits that minimize the overall tree impurity $I(T)$. Take any node $t \in \tilde{T}$ and using a split s, split the node into t_L and t_R. The new tree T' has impurity

$$I(T') = \sum_{\tilde{T}-\{t\}} I(t) + I(t_L) + I(t_R).$$

The decrease in tree impurity is

$$I(T) - I(T') = I(t) - I(t_L) - I(t_R).$$

This depends *only* on the node t and split s. Therefore, maximizing the decrease in tree impurity by splits on t is equivalent to maximizing the expression

$$\Delta I(s, t) = I(t) - I(t_L) - I(t_R). \tag{2.7}$$

Define the proportions p_L, p_R of the node t population that go to t_L and t_R, respectively, by

$$p_L = p(t_L)/p(t), \quad p_R = p(t_R)/p(t).$$

Then

$$p_L + p_R = 1$$

and (2.7) can be written as

$$\begin{aligned} \Delta I(s, t) &= [i(t) - p_L i(t_L) - p_R i(t_R)]p(t) \\ &= \Delta i(s, t)p(t). \end{aligned}$$

Since $\Delta I(s, t)$ differs from $\Delta i(s, t)$ by the factor $p(t)$, the same split s^* maximizes both expressions. Thus, the split selection procedure can be thought of as a repeated attempt to minimize overall tree impurity.

The initial stop-splitting rule was simple (and unsatisfactory). Set a threshold $\beta > 0$, and declare a node t terminal if

$$\max_{s \in S} \Delta I(s, t) < \beta. \tag{2.8}$$

2.4.3 The Class Assignment Rule and Resubstitution Estimates

Suppose a tree T has been constructed and has terminal nodes $\tilde{T}$.

DEFINITION 2.9. *A class assignment rule assigns a class $j \in \{1, \ldots, J\}$ to every terminal node $t \in \tilde{T}$. The class assigned to node $t \in \tilde{T}$ is denoted by $j(t)$.*

In the ship classification example, class assignment was done using the rule—$j(t)$ equals that class for which $p(j|t)$ is largest. If the priors $\{\pi(j)\}$ are estimated by $\{N_j/N\}$, this is just the plurality rule—classify t as that class for which $N_j(t)$ is largest.

For any set of priors and class assignment rule $j(t)$,

$$\sum_{j \neq j(t)} p(j|t)$$

is the resubstitution estimate of the probability of misclassification given that a case falls into node t. We take as our class assignment rule $j^*(t)$ the rule that minimizes this estimate, i.e.,

DEFINITION 2.10. *The class assignment rule $j^*(t)$ is given by: If $p(j|t) = \max_i p(i|t)$, then $j^*(t) = j$. If the maximum is achieved for two or more different classes, assign $j^*(t)$ arbitrarily as any one of the maximizing classes.*

Using this rule, we get

DEFINITION 2.11. *The resubstitution estimate $r(t)$ of the probability of misclassification, given that a case falls into node t, is*

$$r(t) = 1 - \max_j p(j|t).$$

Denote

$$R(t) = r(t)p(t).$$

Then the resubstitution estimate for the overall misclassification rate $R^*(T)$ *of the tree classifier* T *is*

$$R(T) = \sum_{t \in \tilde{T}} R(t).$$

Up to now, the assumption has been tacitly made that the cost or loss in misclassifying a class j object as a class i object was the same for all $i \neq j$. In some classification problems this is not a realistic setting. Therefore, we introduce a set of misclassification costs $C(i|j)$, where

DEFINITION 2.12. $C(i|j)$ *is the cost of misclassifying a class* j *object as a class* i *object and satisfies*

(i) $C(i|j) \geq 0,\ i \neq j,$

(ii) $C(i|j) = 0,\ i = j.$

Given a node t with estimated node probabilities $p(j|t)$, $j = 1, \ldots, J$, if a randomly selected object of unknown class falls into t and is classified as class i, then the estimated expected misclassification cost is

$$\sum_j C(i|j)p(j|t).$$

A natural node assignment rule is to select i to minimize this expression. Therefore,

DEFINITION 2.13. *Put* $j^*(t) = i_0$ *if* i_0 *minimizes* $\sum_j C(i|j)p(j|t)$; *define the resubstitution estimate* $r(t)$ *of the expected misclassification cost, given the node* t, *by*

$$r(t) = \min_i \sum_j C(i|j)p(j|t),$$

and define the resubstitution estimate $R(T)$ *of the misclassification cost of the tree* T *by*

$$R(T) = \sum_{t \in \tilde{T}} r(t)p(t) = \sum_{t \in \tilde{T}} R(t),$$

where $R(t) = r(t)p(t)$.

Note that in the unit misclassification cost case, $C(i|j) = 1$, $i \neq j$,

$$\sum_j C(i|j)p(j|t) = 1 - p(i|t),$$

and the minimum cost rule reduces to the rule given in Definition 2.13.

Henceforth, we take $j^*(t)$ as the class assignment rule with no further worry.

An important property of $R(T)$ is that the more one splits in any way, the smaller $R(T)$ becomes. More precisely, if T' is gotten from T by splitting *in any way* a terminal node of T, then

$$R(T') \leq R(T).$$

Putting this another way:

PROPOSITION 2.14. *For any split of a node t into t_L and t_R,*

$$R(t) \geq R(t_L) + R(t_R).$$

The proof is simple, but to get on with the main story, we defer it until Chapter 4 (Proposition 4.2).

2.5 METHODOLOGICAL DEVELOPMENT

In spite of the attractiveness of tree structured classifiers, it soon became apparent that there were serious deficiencies in the tree growing method used in the ship classification project.

The methodology developed to deal with the deficiencies and to make tree structured classification more flexible and accurate is covered in the next three chapters and, in a more theoretical setting, in Chapters 9 to 12.

As a roadmap, a brief outline follows of the issues and the methods developed to deal with them.

2.5.1 *Growing Right Sized Trees: A Primary Issue*

The most significant difficulty was that the trees often gave dishonest results. For instance, suppose that the stopping rule (2.8) is used, with the threshold set so low that every terminal node has only one data point. Then the $p(j|t)$ are all zero or 1. $R(t) = 0$, and the resubstitution estimate $R(T)$ for the misclassification rate is zero. In general, $R(T)$ decreases as the number of terminal nodes increases. The more you split, the better you think you are doing.

Very few of the trees initially grown held up when a test sample was run through them. The resubstitution estimates $R(T)$ were unrealistically low and the trees larger than the information in the data warranted. This results from the nature of the tree growing process as it continuously optimizes and squeezes boundaries between classes in the learning set.

Using more complicated stopping rules did not help. Depending on the thresholding, the splitting was either stopped too soon at some terminal nodes or continued too far in other parts of the tree. A satisfactory resolution came only after a fundamental shift in focus. Instead of attempting to stop the splitting at the right set of terminal nodes, continue the splitting until all terminal nodes are very small, resulting in a large tree. Selectively prune (recombine) this large tree upward, getting a decreasing sequence of subtrees. Then use cross-validation or test sample estimates to pick out that subtree having the lowest estimated misclassification rate.

This process is a central element in the methodology and is covered in the next chapter.

2.5.2 *Splitting Rules*

Many different criteria can be defined for selecting the best split at each node. As noted, in the ship classification project, the split selected was the split that most reduced the node impurity defined by

$$i(t) = -\sum_j p(j|t) \log [p(j|t)].$$

Chapter 4 discusses various splitting rules and their properties. Variable misclassification costs and prior distributions $\{\pi(j)\}$ are incorporated into the splitting structure. Two splitting rules are singled out for use. One rule uses the *Gini index of diversity* as a measure of node impurity; i.e.,

$$i(t) = \sum_{i \neq j} p(i|t)p(j|t).$$

The other is the *twoing rule*: At a node t, with s splitting t into t_L and t_R, choose the split s that maximizes

$$\frac{p_L p_R}{4}\left[\sum_j \left| p(j|t_L) - p(j|t_R)\right|\right]^2 .$$

This rule is not related to a node impurity measure.

However, the tentative conclusion we have reached is that within a wide range of splitting criteria the properties of the final tree selected are surprisingly insensitive to the choice of splitting rule. The criterion used to prune or recombine upward is much more important.

2.5.3 Variable Combinations

Another deficiency in trees using a standard structure is that all splits are on single variables. Put another way, all splits are perpendicular to the coordinate axes. In situations where the class structure depends on combinations of variables, the standard tree program will do poorly at uncovering the structure. For instance, consider the two-class two-variable data illustrated in Figure 2.10.

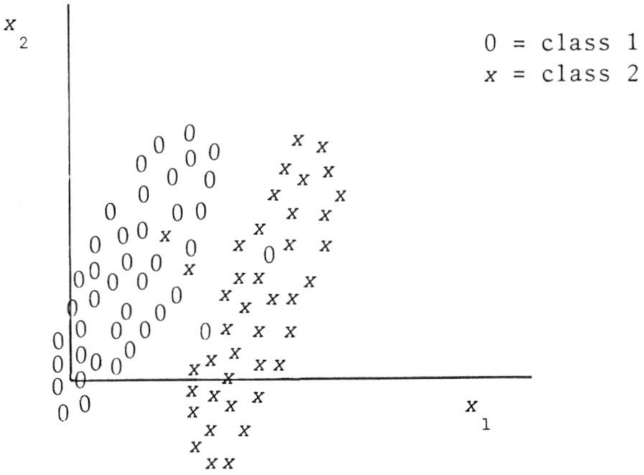

FIGURE 2.10

Discriminant analysis would do very well on this data set. The standard tree program would split many times in an attempt to approximate the separating hyperplane by rectangles. It would be difficult to see the linear structures of the data by examining the tree output.

In problems where linear structure is suspected, the set of allowable splits is extended to include all linear combination splits of the form

$$\text{Is } \sum_m a_m x_m \leq c?$$

An algorithm was developed to search through such splits in an effort to find the one that maximizes the goodness of split criterion.

With linear combination splits incorporated, tree structured classification gives results competitive with or better than linear discriminant analysis on problems such as that shown in Figure 2.10.

In another type of classification problem, variables tend to appear in certain Boolean combinations. For instance, in medical diagnosis, frequently the diagnostic questions are of the type: Does the patient have (symptom *A* and symptom *B*) or symptom *C*?

In problems where a Boolean combination structure is suspected and the measurement space is high dimensional, it is useful to extend the allowable splits to include Boolean combinations of splits. However, since the number of such splits can increase rapidly with the length of the Boolean expressions allowed, a stepwise procedure is devised to make the search computationally feasible.

Both of these variable combination methods are covered in Chapter 5.

2.5.4 *Missing Values*

There are two aspects to the missing value problem. First, some cases in $\mathcal{L}$ may have some missing measurement values (we assume the class label is always present). Second, we may want the completed tree to predict a class label for a measurement vector some of whose values are missing.

The missing value algorithm discussed in Chapter 5 handles both of these problems through the use of *surrogate splits*. The idea is this: Define a measure of similarity between any two splits s, s' of a node t. If the best split of t is the split s on the variable x_m, find the split s' on the variables other than x_m that is most similar to s. Call s' the best surrogate for s. Similarly, define the second best surrogate, third best, and so on.

If a case has x_m missing in its measurement, decide whether it goes to t_L or t_R by using the best surrogate split. If it is missing the variable containing the best surrogate split, use the second best, and so on.

2.5.5 *Tree Interpretation*

Another difficulty in tree structured classification is that the simple structure of the final classification tree can be deceptive, leading to misinterpretation. For instance, if a variable is never split on in the final tree, one interpretation might be that the variable has very little association with class membership. The truth may be that its effect was masked by other variables.

Tree structures may be unstable. If one variable narrowly masks another, then a small change in the priors or learning sample may shift the split from one variable to another. Two splits of a node may be dissimilar but have almost the same goodness of split. Small changes may sometimes favor one, sometimes the other.

Various procedures are suggested in Chapter 5 to assist in tree interpretation. In particular, a method is given for ranking variables in terms of their *potential* effect on the classification. Even though a variable may not appear in a split on the final tree, its ranking may be high, giving an indication of masking.

Often, the analyst may want to explore a range of parameters and/or the effects of adding or deleting variables. Since a number of trees will be grown and used for comparison, the accuracy of full cross-validation may not be necessary and an alternative rapid method for growing exploratory trees is suggested.

2.5.6 Within-Node Misclassification Rate Estimates

In Chapter 4, it is seen that cross-validation gives fairly adequate estimates of $R^*(T)$. However, the cross-validation procedure does not give estimates for the true within-node misclassification rate $r^*(t)$. The resubstitution estimates $r(t)$ are usually optimistic, especially for the smaller terminal nodes.

Since estimates of $r^*(t)$ are often a valuable by-product of the classifier, efforts have been made to get improved estimates. Chapter 5 gives a rather ad hoc method of estimation having some heuristic justification. Its main justification, however, is that in all problems on which it has been tested, its error averaged over the terminal nodes has been less than half of that of $r(t)$.

2.5.7 *Computational Efficiency*

Growing a tree using, say, tenfold cross-validation can be computationally time consuming with a large data set. A node subsampling method is given in Chapter 5 to increase efficiency.

If the sample size in a node t is larger than a fixed maximum, then a subsample from each of the classes represented in t is used to determine the best split. However, the entire data set in t is then sent down the split. Thus, subsampling significantly decreases computation time in the larger nodes without decreasing sample size further down the tree.

2.5.8 *Other Issues*

There are some other problems inherent in the tree procedure. In brief, "end cuts" or splits that tend to separate a node into one small and one large subset are, all other things being equal, favored over "middle cuts." This is discussed in Section 11.8. Also, variable selection is biased in favor of those variables having more values and thus offering more splits.

Finally, another problem frequently mentioned (by others, not by us) is that the tree procedure is only one-step optimal and not overall optimal. That is, suppose the tree growing procedure produced, say, 11 rectangular terminal nodes. If one could search *all* possible partitions of X into 11 rectangles for that partition that minimizes the sum of the node impurities, the two results might be quite different.

This issue is analogous to the familiar question in linear regression of how well the stepwise procedures do as compared with "best subsets" procedures. We do not address this problem. At this stage of computer technology, an overall optimal tree growing procedure does not appear feasible for any reasonably sized data set.

The issue of "honesty" is more critical than "optimality." Constructing a good classifier whose performance will stand up under test samples and that is useful and practical is our first priority.

2.6 TWO RUNNING EXAMPLES

To illustrate various parts of the methodology, two models have been constructed for generating data. The first is a digit recognition model. It has a simple structure and can be analyzed analytically. It is complemented by a more complex waveform recognition model.

2.6.1 Digit Recognition Example

Digits are ordinarily displayed on electronic watches and calculators using seven horizontal and vertical lights in on-off combinations (see Figure 2.11).

FIGURE 2.11

Number the lights as shown in Figure 2.12. Let i denote the ith digit, i = 1, 2, ..., 9, 0, and take $(x_{i1}, \ldots, x_{i7})$ to be a seven-dimensional vector of zeros and ones with $x_{im} = 1$ if the light in the mth position is on for the ith digit, and $x_{im} = 0$ otherwise. The values of x_{im} are given in Table 2.1. Set $C = \{1, \ldots, 10\}$ and let X be the set of all possible 7-tuples $= (x_1, \ldots, x_7)$ of zeros and ones.

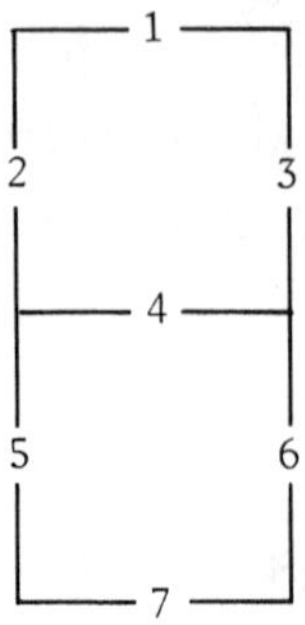

FIGURE 2.12

TABLE 2.1

Digit	x_1	x_2	x_3	x_4	x_5	x_6	x_7
1	0	0	1	0	0	1	0
2	1	0	1	1	1	0	1
3	1	0	1	1	0	1	1
4	0	1	1	1	0	1	0
5	1	1	0	1	0	1	1
6	1	1	0	1	1	1	1
7	1	0	1	0	0	1	0
8	1	1	1	1	1	1	1
9	1	1	1	1	0	1	1
0	1	1	1	0	1	1	1

The data for the example are generated from a faulty calculator. Each of the seven lights has probability .1 of not doing what it is supposed to do. More precisely, the data consist of outcomes from the random vector $(X_1, \ldots, X_7, Y)$ where Y is the class label and assumes the values 1, ..., 10 with equal probability and the $X_1, \ldots, X_7$ are zero-one variables. Given the value of Y, the $X_1, \ldots, X_7$ are each independently equal to the value corresponding to Y in Table 2.1 with probability .9 and are in error with probability .1.

For example,

$$P((X_1, \ldots, X_7) = (1, 1, 0, 1, 0, 1, 1) \mid Y = 5) = (.9)^7$$

and

$$P((X_1, \ldots, X_7) = (1, 1, 0, 1, 0, 1, 1), Y = 5) = (.9)^7 \cdot \frac{1}{10}.$$

Two hundred samples from this distribution constitute the learning sample. That is, every case in $\mathcal{L}$ is of the form $(x_1, \ldots, x_7, j)$ where $j \in C$ is a class label and the measurement vector $x_1, \ldots, x_7$ consists of zeros and ones. The data in $\mathcal{L}$ are displayed in Table 2.2.

To use a tree structured classification construction on $\mathcal{L}$, it is necessary to specify:

First: the set $\mathcal{Q}$ of questions.

(In this example, $\mathcal{Q}$ consisted of the seven questions, Is $x_m = 0$?, $m = 1, \ldots, 7$.)

Second: a rule for selecting the best split.

(The twoing criterion mentioned in Section 2.5.2 was used.)

Third: a criterion for choosing the right sized tree.

(The pruning cross-validation method covered in Chapter 3 was used.)

The resulting tree is shown in Figure 2.13. There are 10 terminal nodes, each corresponding to one class. This is accidental. In general, classification trees may have any number of terminal nodes corresponding to a single class, and occasionally, some class may have no corresponding terminal nodes.

Underneath each intermediate node is the question leading to the split. Those answering yes go left, no to the right.

It is interesting to compare the tree with Table 2.1. For instance, terminal node 8 corresponding to class 1 consists of all measurement vectors such that $x_5 = 0$, $x_4 = 0$, and $x_1 = 0$. If the calculator were not faulty, these three conditions would completely distinguish the digit 1 from the others. Similarly, node 9 cor-

TABLE 2.2

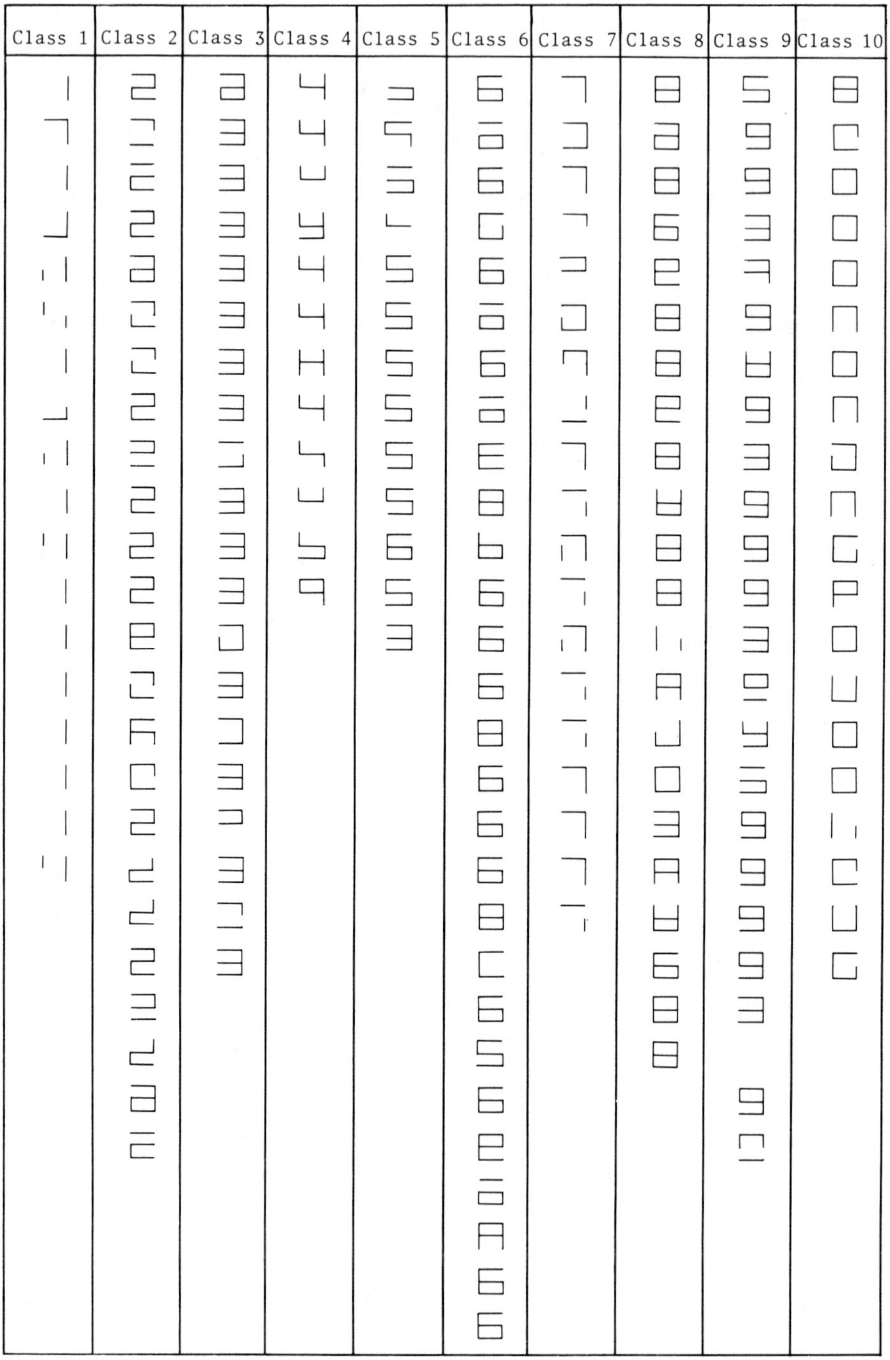
Class 1
Class 2
Class 3
Class 4
Class 5
Class 6
Class 7
Class 8
Class 9
Class 10

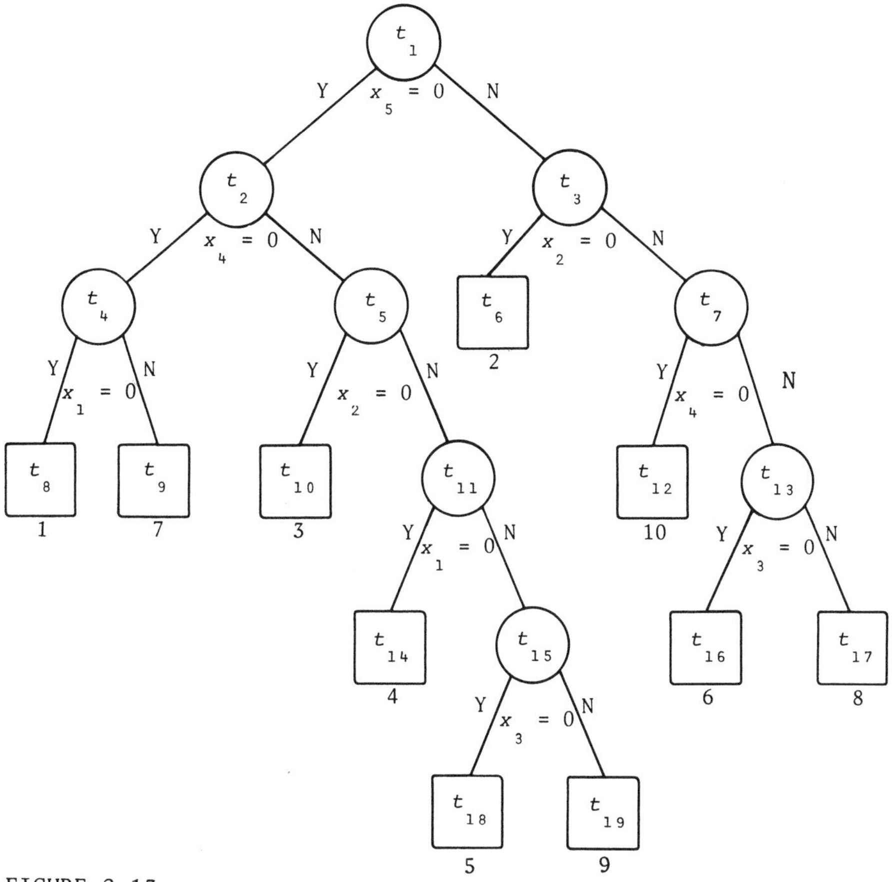

FIGURE 2.13

responding to the digit 7 consists of all vectors such that $x_5 = 0$, $x_4 = 0$, and $x_1 = 1$. Again, for a perfect calculator, the digit 7 is distinguished by these conditions.

For this particular tree, $R^*(T)$ was estimated by using a test sample of size 5000 as .30. The cross-validation estimate, using only the learning sample, is .30. The resubstitution estimate is .29. The Bayes rule can be solved for in this example. The resulting Bayes misclassification rate is .26. In this example, there is not much room for improvement over the tree classifier.

A variant of the digit recognition problem is gotten by adding pure noise variables. Let the grid in Figure 2.12 be replaced by the larger grid in Figure 2.14 with 24 line segments. Take the additional 17 lights to be on or off with probability .5 independently of each other and of the original 7 lights. The 17 lights generate 17 variables $x_8, \ldots, x_{24}$ with $x_m = 0$ or 1 as the mth light is on or off. These variables are pure noise in the sense that they are useless for the purpose of digit recognition.

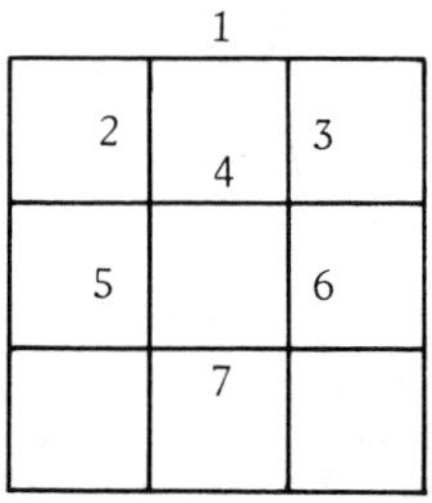

FIGURE 2.14

A learning sample for this problem was constructed by taking each case in the learning sample for the digit recognition problem, sampling 17 measurements from the pure noise distribution, and adding these to the 7 measurements already in the case. The new learning sample, then, consists of 200 cases, each having 24 measurement values.

A classification tree was grown using this learning set, with the same questions, split selection criterion, and pruning cross-validation method as in the previous tree. The tree selected was identical to the tree of Figure 2.13. A test set of size 5000 estimated $R^*(T)$ as .30. The cross-validation estimate is .31 and the resubstitution estimate is .29.

The fact that even with 17 added noise variables the tree selected is identical to the original 7-variable digit recognition tree illustrates an important advantage of tree structured classification over nearest neighbor or other nonparametric methods. Its inherent structure is based on a procedure for distinguishing between those variables useful for classification and those which are not. Of course, the distinction between useful and useless variables is rarely as clearly defined as in the preceding example, and the tree structured approach rarely so obviously triumphant.

2.6.2 Waveform Recognition Problem

Because of the elementary structure of the digit recognition problem, another, more complex, example was constructed. It is a three-class problem based on the waveforms $h_1(t)$, $h_2(t)$, $h_3(t)$ graphed in Figure 2.15.

Each class consists of a random convex combination of two of these waveforms sampled at the integers with noise added. More specifically, the measurement vectors are 21 dimensional: $\mathbf{x} = (x_1, \ldots, x_{21})$. To generate a class 1 vector $\mathbf{x}$, independently generate a uniform random number u and 21 random numbers $\varepsilon_1, \ldots, \varepsilon_{21}$ normally distributed with mean zero and variance 1. Then set

$$x_m = uh_1(m) + (1 - u)h_2(m) + \varepsilon_m, \quad m = 1, \ldots, 21.$$

To generate a class 2 vector, repeat the preceding and set

$$x_m = uh_1(m) + (1 - u)h_3(m) + \varepsilon_m, \quad m = 1, \ldots, 21.$$

Class 3 vectors are generated by

$$x_m = uh_2(m) + (1 - u)h_3(m) + \varepsilon_m, \quad m = 1, \ldots, 21.$$

Three hundred measurement vectors were generated using prior probabilities of $\left(\frac{1}{3}, \frac{1}{3}, \frac{1}{3}\right)$, so there were approximately 100 per class. The data in the first five cases in each class are plotted in Figure 2.16. The tree was constructed using all questions of the form $\{\text{Is } x_m \le c?\}$ for c ranging over all real numbers and $m = 1, \ldots, 21$. The Gini splitting criterion (Section 2.5.2) was used, and the final tree selected by means of pruning and cross-validation.

This tree is shown in Figure 2.17. It has 11 terminal nodes. The resubstitution estimate of misclassification is .14; the cross-validation estimate is .29; and the estimate of $R^*(T)$ gotten by using an independent test set of size 5000 is .28. Within each node in Figure 2.17 is the population, by class, of the node. Underneath each node is the question producing the split.

To illustrate how the splitting mechanism works, examine the first few splits. Figure 2.18 shows the waveforms $h_j(t)$ superimposed. The first split is $x_6 \leq 2.0$. The sixth coordinate is nearly at the peak of the first waveform. Since classes 1 and 2 both contain a random proportion of the first waveform, x_6 will be high for these two, except when the proportion is small or ε_6 is large and negative. This is reflected in the split with 109 out of the 115 class 3's going left (yes) and most of classes 1 and 2 going right (no).

Then the tree mechanism attempts to purify the left node by the split $x_{11} \leq 2.5$. At the eleventh coordinate the third waveform has its peak, so low values of x_{11} would tend to characterize class 1. The split sends 27 out of 36 class 1s left, with most of classes 2 and 3 going right.

The right node resulting from the original split has almost equal proportions of classes 1 and 2 (64, 68). The mechanism attempts to separate these two classes by the split $x_{10} \leq 2.6$ near the peak of the third waveform. We would expect class 1 to be low at x_{10} and class 2 to be high. The resulting split confirms our expectations with most of class 1 going left and most of class 2 going right.

The numbers to the right of each terminal node are the normalized counts, by class, of the 5000 test sample vectors that were run down the tree and ended up in that terminal node. For comparability, the counts are normalized so that the total in the node equals the learning sample total.

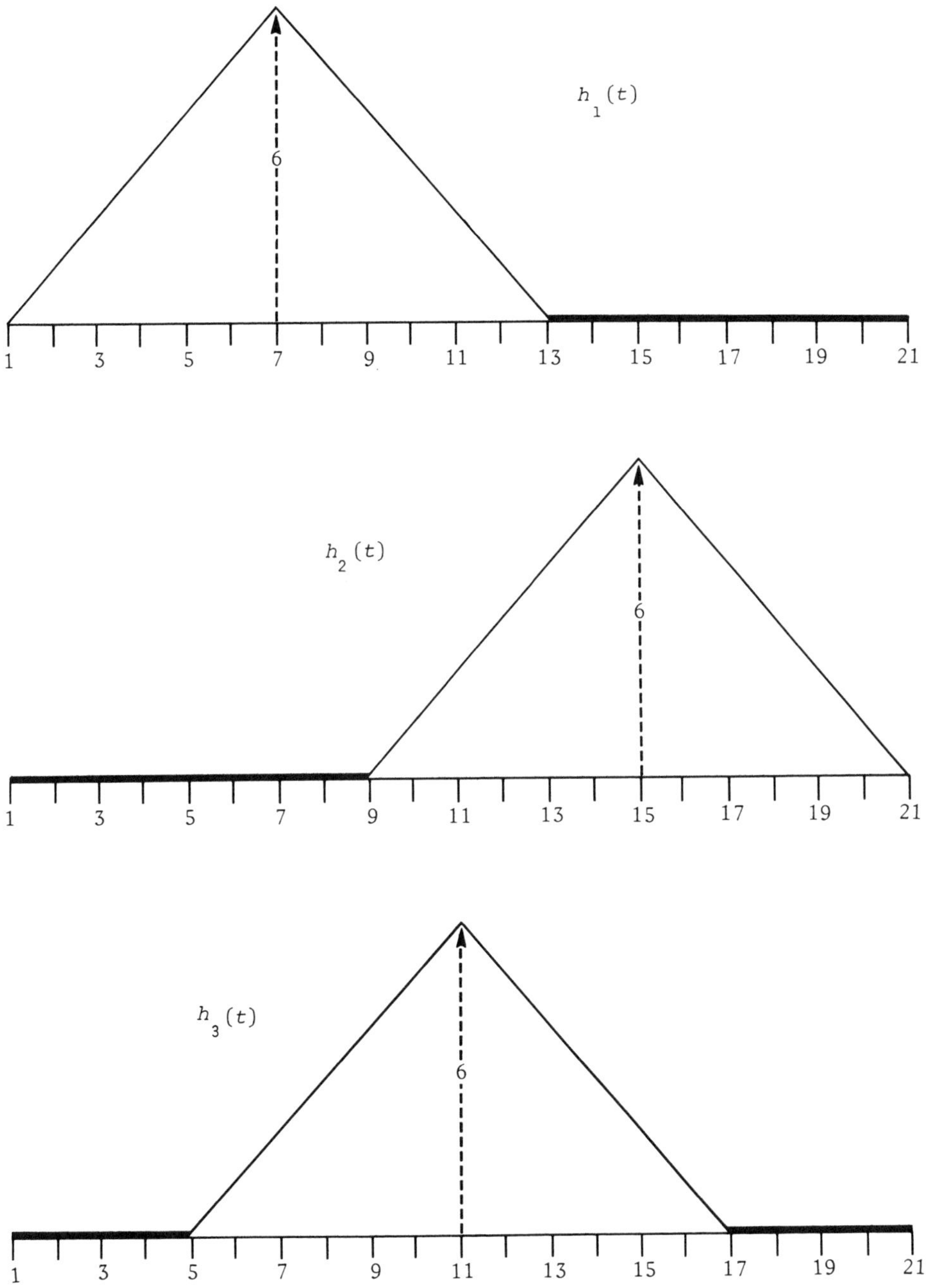

FIGURE 2.15 Waveforms.

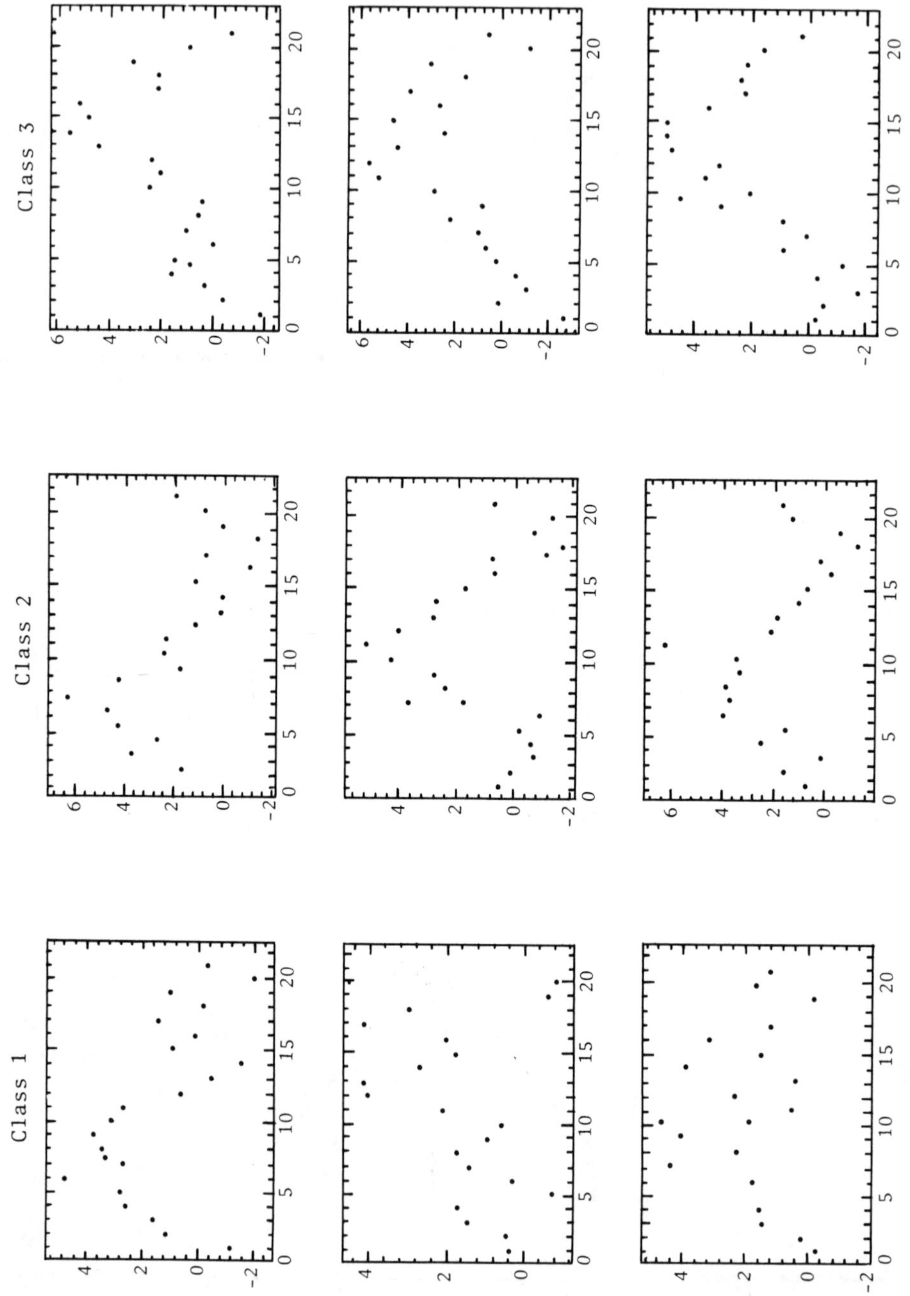
Class 3
Class 2
Class 1

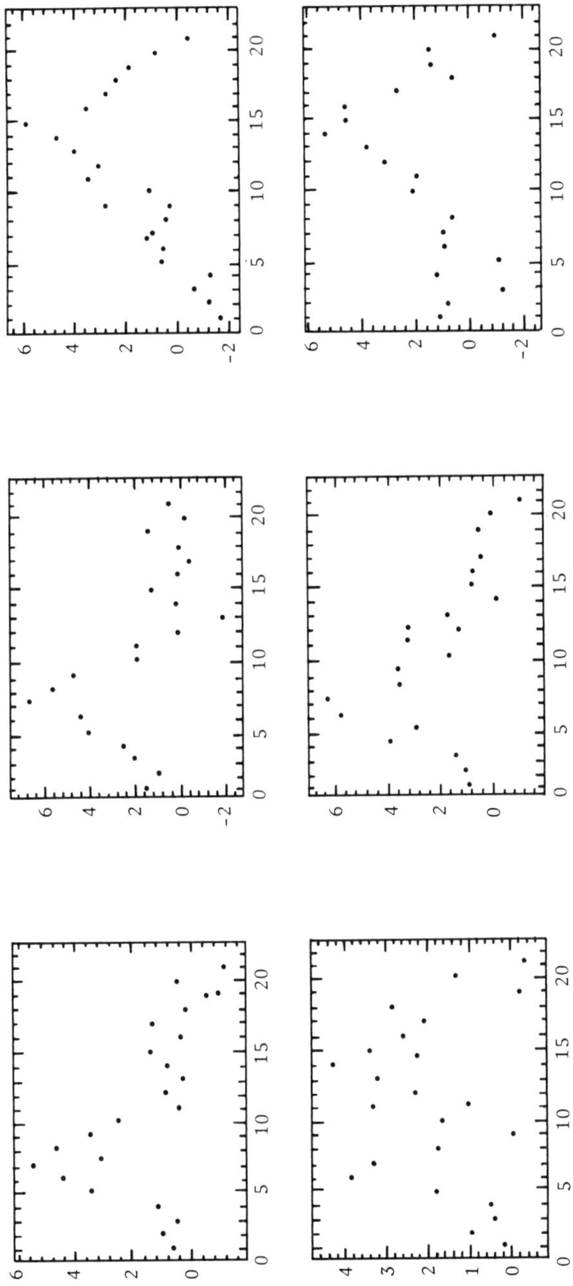

FIGURE 2.16

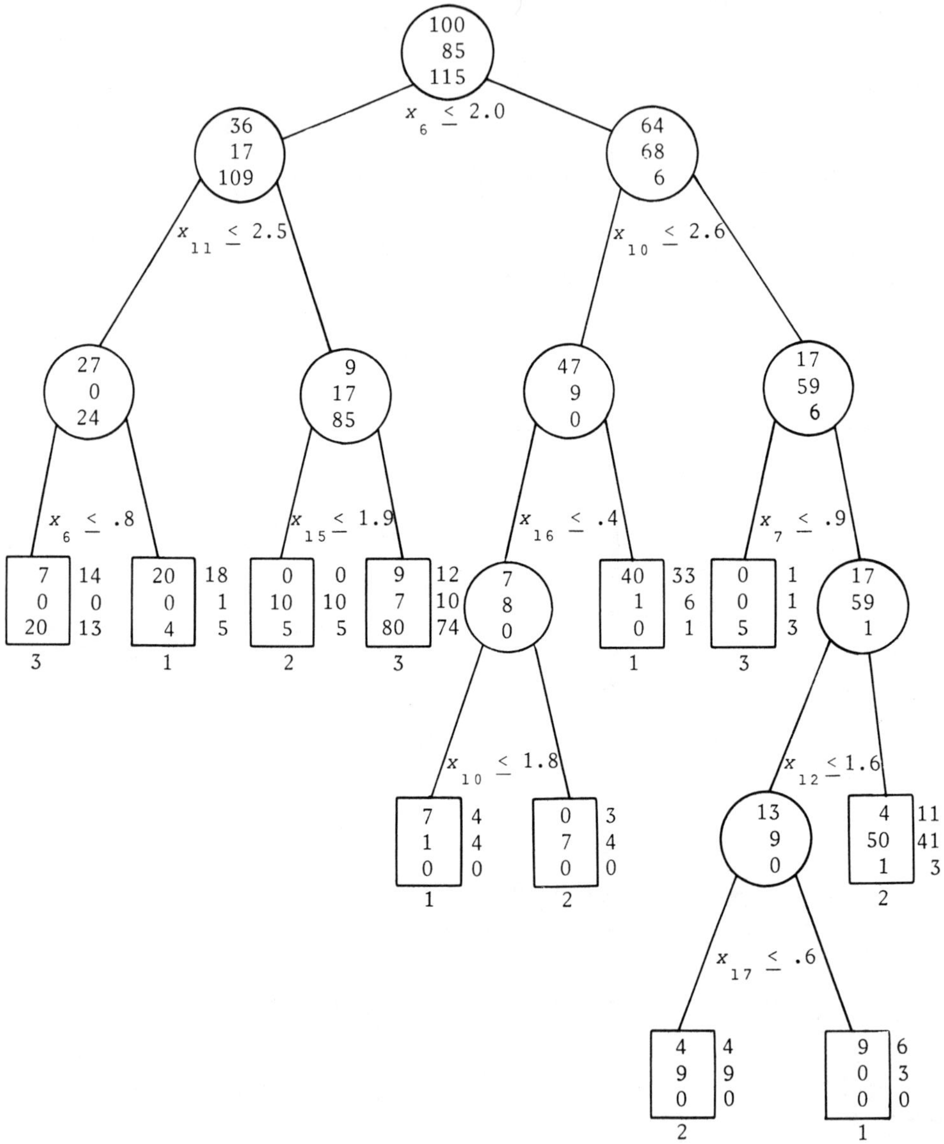

FIGURE 2.17 Waveform tree.

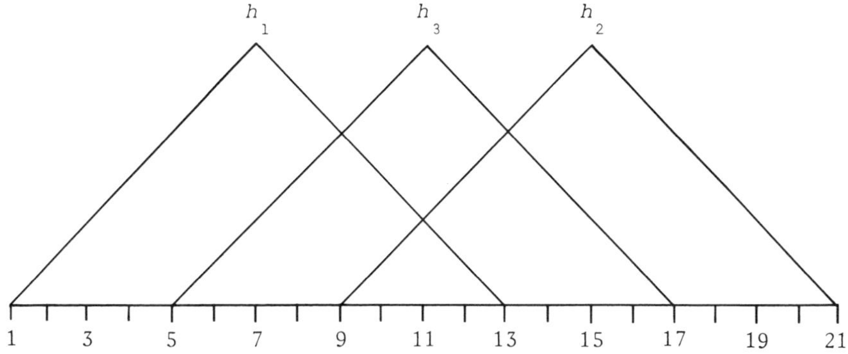

FIGURE 2.18

Comparing these two lists for the various terminal nodes, we can see that the learning sample node proportions can occasionally give quite misleading estimates. This is particularly true for the smaller nodes.

An analytic expression can be derived for the Bayes rule in this problem. Using this rule on a test sample of size 5000 gave an error rate of .14.

This example was selected to illustrate features of the tree construction procedure. That the tree classifier error rate is about twice the Bayes rate is an indication that single coordinate splits are not well suited to extracting the available information from this particular data set. The use of combinations of variables will be looked at in Section 5.2.

2.7 THE ADVANTAGES OF THE TREE STRUCTURED APPROACH

As outlined in the previous sections, tree structured classification is a recursive and iterative procedure that requires the specifications of only a few elements:

1. *The set $\mathcal{Q}$ of questions*
2. *A rule for selecting the best split at any node*
3. *A criterion for choosing the right-sized tree*

It has the potential for being a powerful and flexible classification tool. In particular:

1. It can be applied to any data structure through the appropriate formulation of the set of questions $\mathcal{Q}$. In standard data structures it handles both ordered and categorical variables in a simple and natural way.
2. The final classification has a simple form which can be compactly stored and that efficiently classifies new data.

For example, the classification algorithm for the chlorine problem consists of a list of about 40 questions. The classification process for any new mass spectra is done in fractions of a second.

3. It makes powerful use of conditional information in handling nonhomogeneous relationships.

Once a node t is split into t_L and t_R, then these two nodes are *individually* searched for the most significant split. For example, in the ship problem, for those profiles having a maximum in [.13, .29], the next most significant question was whether there was a maximum in [.24, .53], while for those not having a maximum in [.13, .29], the most significant question was very different, involving a maximum in the range [.54, .76].

4. It does automatic stepwise variable selection and complexity reduction.

By this it is meant that as an intrinsic part of the tree growth structure, a search is made at each node for the most significant split. In this sense it resembles a stepwise procedure rather than a best subsets method. At each stage it attempts to extract the most salient information from the part of the space it is working on.

5. It gives, with no additional effort, not only a classification, but also an estimate of the misclassification probability for the object.

For example, the chlorine classification algorithm outputs not only the predicted classification but also the value $r(t)$ from the terminal node containing the mass spectra of the unknown compound. This is useful in that if $r(t)$, for example, has the value .40 (its maximum is .50 in a two class problem), this indicates that the classification is highly uncertain and that an opinion from a trained mass spectra analyst may be desirable. However, an $r(t)$ value of .03 is quite reassuring. (Actually, the $r(t)$ values given are not the resubstitution estimates, but were estimated using an independent test set. See Chapter 7.)

6. In a standard data structure it is invariant under all monotone transformations of individual ordered variables.

For instance, if x_1 is an ordered variable in a standard problem and the transformation $x_1' = x_1^3$ is made, the optimal split will be the same regardless of whether x_1 or x_1' is used. If the optimal split on x_1 is $x_1 \leq 3$, then the optimal split using x_1' will be $x_1' \leq 27$, and the same cases will go left and right.

7. It is extremely robust with respect to outliers and misclassified points in $\mathcal{L}$.

In the measurement space X, trees have a robustness property similar to medians. A single data case has weight 1 among N data cases. In evaluating any split, one essentially counts how many cases of each class go right and go left.

Another type of frequently occurring error is mislabeling of a few cases in the learning set. This can have a disastrous effect on linear discrimination, as illustrated by Figure 2.19.

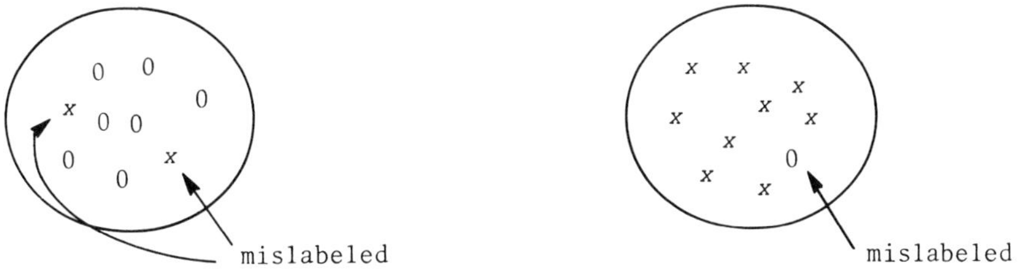

FIGURE 2.19

The tree structured method again weighs each point only as one among N and is not appreciably affected by a few mislabeled points.

8. The tree procedure output gives easily understood and interpreted information regarding the predictive structure of the data.

The tree structured methods have been used in a variety of applications in collaboration with nonstatistically oriented chemists, doctors, meteorologists, physicists, etc. Their reaction, almost universally, has been that the tree classifier provides an illuminating and natural way of understanding the structure of the problem.

3

RIGHT SIZED TREES AND HONEST ESTIMATES

3.1 INTRODUCTION

This chapter is concerned with two main issues: getting the right sized tree T and getting more accurate estimates of the true probability of misclassification or of the true expected misclassification cost $R^*(T)$.

The stepwise tree structure does an optimization at each step over a large number of possible splits of the data. If only resubstitution estimates are used, the usual results are too much splitting, trees that are much larger than the data warrant, and a resubstitution estimate $R(T)$ that is biased downward.

For instance, if the splitting is carried out to the point where each terminal node contains only one data case, then each node is classified by the case it contains, and the resubstitution estimate gives zero misclassification rate.

In general, more splits result in lower values of the resubstitution estimate $R(T)$. In this respect, the tree procedure is similar to stepwise linear regression, in which the estimated R^2 increases with each variable entered, encouraging the entry of variables that have no predictive power when tested on independent samples drawn from the same distribution.

In fact, stepwise regression simulations have shown that past a certain point, entry of additional variables will cause the *true* R^2 to decrease. The situation is similar with trees. Too large a tree will have a higher true misclassification rate than the right sized tree.

On the other hand, too small a tree will not use some of the classification information available in $\mathcal{L}$, again resulting in a higher true misclassification rate than the right sized tree.

These problems are illustrated in Table 3.1 with some output from the digit recognition with noise variables example. Different sized trees were grown. $R(T)$ was computed for each. An independent sample of size 5000 was run down each tree to give a more accurate estimate of $R^*(T)$. This estimate is denoted $R^{ts}(T)$.

TABLE 3.1

No. Terminal Nodes	$R(T)$	$R^{ts}(T)$
71	.00	.42
63	.00	.40
58	.03	.39
40	.10	.32
34	.12	.32
19	.20	.31
10	.29	.30
9	.32	.34
7	.41	.47
6	.46	.54
5	.53	.61
2	.75	.82
1	.86	.91

Notice two important features in Table 3.1:

1. The estimate $R(T)$ becomes increasingly less accurate as the trees grow larger.

2. The estimates R^{ts} indicate that as the trees initially decrease in size, the true misclassification rate decreases. Then it hits a minimum at the tree with 10 terminal nodes and begins to climb again as the trees get too small.

(The standard error of R^{ts} was estimated as around .007.)

The selection of overly large trees and the use of the inaccurate resubstitution estimate have led to much of the past criticism of tree structured procedures.

Trees were produced that seemed to indicate apparent data structure, but the extent to which the splits were "informative" was questionable. Work was centered on finding appropriate *stopping rules*, that is, on finding a criterion for declaring a node terminal. For example, recall that an early stopping rule consisted of setting a threshold β and deciding not to split a node if the maximum decrease in impurity was less than β, that is, if

$$\max_{\mathcal{s}} \Delta I(\mathcal{s}, t) < \beta. \tag{3.1}$$

As it became clear that this rule produced generally unsatisfactory results, other variants were invented and tested. None were generally acceptable. The problem has two aspects, which can be illustrated by rule (3.1). If β is set too low, then there is too much splitting and the tree is too large. Increasing β leads to the following difficulty: There may be nodes t such that $\max_{\mathcal{s}} \Delta I(\mathcal{s}, t)$ is small. But the descendant nodes t_L, t_R of t may have splits with large decreases in impurity. By declaring t terminal, one loses the good splits on t_L or t_R.

Finally, the conclusion was reached that looking for the right stopping rule was the wrong way of looking at the problem. A more satisfactory procedure was found that consisted of two key elements:

1. Prune instead of stopping. Grow a tree that is much too large and prune it upward in the "right way" until you finally cut back to the root node.
2. Use more accurate estimates of $R^*(T)$ to select the right sized tree from among the pruned subtrees.

This new framework leads to two immediate questions: How does one prune upward in the "right way," and how can better estimates of $R^*(T)$ be gotten?

Sections 3.2 and 3.3 discuss a method of pruning upward. The selection of a method for honestly estimating $R^*(T)$ depends on the sample size available. With large sample size, use of an independent test sample is most economical (Section 3.4.1). With small sample sizes, the preferred method, cross-validation (Section 3.4.2), necessitates the growing of auxiliary trees. (See Mabbett, Stone, and Washbrook, 1980, for another way of using cross-validation to select a classification tree.)

It is important to gauge the accuracy of the estimates of $R^*(T)$. Section 3.4.3 gives a brief discussion of approximate standard error formulas for the estimates. Chapter 11 contains a more complete discussion and the derivations.

Combining pruning and honest estimation produces trees that, in our simulated examples, have always been close to the optimal size and produces estimates with satisfactory accuracy. Section 3.4 contains some illustrations and also discusses the effect on accuracy of increasing or decreasing the number of cross-validation trees.

Characteristically, as a tree is pruned upward, the estimated misclassification rate first decreases slowly, reaches a gradual minimum, and then increases rapidly as the number of terminal nodes becomes small. This behavior can be attributed to a tradeoff between bias and variance and is explored in the appendix to this chapter.

3.2 GETTING READY TO PRUNE

To fix the notation, recall that the resubstitution estimate for the overall misclassification cost $R^*(t)$ is given by

$$R(T) = \sum_{t \in \tilde{T}} r(t)p(t)$$
$$= \sum_{t \in \tilde{T}} R(t).$$

We refer to $R(T)$ and $R(t)$ as the tree and node *misclassification costs*.

The first step is to grow a very large tree T_{max} by letting the splitting procedure continue until all terminal nodes are either small or pure or contain only identical measurement vectors. Here, pure means that the node cases are all in one class. With unlimited computer time, the best way of growing this initial tree would be to continue splitting until each terminal node contained exactly one sample case.

The size of the initial tree is not critical as long as it is large enough. Whether one starts with the largest possible tree T'_{max} or with a smaller, but still sufficiently large tree T_{max}, the pruning process will produce the same subtrees in the following sense: If the pruning process starting with T'_{max} produces a subtree contained in T_{max}, then the pruning process starting with T_{max} will produce exactly the same subtree.

The compromise method adopted for growing a sufficiently large initial tree T_{max} specifies a number N_{min} and continues splitting until each terminal node either is pure or satisfies $N(t) \leq N_{min}$ or contains only identical measurement vectors. Generally, N_{min} has been set at 5, occasionally at 1.

Starting with the large tree T_{max} and selectively pruning upward produces a sequence of subtrees of T_{max} eventually collapsing to the tree $\{t_1\}$ consisting of the root node.

To define the pruning process more precisely, call a node t' lower down on the tree a *descendant* of a higher node t if there is a connected path down the tree leading from t to t'. Then also t is called an *ancestor* of t'. Thus, in Figure 3.1a, t_4, t_5, t_8, t_9, t_{10} and t_{11} are all descendants of t_2, but not t_6 and t_7. Similarly, t_4, t_2, and t_1 are ancestors of t_9, but t_3 is not an ancestor of t_9.

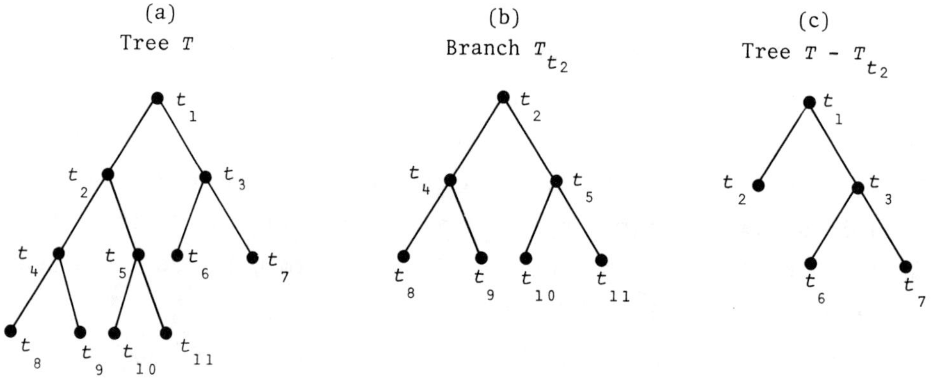

FIGURE 3.1

DEFINITION 3.2. *A branch T_t of T with root note $t \in T$ consists of the node t and all descendants of t in T.*

The branch T_{t_2} is illustrated in Figure 3.1b.

DEFINITION 3.3. *Pruning a branch T_t from a tree T consists of deleting from T all descendants of t, that is, cutting off all of T_t except its root node. The tree pruned this way will be denoted by $T - T_t$.*

The pruned tree $T - T_{t_2}$ is shown in Figure 3.1c.

DEFINITION 3.4. *If T' is gotten from T by successively pruning off branches, then T' is called a pruned subtree of T and denoted by $T' \prec T$. (Note that T' and T have the same root node.)*

Even for a moderate sized T_{max} containing, say, 30 to 40 nodes, there is an extremely large number of subtrees and an even larger number of distinct ways of pruning up to $\{t_1\}$. A "selective" pruning procedure is necessary, that is, a selection of a reasonable number of subtrees, decreasing in size, such that roughly speaking, each subtree selected is the "best" subtree in its size range.

The word *best* indicates the use of some criterion for judging how good a subtree T is. Even though $R(T)$ lacks accuracy as an estimate of $R^*(T)$, it is the most natural criterion to use in *comparing* different subtrees of the same size.

Regardless of how T_{max} was constructed, what splitting criterion was used, and so on, the selective pruning process starts with the given initial tree T_{max}, computes $R(t)$ for each node $t \in T_{max}$, and progressively prunes T_{max} upward to its root node such that at each stage of pruning, $R(T)$ is as small as possible.

Here is a simple example of such a selective pruning process. Suppose that T_{max} has L terminal nodes. Then construct a sequence of smaller and smaller trees

$$T_{max}, T_1, T_2, \ldots, \{t_1\}$$

as follows: For every value of H, $1 \leq H < L$. consider the class $\mathcal{T}_H$ of all subtrees of T_{max} having $L - H$ terminal nodes. *Select* T_H *as that subtree in* $\mathcal{T}_H$ *which minimizes* $R(T)$; that is,

$$R(T_H) = \min_{T \in \mathcal{T}_H} R(T).$$

Put another way, T_H is the minimal cost subtree having $L - H$ nodes.

This is an intuitively appealing procedure and can be efficiently implemented by a backward dynamic programming algorithm. However, it has some drawbacks. Perhaps the most important is that the sequence of subtrees is not nested, that is, T_{H+1} is not necessarily a subtree of T_H. As we go through the sequence, nodes may reappear that were previously cut off. In short, the sequence of subtrees is not formed by a progressive upward pruning.

Instead, we have adopted another selection method discussed in the next section. A preliminary version of this method was described in Breiman and Stone (1978).

3.3 MINIMAL COST-COMPLEXITY PRUNING

The idea behind minimal cost-complexity pruning is this:

DEFINITION 3.5. *For any subtree* $T \preccurlyeq T_{max}$, *define its complexity as* $|\widetilde{T}|$, *the number of terminal nodes in* T. *Let* $\alpha \geq 0$ *be a real number called the complexity parameter and define the cost-complexity measure* $R_\alpha(T)$ *as*

$$R_\alpha(T) = R(T) + \alpha|\widetilde{T}|.$$

Thus, $R_\alpha(T)$ is a linear combination of the cost of the tree and its complexity. If we think of α as the complexity cost per terminal node, $R_\alpha(T)$ is formed by adding to the misclassification cost of the tree a cost penalty for complexity.

Now, for each value of α, find that subtree $T(\alpha) \preccurlyeq T_{max}$ which minimizes $R_\alpha(T)$, i.e.,

$$R_\alpha\big(T(\alpha)\big) = \min_{T \preccurlyeq T_{max}} R_\alpha(T).$$

If α is small, the penalty for having a large number of terminal nodes is small and $T(\alpha)$ will be large. For instance, if T_{max} is so large that each terminal node contains only one case, then every case is classified correctly; $R(T_{max}) = 0$, so that T_{max} minimizes $R_0(T)$. As the penalty α per terminal node increases, the minimizing subtrees $T(\alpha)$ will have fewer terminal nodes. Finally, for α sufficiently large, the minimizing subtree $T(\alpha)$ will consist of the root node only, and the tree T_{max} will have been completely pruned.

Although α runs through a continuum of values, there are at most a finite number of subtrees of T_{max}. Thus, the pruning process produces a finite sequence of subtrees $T_1, T_2, T_3, \ldots$ with progressively fewer terminal nodes. Because of the finiteness, what happens is that if $T(\alpha)$ is the minimizing tree for a given value of α, then it continues to be minimizing as α increases until a jump point α' is reached, and a new tree $T(\alpha')$ becomes minimizing and continues to be the minimizer until the next jump point α''.

Though the pruning process is not difficult to describe, certain critical questions have been left open. For instance:

Is there a unique subtree $T \preceq T_{max}$ which minimizes $R_\alpha(T)$?

In the minimizing sequence of trees $T_1, T_2, \ldots$, is each subtree gotten by pruning upward from the previous subtree, i.e., does the nesting $T_1 \succ T_2 \succ \cdots \succ \{t_1\}$ hold?

More practically, perhaps the most important problem is that of finding an effective algorithm for implementing the pruning process. Clearly, a direct search through all possible subtrees to find the minimizer of $R_\alpha(T)$ is computationally expensive.

This section outlines the resolution of these problems. The uniqueness problem centers around an appropriate definition and a proof that the object defined really exists. The inclusion and effective implementation then both follow from a closer examination of the mechanism of minimal cost-complexity pruning.

Begin with:

DEFINITION 3.6. *The smallest minimizing subtree $T(\alpha)$ for complexity parameter α is defined by the conditions*

(i) $R_\alpha(T(\alpha)) = \min_{T \preceq T_{max}} R_\alpha(T)$

(ii) *If $R_\alpha(T) = R_\alpha(T(\alpha))$, then $T(\alpha) \preceq T$.*

This definition breaks ties in minimal cost-complexity by selecting the smallest minimizer of R_α. Obviously, if such a subtree exists, then it must be unique. The question is existence. For instance, suppose there are exactly two minimizers T, T' of R_α, but neither contains the other. Then $T(\alpha)$, as defined in (3.6), does not exist. However,

PROPOSITION 3.7. *For every value of* α, *there exists a smallest minimizing subtree as defined by (3.6).*

The proof of this is contained in Chapter 10 (Theorem 10.9). It is not difficult.

The jumping-off point for the pruning is not $T_{\max}$ but rather $T_1 = T(0)$. That is, T_1 is the smallest subtree of $T_{\max}$ satisfying

$$R(T_1) = R(T_{\max}).$$

To get T_1 from $T_{\max}$, let t_L, t_R be any two terminal nodes in $T_{\max}$ resulting from a split of the immediate ancestor node t. Recall from Proposition 2.14 that $R(t) \geq R(t_L) + R(t_R)$. If $R(t) = R(t_L) + R(t_R)$, then prune off t_L and t_R. Continue this process until no more pruning is possible. The resulting tree is T_1 (see Section 10.2 for the proof).

For T_t any branch of T_1, define $R(T_t)$ by

$$R(T_t) = \sum_{t' \in \tilde{T}_t} R(t'),$$

where $\tilde{T}_t$ is the set of terminal nodes of T_t. In Section 10.2 (Theorem 10.11) we show that the tree T_1 has the following property:

PROPOSITION 3.8. *For* t *any nonterminal node of* T_1,

$$R(t) > R(T_t).$$

Starting with T_1, the heart of minimal cost-complexity pruning lies in understanding that it works by *weakest-link cutting*. For any node $t \in T_1$, denote by $\{t\}$ the subbranch of T_t consisting of the single node $\{t\}$.

Set

$$R_\alpha(\{t\}) = R(t) + \alpha.$$

For any branch T_t, define

$$R_\alpha(T_t) = R(T_t) + \alpha|\widetilde{T}_t|.$$

As long as

$$R_\alpha(T_t) < R_\alpha(\{t\}),$$

the branch T_t has a smaller cost-complexity than the single node $\{t\}$. But at some critical value of α, the two cost-complexities become equal. At this point the subbranch $\{t\}$ is smaller than T_t, has the same cost-complexity, and is therefore preferable. To find this critical value of α, solve the inequality

$$R_\alpha(T_t) < R_\alpha(\{t\}),$$

getting

$$\alpha < \frac{R(t) - R(T_t)}{|\widetilde{T}_t| - 1} . \tag{3.9}$$

By (3.8) the critical value on the right of (3.9) is positive.

Define a function $g_1(t)$, $t \in T_1$, by

$$g_1(t) = \begin{cases} \dfrac{R(t) - R(T_t)}{|\widetilde{T}_t| - 1}, & t \notin \widetilde{T}_1 \\ +\infty, & t \in \widetilde{T}_1. \end{cases}$$

Then define the *weakest link* $\bar{t}_1$ *in* T_1 as the node such that

$$g_1(\bar{t}_1) = \min_{t \in T_1} g_1(t)$$

and put

$$\alpha_2 = g_1(\bar{t}_1).$$

The node $\bar{t}_1$ is the weakest link in the sense that as the parameter α increases, it is the first node such that $R_\alpha(\{t\})$ becomes equal to $R_\alpha(T_t)$. Then $\{\bar{t}_1\}$ becomes preferable to $T_{\bar{t}_1}$, and α_2 is the value of α at which equality occurs.

Define a new tree $T_2 \prec T_1$ by pruning away the branch $T_{\bar{t}_1}$, that is,

$$T_2 = T_1 - T_{\bar{t}_1}.$$

Now, *using T_2 instead of T_1, find the weakest link in T_2*. More precisely, letting T_{2t} be that part of the branch T_t which is contained in T_2, define

$$g_2(t) = \begin{cases} \dfrac{R(t) - R(T_{2t})}{|\widetilde{T}_{2t}| - 1}, & t \in T_2,\ t \notin \widetilde{T}_2 \\ +\infty, & t \in \widetilde{T}_2 \end{cases}$$

and $\bar{t}_2 \in T_2$, α_3 by

$$g_2(\bar{t}_2) = \min_{t \in T_2} g_2(t)$$
$$\alpha_3 = g_2(\bar{t}_2).$$

Repeat the procedure by defining

$$T_3 = T_2 - T_{\bar{t}_2}$$

and finding the weakest link $\bar{t}_3$ in T_3 and the corresponding parameter value α_4; now form T_4 and repeat again.

If at any stage there is a multiplicity of weakest links, for instance, if

$$g_k(\bar{t}_k) = g_k(\bar{t}'_k),$$

then define

$$T_{k+1} = T_k - T_{\bar{t}_k} - T_{\bar{t}'_k}.$$

Continuing this way, we get a decreasing sequence of subtrees

$$T_1 \succ T_2 \succ T_3 \succ \cdots \succ \{t_1\}.$$

The connection to minimal cost-complexity pruning is given by:

THEOREM 3.10. *The* $\{\alpha_k\}$ *are an increasing sequence, that is,* $\alpha_k < \alpha_{k+1}$, $k \geq 1$, *where* $\alpha_1 = 0$. *For* $k \geq 1$, $\alpha_k \leq \alpha < \alpha_{k+1}$, $T(\alpha) = T(\alpha_k) = T_k$.

This theorem spells out how minimal cost-complexity pruning works. It starts with T_1, finds the weakest-link branch $T_{\bar{t}_1}$, and prunes it to get T_2 when α reaches α_2. Now it finds the weakest-link branch $T_{\bar{t}_2}$ in T_2 and prunes it to get T_3 when α reaches α_3; and so on. These recursive pruning steps are computationally rapid and require only a small fraction of the total tree construction time. The proof of Theorem 3.10 is given in Section 10.2.

Starting with T_1, the algorithm initially tends to prune off large subbranches with many terminal nodes. As the trees get smaller, it tends to cut off fewer at a time. This is illustrated in Table 3.2 for the digit recognition with noise variables example.

TABLE 3.2

Tree	T_1	T_2	T_3	T_4	T_5	T_6	T_7	T_8	T_9	T_{10}	T_{11}	T_{12}	T_{13}
$\lvert\tilde{T}_k\rvert$	71	63	58	40	34	19	10	9	7	6	5	2	1

Finally, we remark that the sequence of minimal cost-complexity trees is a subsequence of the sequence of subtrees constructed by finding the minimum cost subtree for a given number of terminal nodes. For instance, if $T(\alpha)$ has seven terminal nodes, there is no other subtree T having seven terminal nodes with smaller $R(T)$. If so,

$$R_\alpha(T) = R(T) + 7\alpha < R_\alpha(T(\alpha)),$$

which, by definition, is impossible. (See Stone, 1981, for a brief survey of the statistical literature involving complexity costs.)

3.4 THE BEST PRUNED SUBTREE: AN ESTIMATION PROBLEM

The method of pruning discussed in the previous section results in a decreasing sequence of subtrees $T_1 \succ T_2 \succ \cdots \succ \{t_1\}$, where $T_k = T(\alpha_k)$, $\alpha_1 = 0$. The problem is now reduced to selecting one of these as the optimum-sized tree.

If the resubstitution estimate $R(T_k)$ is used as a criterion, the largest tree T_1 would be selected. But if one had an "honest" estimate $\hat{R}(T_k)$ of the misclassification cost, then the best subtree T_{k_0} could be defined as the subtree that minimizes $\hat{R}(T_k)$; i.e.,

$$\hat{R}(T_{k_0}) = \min_k \hat{R}(T_k). \tag{3.11}$$

The issue discussed in this section is the construction of relatively unbiased estimates of the true misclassification cost $R^*(T_k)$. Two methods of estimation are discussed: Use of an independent test sample and cross-validation. Of the two, use of an independent test sample is computationally more efficient and is preferred when the learning sample contains a large number of cases. As a useful by-product it gives relatively unbiased estimates of the node misclassification costs. Cross-validation is computationally more expensive, but makes more effective use of all cases and gives useful information regarding the stability of the tree structure.

To study the bias or standard error of an estimate, a probability model is necessary. Assume in this section the model used previously: The cases in $\mathcal{L}$ are N independent draws from the probability distribution $P(A, j)$ on $X \times C$, and $(\mathbf{X}, Y)$ is random with distribution $P(A, j)$, independent of $\mathcal{L}$. If there are no variable misclassification costs, recall that $R^*(d)$ is defined as

$$R^*(d) = P(d(\mathbf{X}) \neq Y).$$

In the general case, with variable misclassification costs $C(i|j)$,

DEFINITION 3.12. *Define*

(i) $Q^*(i|j) = P(d(\mathbf{X}) = i|Y = j)$

so that $Q^*(i|j)$ *is the probability that a case in* j *is classified into* i *by* d. *Define*

(ii) $R^*(j) = \sum_i C(i|j)Q^*(i|j)$

so that $R^*(j)$ *is the expected cost of misclassification for class* j *items. Define*

(iii) $R^*(d) = \sum_j R^*(j)\pi(j)$

as the expected misclassification cost for the classifier d.

Both test sample and cross-validation provide estimates of $Q^*(i|j)$ and $R^*(j)$, as well as $R^*(d)$. These are useful outputs of the tree program. The basic idea in both procedures is that $Q^*(i|j)$ can be estimated using simple counts of cases misclassified. Then $R^*(j)$, $R^*(T_k)$ are estimated through Definitions 3.12(ii) and (iii). Furthermore, standard errors can be computed by assuming a simple binomial model for the estimate of $Q^*(i|j)$.

3.4.1 *Test Sample Estimates*

Select a fixed number $N^{(2)}$ of cases at random from $\mathcal{L}$ to form the test sample $\mathcal{L}_2$. The remainder $\mathcal{L}_1$ form the new learning sample.

The tree $T_{\max}$ is grown using only $\mathcal{L}_1$ and pruned upward to give the sequence $T_1 > T_2 > \cdots > \{t_1\}$. That is, the $\{T_k\}$ sequence of trees is constructed and the terminal nodes assigned a classification without ever seeing any of the cases in $\mathcal{L}_2$.

Now take the cases in $\mathcal{L}_2$ and drop them through T_1. Each tree T_k assigns a predicted classification to each case in $\mathcal{L}_2$. Since the true class of each case in $\mathcal{L}_2$ is known, the misclassification cost of T_k operating on $\mathcal{L}_2$ can be computed. This produces the estimate $R^{ts}(T_k)$.

In more detail, denote by $N_j^{(2)}$ the number of class j cases in $\mathcal{L}_2$. For T any one of the trees T_1, T_2, ..., take $N_{ij}^{(2)}$ to be the number of class j cases in $\mathcal{L}_2$ whose predicted classification by T is class i.

The basic estimate is gotten by setting

$$Q^{ts}(i|j) = N_{ij}^{(2)}/N_j^{(2)}.$$

That is, $Q^*(i|j)$ is estimated as the proportion of test sample class j cases that the tree T classifies as i (set $Q^{ts}(i|j) = 0$ if $N_j^{(2)} = 0$).

Using Definition 3.12(ii) gives the estimate

$$R^{ts}(j) = \sum_i C(i|j)Q^{ts}(i|j).$$

For the priors $\{\pi(j)\}$ either given or estimated, Definition 3.12 (iii) indicates the estimate

$$R^{ts}(T) = \sum_j R^{ts}(j)\pi(j). \tag{3.13}$$

If the priors are data estimated, use $\mathcal{L}_2$ to estimate them as $\pi(j) = N_j^{(2)}/N^{(2)}$. In this case, (3.13) simplifies to

$$R^{ts}(T) = \frac{1}{N^{(2)}} \sum_{i,j} C(i|j)N_{ij}^{(2)}. \tag{3.14}$$

This last expression (3.14) has a simple interpretation. Compute the misclassification cost for every case in $\mathcal{L}_2$ dropped through T and then take the average.

In the unit cost case, $R^{ts}(j)$ is the proportion of class j test cases misclassified, and with estimated priors $R^{ts}(T)$ is the total proportion of test cases misclassified by T.

Using the assumed probability model, it is easy to show that the estimates $Q^{ts}(i|j)$ are biased only if $N_j^{(2)} = 0$. For any reasonable distribution of sample sizes, the probability that $N_j^{(2)} = 0$ is so small that these estimates may be taken as unbiased. In consequence, so are the estimators $R^{ts}(T)$. In fact, in the estimated prior case, there is cancellation and $R^{ts}(T)$ is exactly unbiased.

The test sample estimates can be used to select the right sized tree T_{k_0} by the rule

$$R^{ts}(T_{k_0}) = \min_k R^{ts}(T_k).$$

In a later section, a slight modification of this rule is suggested. After selection of T_{k_0}, $R^{ts}(T_{k_0})$ is used as an estimate of its expected misclassification cost.

3.4.2 *Cross-Validation Estimates*

Unless the sample size in $\mathcal{L}$ is quite large, cross-validation is the preferred estimation method. In fact, the only time we have used test sample estimation recently is in the various mass spectra projects where the minimum sample size in any class was about 900. (However, see Section 6.2, where a different kind of test sample procedure is used in a heart attack diagnosis project.)

In V-fold cross-validation, the original learning sample $\mathcal{L}$ is divided by random selection into V subsets, $\mathcal{L}_v$, $v = 1, \ldots, V$, each containing the same number of cases (as nearly as possible). The vth learning sample is

$$\mathcal{L}^{(v)} = \mathcal{L} - \mathcal{L}_v, \quad v = 1, \ldots, V,$$

so that $\mathcal{L}^{(v)}$ contains the fraction $(V - 1)/V$ of the total data cases. Think of V as being reasonably large. Usually V is taken as 10, so each learning sample $\mathcal{L}^{(v)}$ contains 9/10 of the cases.

In V-fold cross-validation, V auxiliary trees are grown together with the main tree grown on $\mathcal{L}$. The vth auxiliary tree is grown using the learning sample $\mathcal{L}^{(v)}$. Start by growing V overly large trees $T^{(v)}_{\max}$, $v = 1, \ldots, V$, as well as $T_{\max}$, using the criterion that the splitting continues until nodes are pure or have fewer cases than $N_{\min}$.

For each value of the complexity parameter α, let $T(\alpha)$, $T^{(v)}(\alpha)$, $v = 1, \ldots, V$, be the corresponding minimal cost-complexity subtrees of T_{max}, $T^{(v)}_{max}$. For each v, the trees $T^{(v)}_{max}$, $T^{(v)}(\alpha)$ have been constructed without ever seeing the cases in $\mathcal{L}_v$. Thus, the cases in $\mathcal{L}_v$ can serve as an independent test sample for the tree $T^{(v)}(\alpha)$.

Put $\mathcal{L}_v$ down the tree $T^{(v)}_{max}$, $v = 1, \ldots, V$. Fix the value of the complexity parameter α. For every value of v, i, j, define

$$N^{(v)}_{ij} = \text{the number of class } j \text{ cases in } \mathcal{L}_v \text{ classified as } i \text{ by } T^{(v)}(\alpha),$$

and set

$$N_{ij} = \sum_v N^{(v)}_{ij},$$

so N_{ij} is the total number of class j test cases classified as i. Each case in $\mathcal{L}$ appears in one and only one test sample $\mathcal{L}_v$. Therefore, the total number of class j cases in all test samples is N_j, the number of class j cases in $\mathcal{L}$.

The idea now is that for V large, $T^{(v)}(\alpha)$ should have about the same classification accuracy as $T(\alpha)$. Hence, we make the fundamental step of estimating $Q^*(i|j)$ for $T(\alpha)$ as

$$Q^{CV}(i|j) = N_{ij}/N_j. \tag{3.15}$$

For the $\{\pi(j)\}$ given or estimated, set

$$R^{CV}(j) = \sum_i C(i|j)Q^{CV}(i|j)$$

and put

$$R^{CV}(T(\alpha)) = \sum_j R^{CV}(j)\pi(j). \tag{3.16}$$

If the priors are data estimated, set $\pi(j) = N_j/N$. Then (3.16) becomes

$$R^{CV}(T(\alpha)) = \frac{1}{N} \sum_{i,j} C(i|j)N_{ij}. \tag{3.17}$$

In the unit cost case, (3.17) is the proportion of test set cases misclassified.

The implementation is simplified by the fact that although α may vary continuously, the minimal cost-complexity trees *grown on* $\mathcal{L}$ are equal to T_k for $\alpha_k \leq \alpha < \alpha_{k+1}$. Put

$$\alpha'_k = \sqrt{\alpha_k \alpha_{k+1}}$$

so that α'_k is the geometric midpoint of the interval such that $T(\alpha) = T_k$. Then put

$$R^{CV}(T_k) = R^{CV}(T(\alpha'_k)),$$

where the right-hand side is defined by (3.16). That is, $R^{CV}(T_k)$ is the estimate gotten by putting the test samples $\mathcal{L}_v$ through the trees $T^{(v)}(\alpha'_k)$. For the root node tree $\{t_1\}$, $R^{CV}(\{t_1\})$ is set equal to the resubstitution cost $R(\{t_1\})$.

Now the rule for selecting the right sized tree (modified in Section 3.4.3) is: Select the tree T_{k_0} such that

$$R^{CV}(T_{k_0}) = \min_k R^{CV}(T_k).$$

Then use $R^{CV}(T_{k_0})$ as an estimate of the misclassification cost.

Questions concerning the exact bias of R^{CV} are difficult to settle, since the probabilistic formulation is very complex. One point is fairly clear. Cross-validation estimates misclassification costs for trees grown on a fraction $(V - 1)/V$ of the data. Since the auxiliary cross-validation trees are grown on smaller samples, they tend to be less accurate. The cross-validation estimates, then, tend to be conservative in the direction of overestimating misclassification costs.

In all of our simulated work, the cross-validation estimates have selected trees that were close to optimal in terms of minimizing $R^*(T_k)$. Some examples will be looked at in Section 3.5.

An important practical question is how large to take V. The major computational burden in tree construction is growing the initial large tree. Computing time in cross-validation goes up about linearly in V. But also, the larger V, the more accurate the estimates R^{CV} should be. This issue is explored in Section 3.5.3 through the use of simulated examples.

3.4.3 *Standard Errors and the 1 SE Rule*

It is a good statistical practice to gauge the uncertainties in the estimates $R^{ts}(T)$ and $R^{CV}(T)$ by estimating their standard errors. Chapter 11 derives expressions for these standard errors as well as for the standard errors in the estimates of $Q^*(i|j)$ and $R^*(j)$. The derivations are standard statistics for the test sample estimates, but are heuristic when cross-validation is used.

As an illustration, we derive the expression for the standard error in $R^{ts}(T)$ when the priors are estimated from the data in the unit cost case.

The learning sample $\mathcal{L}_1$ is used to construct T. Take the test sample $\mathcal{L}_2$ to be independently drawn from the same underlying distribution as $\mathcal{L}_1$ but independent of $\mathcal{L}_1$. The estimate $R^{ts}(T)$ is the proportion of cases in $\mathcal{L}_2$ misclassified by T (see Section 3.4.1).

Now drop the N_2 cases in $\mathcal{L}_2$ through T. The probability p^* that any single case is misclassified is $R^*(T)$. Thus, we have the binomial situation of N_2 independent trials with probability p^* of success at each trial, where p^* is estimated as the proportion p of successes. Clearly, $Ep = p^*$, so p is unbiased. Further,

$$\text{Var}(p) = p^*(1 - p^*)/N_2.$$

Thus, the standard error of the estimate p is estimated by $[p(1 - p)/N_2]^{1/2}$, leading to the standard error estimate for $R^{ts}(T)$ as $\text{SE}(R^{ts}(T)) = [R^{ts}(T)(1 - R^{ts}(T))/N_2]^{1/2}$.

In most of the examples we have worked with, the estimates $\hat{R}(T_k)$ ($R^{ts}(T_k)$ or $R^{CV}(T_k)$) as a function of the number of terminal nodes $|\tilde{T}_k|$ look something like Figure 3.2. The characteristics are

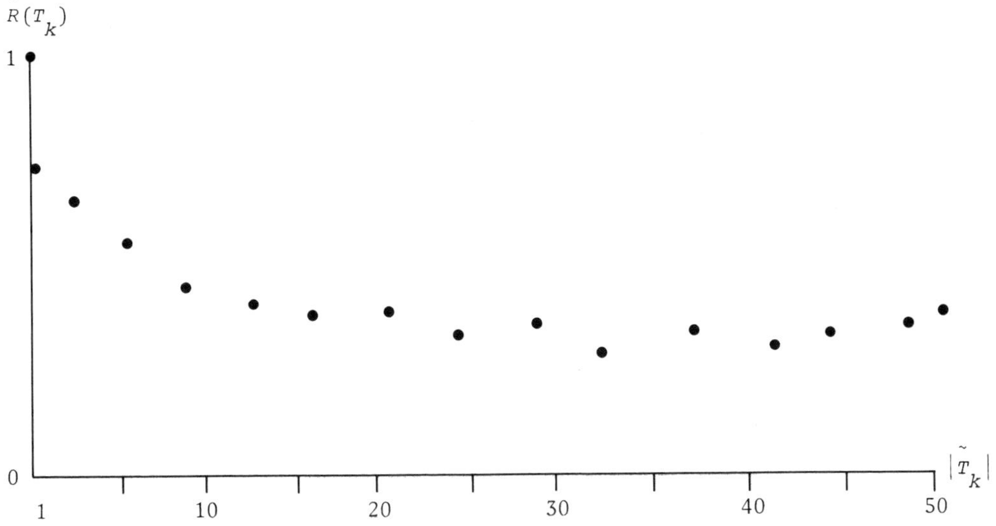

FIGURE 3.2

a fairly rapid initial decrease followed by a long, flat valley and then a gradual increase for larger $|\tilde{T}_k|$. The minimum occurs somewhere in the long, flat valley region, where $\hat{R}(T_k)$ is almost constant except for up-down changes well within the ±1 SE range.

The position of the minimum of $\hat{R}(T_k)$ within this valley may be unstable. Small changes in parameter values or even in the seed of the random number generator used to separate $\mathcal{L}$ into V test sets may cause large changes in $|\tilde{T}_k|$ for the tree that minimizes $\hat{R}$.

The 1 SE rule for selecting the right sized tree was created to

1. Reduce the instability noted here
2. Choose the simplest tree whose accuracy is comparable to $\min_k \hat{R}(T_k)$

When (2) is interpreted as being within one standard error, the resulting rule is

DEFINITION 3.18. *Define* T_{k_0} *by*

$$\hat{R}(T_{k_0}) = \min_k \hat{R}(T_k).$$

Then the tree selected is T_{k_1}*, where* k_1 *is the* maximum k *satisfying*

$$\hat{R}(T_{k_1}) \leq \hat{R}(T_{k_0}) + \text{SE}(\hat{R}(T_{k_0})).$$

This is called the 1 SE rule.

Although the 1 SE rule is consistently used on the simulated examples in this book, the analyst has the liberty of inspecting the sequence of trees, T_1, T_2, ... and the corresponding estimates $\hat{R}(T_1)$, $\hat{R}(T_2)$, ... The context of the problem and prior knowledge of the variables involved may suggest an alternative selection. This is illustrated in Chapter 6, and a different selection rule is mentioned in Section 11.6.

3.4.4 *Other Estimation Issues*

There are two minor issues concerning the estimation of $R^*(T)$ that are more or less unresolved.

The first is this: Recall that in the selection of the test sample a fraction $f = N_2/N$ was selected at random from $\mathcal{L}$. The exact proportions of the classes in $\mathcal{L}_2$ are left to a random device. An alternative strategy is to select a fixed fraction f_j from the N_j cases in class j in $\mathcal{L}$. The question is: When is the latter strategy preferable to the former?

Similarly, in V-fold cross-validation, $\mathcal{L}$ is randomly divided into V nearly equal sets $\mathcal{L}_v$ of cases without regard for class membership. When is it better to constrain the random selection such that each class is equally spread among the $\mathcal{L}_v$, $v = 1, \ldots, V$?

Our most recent and tentative thinking, based on a simplified analytic model and simulation results, is that constructing the test sample (or samples) to contain fixed fractions of the classes may produce more accurate estimates. Chapter 8 contains further discussion of this problem in the regression context.

Another problem is that when the tree T_{k_0} is selected by the rule

$$\hat{R}(T_{k_0}) = \min_k \hat{R}(T_k),$$

where ($\hat{R} = R^{ts}$ or R^{cv}), then $\hat{R}(T_{k_0})$ is used as the estimate of $R^*(T_{k_0})$. The potential difficulty here is reminiscent of the problem with the resubstitution estimate $R(T)$. When a procedure tries to minimize some criterion, then using the minimum value of the criterion as an estimate of performance can lead to biased results.

For this reason, suggestions have been made that once k_0 has been selected, another cross-validation be carried out to get a less biased estimate of $R^*(T_{k_0})$. However, the outputs of the many simulated examples that have been run do not show a tendency for $\hat{R}(T_{k_0})$ to be biased low. Further, since an extra stage of cross-validation would significantly increase the computational burden, the suggestion has not been adopted.

The bootstrap method, due to Efron (1979), was also considered as a possible way of estimating the bias in the resubstitution estimate of $R^*(T)$. As it turns out, this method does not always work well for tree structured classifiers. Section 11.7 has a discussion of this issue.

3.5 SOME EXAMPLES

In this section we focus on the performance of cost-complexity pruning combined with cross-validation estimation. Test sample estimates using 5000 additional samples from the same distribution are used to get accurate estimates of $R^*(T)$. In the two examples discussed earlier, the estimated SE's of the test sample estimates are always less than .007. However, test samples are not explored as an internal method of estimation. There are two reasons for this. First, the distribution of test sample estimates can be explicitly derived from the probability model, but the distribution of cross-validation estimates is complex and unknown. The second reason is that in applications, cross-validation has been used much more frequently than test sample estimation.

3.5.1 Digit Recognition

In the digit recognition problem described in Section 2.6, ten-fold cross-validation was used and the priors estimated from the data. The tree T_{max} was grown with $N_{min} = 1$. For the sequence of trees resulting from cost-complexity pruning, Table 3.3 gives the corresponding values of the resubstitution estimate $R(T_k)$, the number of terminal nodes $|\widetilde{T}_k|$, the cross-validation estimate $R^{CV}(T_k)$, and the estimate $R^{ts}(T_k)$ gotten from the 5000-case test sample. The plus-minus values on the cross-validation estimates are ±SE.

TABLE 3.3

k	$\lvert\widetilde{T}_k\rvert$	$R(T_k)$	$R^{CV}(T_k)$	$R^{ts}(T_k)$
1	31	.17	.30 ± .03	.29
2**	23	.19	.27 ± .03	.29
3	17	.22	.30 ± .03	.29
4	15	.23	.30 ± .03	.28
5	14	.24	.31 ± .03	.28
6*	10	.29	.30 ± .03	.30
7	9	.32	.41 ± .04	.34
8	7	.41	.51 ± .04	.47
9	6	.46	.53 ± .04	.54
10	5	.53	.61 ± .04	.61
11	2	.75	.75 ± .03	.82
12	1	.86	.86 ± .03	.90

Notice that

1. The minimum R^{CV} tree is T_2 with 23 terminal nodes (indicated by **). The 1 SE rule selects T_6 with 10 terminal nodes (indicated by *).

2. In 9 of the 12 trees, R^{ts} is in the range R^{CV} ± SE. In 11 trees, R^{ts} is in the R^{CV} ± 2 SE range.
3. The estimates R^{CV} are higher than R^{ts} for the larger trees (except for T_2), are about equal for T_9 and T_{10}, and are low for the two smallest trees T_{11} and T_{12}.

In the digit recognition problem with 17 noise variables added and the priors estimated from data, cost-complexity pruning and tenfold cross-validation produced the results shown in Table 3.4. Recall that N = 200. To grow $T_{\max}$, $N_{\min}$ = 1 was used.

TABLE 3.4

k	$\lvert\widetilde{T}_k\rvert$	$R(T_k)$	$R^{CV}(T_k)$	$R^{ts}(T_k)$
1	71	.00	.46 ± .04	.42
2	63	.00	.45 ± .04	.40
3	58	.04	.43 ± .04	.39
4	40	.10	.38 ± .03	.32
5	34	.12	.38 ± .03	.32
6	19	.20	.32 ± .03	.31
7*,**	10	.29	.31 ± .03	.30
8	9	.32	.39 ± .03	.34
9	7	.41	.47 ± .04	.47
10	6	.46	.53 ± .04	.54
11	5	.53	.64 ± .03	.61
12	2	.75	.78 ± .03	.82
13	1	.86	.86 ± .03	.91

The results differ slightly from the first example.

1. Tree T_7 is selected by both the minimum R^{CV} and the 1 SE rule. It also minimizes R^{ts}.
2. Only 5 of the 13 trees have R^{ts} in the R^{CV} ± SE range. But all 13 are in the R^{CV} ± 2 SE range.

3. The R^{cv} estimates are consistently high for the larger trees, about equal to R^{ts} in the midrange (T_6-T_{11}), and low for the two smallest trees.

As another check on the cross-validation estimates, four data sets were generated replicating this example but each with a different random number seed. The 1 SE trees were selected for each and a test sample of 5000 used to get the R^{ts} estimates. The results appear in Table 3.5.

TABLE 3.5

R^{cv}	R^{ts}
.34 ± .03	.31
.31 ± .03	.30
.36 ± .03	.30
.33 ± .03	.31

3.5.2 Waveform Classification

For the waveform problem, with N = 300 and N_{min} = 1, the results are given in Table 3.6.

Tree T_4 is the minimizing tree, and tree T_6 is the 1 SE tree. These two trees also have the lowest R^{ts}. The cross-validation estimates are consistently above the test sample estimates. In 8 out of 12 trees, R^{ts} is in the range R^{cv} ± SE. In 3 of the remaining cases, it is in the ±2 SE range. Note that in all three examples, the most marked lack of accuracy is in the very small trees. This phenomenon is discussed in the regression context in Section 8.7.

Again four replicate data sets were generated using different seeds, the 1 SE trees grown, and 5000 cases used to get R^{ts}. The results are given in Table 3.7.

TABLE 3.6

k	$\lvert\widetilde{T}_k\rvert$	$R(T_k)$	$R^{cv}(T_k)$	$R^{ts}(T_k)$
1	41	.00	.31 ± .03	.31
2	39	.00	.31 ± .03	.31
3	23	.06	.30 ± .03	.29
4*	21	.07	.28 ± .03	.28
5	14	.11	.29 ± .03	.29
6**	11	.14	.29 ± .03	.28
7	9	.17	.31 ± .03	.29
8	7	.20	.35 ± .03	.30
9	5	.23	.35 ± .03	.33
10	3	.28	.39 ± .03	.36
11	2	.41	.52 ± .03	.44
12	1	.62	.62 ± .03	.66

TABLE 3.7

R^{cv}	R^{ts}
.32 ± .03	.29
.32 ± .03	.34
.29 ± .03	.33
.25 ± .03	.29

3.5.3 *How Many Folds in the Cross-Validation?*

In all the simulation examples we have run, taking $V = 10$ gave adequate accuracy. In some examples, smaller values of V also gave sufficient accuracy. But we have not come across any situations where taking V larger than 10 gave a significant improvement in accuracy for the tree selected.

This is illustrated in Tables 3.8 and 3.9. The waveform recognition example and the digit recognition example were run using $V = 2, 5, 10, 25$. A test sample of size 5000 was used to give the estimates R^{ts}.

TABLE 3.8 Waveform Recognition

k	$\|\tilde{T}\|$	R^{ts}	R^{CV} $V = 2$	$V = 5$	$V = 10$	$V = 25$
1	41	.31	.34	.33	.31	.32
2	39	.31	.34	.33	.31	.31
3	23	.29	.35	.33	.30	.29
4	21	.28	.35	.32	.28	.28
5	14	.29	.33	.33	.29	.30
6	11	.28	.33	.34	.29*	.31
7	9	.29	.33	.33	.31	.30*
8	7	.30	.31	.33*	.35	.34
9	5	.33	.30*	.36	.35	.34
10	3	.36	.33	.39	.39	.36
11	2	.44	.41	.51	.52	.47
12	1	.66	.62	.62	.62	.62

TABLE 3.9 Digit Recognition

k	$\|\tilde{T}\|$	R^{ts}	R^{CV} $V = 2$	$V = 5$	$V = 10$	$V = 25$
1	31	.29	.33	.31	.30	.27
2	23	.29	.34	.28	.27	.26
3	17	.29	.34	.28	.30	.29
4	15	.28	.34	.31	.30	.29
5	14	.28	.36	.31	.31	.32
6	10	.30	.36*	.31*	.30*	.30*
7	9	.34	.38	.38	.41	.36
8	7	.47	.47	.48	.51	.49
9	6	.54	.51	.52	.53	.49
10	5	.61	.51	.71	.61	.61
11	2	.82	.75	.84	.75	.75
12	1	.91	.86	.86	.86	.86

In the waveform example, $V = 25$ gives a significantly better estimate than $V = 10$ only for the small trees T_9 and T_{10}. For tree T_6 on up, $V = 5$ and $V = 2$ give estimates that are too large. The starred entries indicate the trees selected by the 1 SE rule. It seems clear that $V = 10$ does as well as $V = 25$ and that at $V = 5$ and $V = 2$, performance is degraded.

The situation in the digit recognition problem differs a bit. The twofold cross-validation gives less accuracy. But for $V = 5$, 10, 25, the accuracy is roughly comparable.

APPENDIX

Heuristics of Bias Versus Variance

In those examples where $|\widetilde{T}_1|$ is large, when the cross-validated or test sample estimates $\hat{R}(T_k)$ are graphed as a function of $|\widetilde{T}_k|$, similar-shaped curves result. A typical graph, including the resubstitution estimate $R(T_k)$, is shown in Figure 3.3.

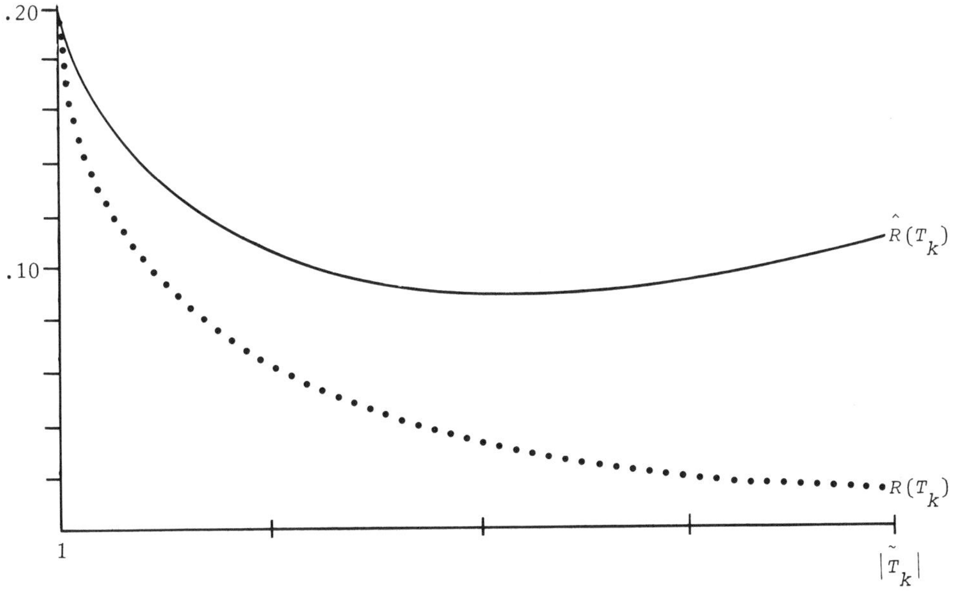

FIGURE 3.3

The graph of $\hat{R}(T_k)$ starts high for $|\tilde{T}_k| = 1$, decreases as $|\tilde{T}_k|$ increases, reaches a shallow minimum region, and then increases slowly to the misclassification rate corresponding to the largest tree T_1. Another feature is that $\hat{R}(T_1)$ is invariably less than twice the minimum misclassification rate $\min_k \hat{R}(T_k)$.

The remainder of this appendix is a heuristic attempt to understand the mechanism leading to the preceding characteristics. The discussion is not a part of the methodological development, and readers can, without loss of generality, skip to the next chapter.

The tree structured procedure attempts to fit the classification surface by a surface that is constant over multidimensional rectangles. When these rectangles are too large, that is, too little splitting and small $|\tilde{T}_k|$, the fit is poor. We refer to this lack of fitting of the surface as *bias*. When $|\tilde{T}_k|$ is large and the space X is split into many small rectangles, the bias is small. On the other hand, these small rectangles are more likely to have a plurality of the wrong class. This latter type of error is referred to as *variance*,

Since the trade-off between bias and variance is an important characteristic of the tree structure, we illustrate it by a simple model. Take a two-class situation with priors $\pi(1)$, $\pi(2)$ and M-dimensional data with class j sampled from the density $f_j(\mathbf{x})$, j = 1, 2. The Bayes optimal misclassification rate is

$$R^* = 1 - \int \max_j \left(\pi(j) f_j(\mathbf{x})\right) d\mathbf{x}.$$

With the space X divided up into L rectangles, $S_1, \ldots, S_L$, let the classification assigned to S_ℓ by $\mathcal{L}$ be denoted by y_ℓ. Take $(\mathbf{X}, Y)$ to be independent of $\mathcal{L}$ with the same distribution. By definition, the true misclassification rate $R^*(L)$ for the partition $S_1, \ldots, S_L$ is

$$R^*(L) = \sum_\ell P(\mathbf{X} \in s_\ell,\ Y \neq y_\ell) \\ = 1 - \sum_\ell P(\mathbf{X} \in s_\ell,\ Y = y_\ell). \tag{A.1}$$

Using $\chi(\ )$ as the indicator function, (A.1) becomes

$$R^*(L) = 1 - \sum_{\ell,j} \chi(y_\ell = j) P(\mathbf{X} \in s_\ell,\ Y = j). \tag{A.2}$$

Define

$$y_\ell^* = \begin{cases} 1, & P(\mathbf{X} \in s_\ell,\ Y = 1) > P(\mathbf{X} \in s_\ell,\ Y = 2) \\ 2 & \text{otherwise.} \end{cases}$$

Then, (A.2) can be written as

$$R^*(L) = 1 - \sum_\ell \max_j P(\mathbf{X} \in s_\ell,\ j) \\ + \sum_\ell \chi(y_\ell \neq y_\ell^*) \left| P(\mathbf{X} \in s_\ell,\ 1) - P(\mathbf{X} \in s_\ell,\ 2) \right|, \tag{A.3}$$

where $P(\mathbf{X} \in s_\ell,\ j) = P(\mathbf{X} \in s_\ell,\ Y = j)$.

The first two terms in (A.3), namely,

$$R_1^*(L) = 1 - \sum_\ell \max_j P(\mathbf{X} \in s_\ell,\ j),$$

form an approximation to the Bayes rate constructed by averaging the densities over the rectangles in the partition. The bias $B(L)$ is defined by

$$B(L) = R_1^*(L) - R^*.$$

The last term of (A.3), denoted by $R^*(L)$, is a variance-like term formed by contributions of these rectangles in which $y_\ell \neq y_\ell^*$. This inequality occurs when the class majority of $\mathcal{L}$ in s_ℓ goes in the opposite direction from the true class probabilities in s_ℓ.

By assuming an increasing sequence of partitions $\{s_1, \ldots, s_L)$ such that the partitioning does not depend too strongly on $\mathcal{L}$, heuristic arguments can be given to show that

1. The bias term $B(L)$ decreases rapidly for small L, more slowly as L increases, and eventually decreases to zero.
2. The variance term increases slowly as L increases and is bounded by a slow growth factor in L.

3. For $L = N$ with each S_ℓ containing one case in $\mathcal{L}$, the variance term is bounded by the Bayes rate R^*.

Thus, we reach the somewhat surprising conclusion that the largest tree possible has a misclassification rate not larger than twice the Bayes rate. This is similar to Cover and Hart's result (1967) that the first nearest neighbor classification algorithm has, asymptotically, a misclassification rate bounded by twice the Bayes rate. When the partition is so small that every rectangle contains only one case in $\mathcal{L}$, then the classification rule becomes similar to first nearest neighbor classification, and Cover and Hart's result becomes relevant.

To illustrate the behavior of the bias term, write

$$R^* = 1 - \sum_\ell \int_{S_\ell} \max_j(\pi(j)f_j)d\mathbf{x}.$$

Then

$$B(L) = \sum_\ell \left[\int_{S_\ell} \max_j(\pi(j)f_j)d\mathbf{x} - \max_j\left(\int_{S_\ell} \pi(j)f_j\, d\mathbf{x}\right)\right]. \qquad \text{(A.4)}$$

If f_1, f_2 are continuous and $\pi(1)f_1(\mathbf{x}) \neq \pi(2)f_2(\mathbf{x})$ for $\mathbf{x} \in S_\ell$, then the corresponding term in the sum (A.4) is zero. For f_1, f_2 smooth and nonconstant, the hypersurface $H \subset X$ defined by

$$\pi(1)f_1(\mathbf{x}) = \pi(2)f_2(\mathbf{x})$$

is generally a smooth surface of lower dimension than X. Putting $f(\mathbf{x}) = \pi(1)f_1(\mathbf{x}) + \pi(2)f_2(\mathbf{x})$, it follows from (A.4) that

$$B(L) \leq \sum_{\ell'} \int_{S_{\ell'}} f(\mathbf{x})d\mathbf{x}, \qquad \text{(A.5)}$$

where the sum is over all rectangles $S_{\ell'}$ containing points in H. Clearly, then, $B(L) \to 0$ as we make the rectangles in the partition smaller.

A better bound on $B(L)$ can be gotten using the fact that if S_ℓ is small and $\pi(1)f_1 = \pi(2)f_2$ somewhere in S_ℓ, then even the nonzero terms in (A.4) are small. In fact, if $P(1|\mathbf{X} = \mathbf{x}) - P(2|\mathbf{X} = \mathbf{x})$ is assumed to have bounded first partial derivatives, if $f(\mathbf{x})$ is zero outside a sufficiently large rectangle, and if the partition is regular enough, then it can be shown that

$$B(L) \leq \frac{C}{L^{2/M}}, \tag{A.6}$$

where M is the dimension of X.

The inequality (A.6) is indicative of the rapid decrease in $B(L)$ for small values of L and slower decrease for larger L. But it also shows the strong effect of dimensionality on bias. The number of nodes needed to reduce bias by 50 percent, say, goes up exponentially with the dimension.

Another interesting facet coming out of this argument is that as L gets larger, virtually all of the bias is contributed by the region near the hypersurface where $\pi(1)f_1(\mathbf{x}) = \pi(2)f_2(\mathbf{x})$. If the classes are well separated in the sense that $f(\mathbf{x})$ is small near H, the bias will be correspondingly small.

In the second term of (A.3), assume that there are n_ℓ cases of $\mathcal{L}$ in S_ℓ. Put

$$p_\ell = \max_j(P(j|\mathbf{X} \in S_\ell)), \; q_\ell = 1 - p_\ell.$$

For L large, we can approximate the second term by its expectation over $\mathcal{L}$, that is, put

$$R_2^*(L) = \sum_\ell P(y_\ell \neq y_\ell^*)|P(\mathbf{X} \in S_\ell, 1) - P(\mathbf{X} \in S_\ell, 2)|.$$

Compute $P(y_\ell^* \neq y_\ell|n_\ell)$ by assuming that the distribution of class 1 and class 2 cases is given by n_ℓ independent trials with probabilities (p_ℓ, q_ℓ). Call p_ℓ the probability of heads, and let H be the random number of heads. Then

$$\begin{aligned} P(y_\ell \neq y_\ell^*|n_\ell) &= P(H \leq [n_\ell/2]) \\ &= \sum_{j=1}^{[n_\ell/2]} \binom{n_\ell}{j} p_\ell^j q_\ell^{n_\ell - j}. \end{aligned} \tag{A.7}$$

Our conclusions can be reached using (A.7). For instance, some elementary inequalities lead to

$$\max_{p_\ell \geq q_\ell} |p_\ell - q_\ell| P(y_\ell \neq y_\ell^* | n_\ell) \leq \frac{1}{\sqrt{n_\ell}} .$$

Since $n_\ell \simeq NP(\mathbf{X} \in S_\ell)$, we get

$$R_2^*(L) \leq \sqrt{\frac{L}{N}} . \tag{A.8}$$

This is the slow growth bound referred to in point 2 earlier.

Finally, for $n_\ell = 1$, $P(y_\ell \neq y_\ell^* | n_\ell) = q_\ell$. Then if $n_\ell = 1$, for all ℓ,

$$R_2^*(L) \simeq \sum_\ell q_\ell |P(\mathbf{X} \in S_\ell, 1) - P(\mathbf{X} \in S_\ell, 2)|$$

$$\leq \sum_\ell q_\ell P(\mathbf{X} \in S_\ell)$$

$$= \sum_\ell [1 - \max_j (P(j | \mathbf{X} \in S_\ell))] P(\mathbf{X} \in S_\ell).$$

Thus,

$$R_2^*(L) \leq 1 - \sum_\ell \max_j (P(j, \mathbf{X} \in S_\ell)) \simeq R^*,$$

assuming, of course, that the partition is fairly fine. This last equation summarizes our argument that the variance term is bounded by the Bayes rate for L large, that is, for $L \simeq N$.

A transition occurs between the situation at which L is large but $N/L \gg 1$ and the limiting $L = N$. If n_ℓ is moderately large, then $P(y_\ell \neq y_\ell^* | n_\ell)$ is small unless $p_\ell \simeq q_\ell$. Therefore, the major contribution to the variance term is from those rectangles S_ℓ near the hypersurface on which $\pi(1)f_1 = \pi(2)f_2$.

As L becomes a sizable fraction of N, the variance contributions become spread out among all rectangles S_ℓ for which n_ℓ is small.

Since detailed proofs of parts of a heuristic argument are an overembellishment, they have been omitted here. Although the lack of rigor is obvious, the preceding derivations and inequalities may help explain the balance involved when the optimal tree is chosen near the bottom of the $\hat{R}$ curve.

4

SPLITTING RULES

In the previous chapter, a method was given for selecting the right sized tree assuming that a large tree T_{max} had already been grown. Assuming that a set of questions $\mathcal{Q}$ or, equivalently, a set S of splits at every node t has been specified, then the fundamental ingredient in growing T_{max} is a splitting rule. Splitting rules are defined by specifying a goodness of split function $\Phi(s, t)$ defined for every $s \in S$ and node t. At every t, the split adopted is the split s^* which maximizes $\Phi(s, t)$.

A natural goodness of split criterion is to take that split at any node which most reduces the resubstitution estimate of tree misclassification cost. Unfortunately, this criterion has serious deficiencies, which are discussed in Section 4.1. In the two-class problem, a class of splitting criteria is introduced which remedies the deficiencies, and the simplest of these is adopted for use (Section 4.2). In Section 4.3 this criterion is generalized in two directions for the multiclass problem resulting in the Gini and twoing criteria.

Section 4.4 deals with the introduction of variable misclassification costs into the splitting criterion in two ways: through a generalization of the Gini criterion and by alteration of the priors. Examples are given in Section 4.5.

After looking at the outputs of a number of simulated examples, we have come to these tentative conclusions:

1. The overall misclassification rate of the tree constructed is not sensitive to the choice of a splitting rule, as long as it is within a reasonable class of rules.
2. The method of incorporation of variable misclassification costs into the splitting rule has less effect than their incorporation into the pruning criterion, that is, into $R(T)$.

In Section 4.6, we look at the problem of growing and pruning trees in terms of a different accuracy criterion. Given a measurement vector of unknown class, suppose that instead of assigning it to a single class, we are more interested in estimating the probabilities that it belongs to class 1, ..., class J. What splitting and pruning criteria should be used to maximize the accuracy of these class probability estimates? In the formulation used, the answer is intimately connected with the Gini criterion.

For other discussions of splitting rules in classification, see Belson (1959), Anderson (1966), Hills (1967), Henrichon and Fu (1969), Messenger and Mandell (1972), Meisel (1972), Meisel and Michalpoulos (1973), Morgan and Messenger (1973), Friedman (1977), Gordon and Olshen (1978), and Rounds (1980).

4.1 REDUCING MISCLASSIFICATION COST

In Section 2.4.2 a framework was given for generating splitting rules. The idea was to define an impurity function $\phi(p_1, \ldots, p_J)$ having certain desirable properties (Definition 2.5). Define the node impurity function $i(t)$ as $\phi(p(1|t), \ldots, p(J|t))$, set $I(t) = i(t)p(t)$, and define the tree impurity $I(T)$ as

$$I(T) = \sum_{t \in \tilde{T}} I(t).$$

Then at any current terminal node, choose that split which most reduces $I(T)$ or, equivalently, maximizes

$$\Delta I(s, t) = I(t) - I(t_L) - I(t_R)$$

or

$$\Delta i(s, t) = i(t) - p_L i(t_L) - p_R i(t_R).$$

Within this framework it seems most natural to take the tree impurity as $R(T)$, the resubstitution estimate for the expected misclassification cost. The best splits would then be those that most reduced the estimated misclassification rate. This is equivalent to defining $i(t)$ as equal to $r(t)$, where

$$\begin{aligned} r(t) &= \min_i \textstyle\sum_j C(i|j)p(j|t) \\ &= 1 - \max_j p(j|t) \quad \text{(unit cost case).} \end{aligned}$$

Then the best split of t maximizes

$$r(t) - p_L r(t_L) - p_R r(t_R) \tag{4.1a}$$

or, equivalently, maximizes

$$R(t) - R(t_L) - R(t_R). \tag{4.1b}$$

The corresponding node impurity function (unit cost) is

$$\phi(p_1, \ldots, p_J) = 1 - \max_j p_j.$$

This function has all the desirable properties listed in Definition 2.5.

Still, in spite of its natural attractiveness, selecting splits to maximize the reduction in $R(T)$ has two serious defects. The first is that the criterion (4.1) may be zero for all splits in S.

PROPOSITION 4.2. *For any split of t into t_L and t_R,*

$$R(t) \geq R(t_L) + R(t_R)$$

with equality if $j^*(t) = j^*(t_L) = j^*(t_R)$.

PROOF. Note that

$$\begin{aligned} R(t) &= \sum_j c(j^*(t)|j)p(j, t) \\ &= \sum_j c(j^*(t)|j)[p(j, t_L) + p(j, t_R)] \end{aligned}$$

or

$$\begin{aligned} &R(t) - R(t_L) - R(t_R) \\ &\quad = \sum_j c(j^*(t)|j)p(j, t_L) - \min_i \sum_j c(i|j)p(j, t_L) \\ &\qquad + \sum_j c(j^*(t)|j)p(j|t_R) - \min_i \sum_j c(i|j)p(j, t_R). \end{aligned}$$

The right hand side is certainly nonnegative and equals zero under the conditions $j^*(t) = j^*(t_L) = j^*(t_R)$.

Now suppose we have a two-class problem with equal priors and are at a node t which has a preponderance of class 1 cases. It is conceivable that every split of t produces nodes t_L, t_R where both have class 1 majorities. Then $R(t) - R(t_L) - R(t_R) = 0$ for all splits in S and there is no single or small number of best splits.

The second defect is more difficult to quantify. In summary, it is that reducing the misclassification rate does not appear to be a good criterion for the overall multistep tree growing procedure.

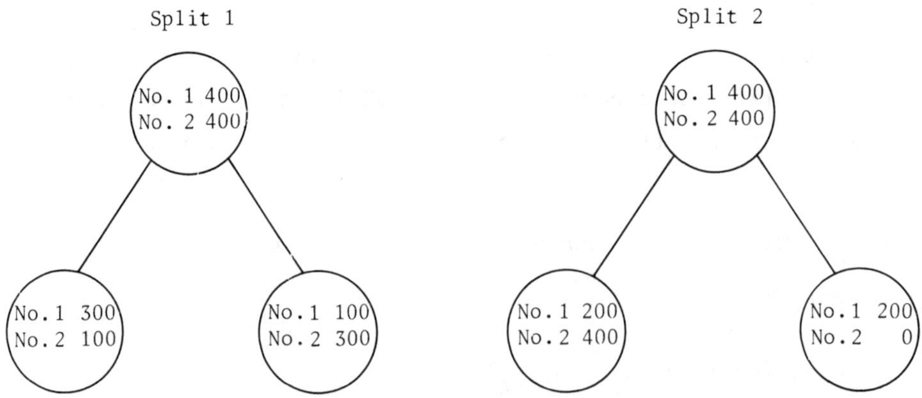

FIGURE 4.1

Look at the example in Figure 4.1. Suppose that the top node is the root node and assume equal priors. The first split leads to a tree in which 200 cases are misclassified and $R(T) = 200/800 = .25$. In the second split, 200 cases are also misclassified and $R(T) = .25$.

Even though both splits are given equal ratings by the $R(T)$ criterion, the second split is probably more desirable in terms of the future growth of the tree. The first split gives two nodes, both having $r(t) = .25$. Both of these nodes will probably need more splitting to get a tree with a lower value of $R(T)$. The second tree has one node with $r(t) = .33$, which will have to be split again on the same grounds. But it also has one node with 1/4 of the cases in it for which $r(t) = 0$. This node is terminal, no more work has to be done on it and it gives, at least on $\mathcal{L}$, perfect classification accuracy.

Besides the degeneracy problem, then, the $R(T)$ criterion does not seem to appropriately reward splits that are more desirable in the context of the continued growth of the tree. Other examples show that this behavior can be even more pronounced as the number of classes gets larger.

This problem is largely caused by the fact that our tree growing structure is based on a one-step optimization procedure. For example, in bridge, the team goal in every deal is to take as many tricks as possible. But a team that attempts to take the trick every time a card is led will almost invariably lose against experienced players.

A better single-play criterion would take into account overall improvement in strategic position as well as the possibility of taking the current trick. In other words, a good single-play criterion would incorporate the fact that any play currently selected has implications for the future course of the game.

Similarly, while the tree T finally constructed should have as small as possible "true" probability of misclassification, the one-step minimization of $R(T)$ is not desirable either from a practical point of view (degeneracy) or from the overall strategic viewpoint.

4.2 THE TWO-CLASS PROBLEM

4.2.1 A Class of Splitting Criteria for the Two-Class Problem

In the two-class unit cost situation, the node impurity function corresponding to the node misclassification rate is

$$\phi(p_1, p_2) = 1 - \max(p_1, p_2) = \min(p_1, p_2)$$
$$= \min(p_1, 1 - p_1).$$

Its graph as a function of p_1 is given in Figure 4.2.

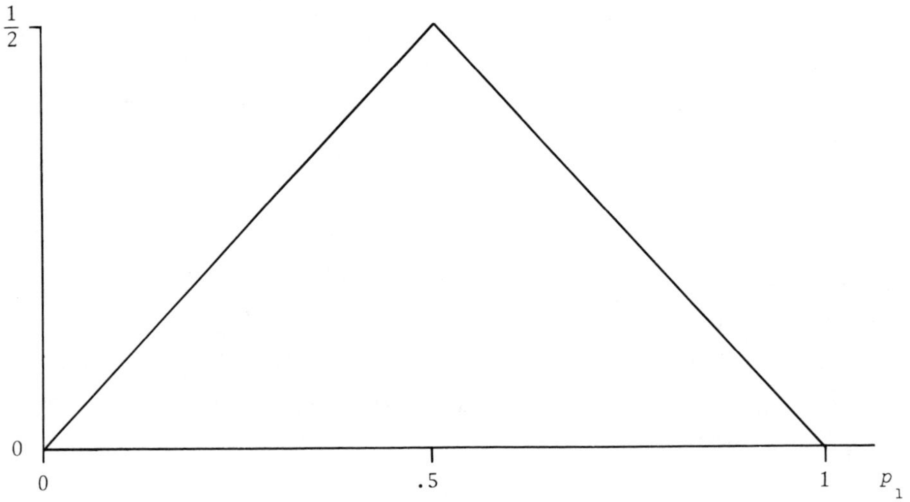

FIGURE 4.2

This section defines a class of node impurity functions $i(t) = \phi(p_1)$ which do not have the undesirable properties of the $\min(p_1, p_2)$ function. We retain the intuitively reasonable requirements that

(i) $\phi(0) = \phi(1) = 0$.

(ii) $\phi(1/2)$ = maximum.

In addition, since $i(t)$ reasonably should be symmetric in p_1, p_2, we require that $\phi(p_1) = \phi(1 - p_1)$.

The example illustrated in Figure 4.1 indicated that $\phi(p_1) = \min(p_1, 1 - p_1)$ did not sufficiently reward purer nodes. Suppose $p_1 > 1/2$; then the problem is that $\phi(p_1) = 1 - p_1$ decreases only linearly in p_1. To construct a class of criteria that select the second split of the example in Figure 4.1 as being more desirable, we will require that $\phi(p_1)$ *decrease faster than linearly* as p_1 increases.

This is formulated by insisting that if $p_1'' > p_1'$, then $\phi(p_1'')$ is less than the corresponding point on the tangent line at p_1' (see Figure 4.3). Equivalently, this requires that ϕ be strictly

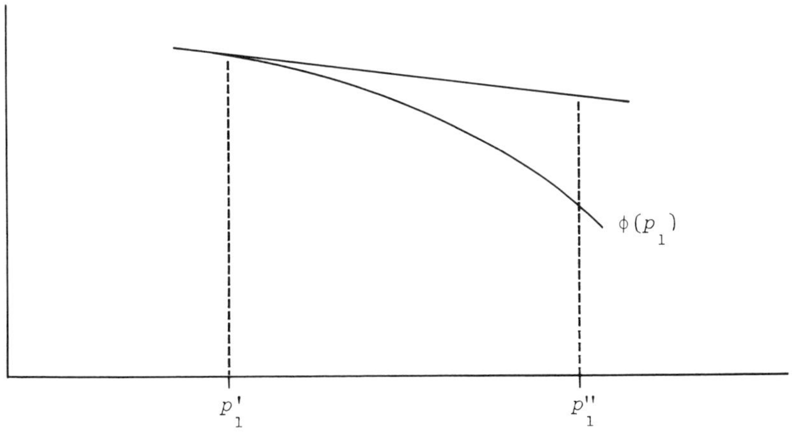

FIGURE 4.3

concave. If ϕ has a continuous second derivative on [0, 1], then the strict concavity translates into $\phi''(p_1) < 0$, $0 < p_1 < 1$.

The class of node impurity functions $\phi(p_1)$ which seems natural to this context is therefore defined as the class F of functions

$\phi(p_1)$, $0 \leq p_1 \leq 1$, with continuous second derivatives on $0 \leq p_1 \leq 1$ satisfying

$$\begin{aligned} &\text{(i)}\ \phi(0) = \phi(1) = 0, \\ &\text{(ii)}\ \phi(p_1) = \phi(1 - p_1), \\ &\text{(iii)}\ \phi''(p_1) < 0,\ 0 < p_1 < 1. \end{aligned} \tag{4.3}$$

The functions of class F generally look like Figure 4.4. There are

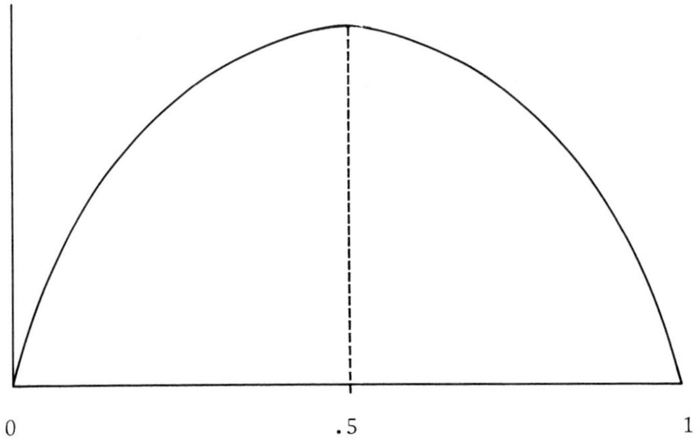

FIGURE 4.4

some general properties common to all functions in F. To begin with, let $i(t) = \phi(p(1|t))$, $\phi \in F$. Then

PROPOSITION 4.4. *For any node t and split s,*

$$\Delta i(s, t) \geq 0$$

with equality if, and only if, $p(j|t_L) = p(j|t_R) = p(j|t)$, $j = 1, 2$.

The proof is in the appendix to this chapter.

The impurity is never increased by splitting. It stays the same only under rare circumstances. The condition for equality in Proposition 4.4 is generally much more difficult to satisfy than the corresponding condition using $R(t)$ in Proposition 4.2. The requirement that $\phi(p_1)$ be strictly concave largely removes the degeneracy. It can still happen that a few splits simultaneously achieve the maximum values of $\Delta i(s, t)$, but in practice, we have found that multiplicity is exceptional.

4.2.2 *The Criteria Class and Categorical Variables*

Perhaps the most interesting evidence that the class F is a natural class of node impurity functions comes from a different direction. Suppose that a measurement variable x is a categorical variable taking values in the set $\{b_1, \ldots, b_L\}$. At node t, the tree construction proceeds by searching for a subset

$$B^* = \{b_{i_1}, \ldots\} \subset \{b_1, \ldots, b_L\}$$

such that the split s^*, Is $x \in B^*$?, maximizes $\Delta i(s,t)$. Call any such subset B^* *maximizing*. (There could be more than one.)

Let $N_{j,\ell}(t)$ equal the number of cases in t in class j such that $x = b_\ell$, and denote

$$p(j|x = b_\ell) = \pi(j)N_{j,\ell}(t) / \sum_{j=1,2} \pi(j)N_{j,\ell}(t).$$

That is, $p(j|x = b_\ell)$ can be interpreted as the estimated probability of being in class j given that the object is in node t and its x category is b_ℓ. Then this result holds.

THEOREM 4.5. *Order the* $p(1|x = b_\ell)$, *that is,*

$$p(1|x = b_{\ell_1}) \leq p(1|x = b_{\ell_2}) \leq \cdots \leq p(1|x = b_{\ell_L}).$$

If ϕ *is in the class* F, *then one of the* L *subsets*

$$\{b_{\ell_1}, \ldots, b_{\ell_h}\}, \; h = 1, \ldots, L,$$

is maximizing.

For a categorical variable with L large, this result provides considerable improvement in computational efficiency. The search is reduced from looking at 2^{L-1} subsets to L subsets. The intuitive content is clear. The best split should put all those categories leading to high probabilities of being in class 1 into one node and the categories leading to lower class 1 probabilities in the other.

The proof is a generalization of a result due to Fisher (1958) and is given in Section 9.4.

4.2.3 Selection of a Single Criterion

The simplest polynomial in the class F is quadratic, say,

$$\phi(x) = a + bx + cx^2.$$

Condition (4.3)(i) gives $a = 0$, $b + c = 0$, so

$$\phi(x) = b(x - x^2),$$

and (4.3)(ii) implies that $b > 0$. Without loss of generality we take $b = 1$, giving the criterion

$$i(t) = p(1|t)p(2|t). \tag{4.6}$$

The criterion

$$i(t) = -p(1|t) \log p(1|t) - p(2|t) \log p(2|t) \tag{4.7}$$

also belongs to F.

The function $p(1|t)p(2|t)$ is simple and quickly computed. It has a familiar interpretation. Suppose all class 1 objects in a node t are given the numerical value 1 and class 2 objects the value 0. Then if $p(1|t)$ and $p(2|t)$ are the proportions of the two classes in the node, the sample variance of the numerical values in the node is equal to $p(1|t)p(2|t)$.

Since we could not think of any intrinsic reason why one function in the class F should be preferred to any other, and since preliminary tests indicated that both (4.6) and (4.7) gave very similar results, the principle of simplicity was appealed to and $p(1|t)p(2|t)$ selected as the node impurity function in the two-class problem.

4.3 THE MULTICLASS PROBLEM: UNIT COSTS

Two different criteria have been adopted for use in the multiclass problem with unit costs. These come from two different approaches toward the generalization of the two-class criterion and are called the

Gini criterion
Twoing criterion

4.3.1 The Gini Criterion

The concept of a criterion depending on a node impurity measure has already been introduced. Given a node t with estimated class probabilities $p(j|t)$, $j = 1, \ldots, J$, a measure of node impurity given t

$$i(t) = \phi(p(1|t), \ldots, p(J|t))$$

is defined and a search made for the split that most reduces node, or equivalently tree, impurity. As remarked earlier, the original function selected was

$$\phi(p_1, \ldots, p_J) = -\sum_j p_j \log p_j.$$

In later work the Gini diversity index was adopted. This has the form

$$i(t) = \sum_{j \neq i} p(j|t)p(i|t) \tag{4.8}$$

and can also be written as

$$i(t) = \Big(\sum_j p(j|t)\Big)^2 - \sum_j p^2(j|t) = 1 - \sum_j p^2(j|t). \tag{4.9}$$

In the two-class problem, the index reduces to

$$i(t) = 2p(1|t)p(2|t),$$

equivalent to the two-class criterion selected previously.

The Gini index has an interesting interpretation. Instead of using the plurality rule to classify objects in a node t, use the rule that assigns an object selected at random from the node to class i with probability $p(i|t)$. The estimated probability that the item is actually in class j is $p(j|t)$. Therefore, the estimated probability of misclassification under this rule is the Gini index

$$\sum_{j \neq i} p(i|t)p(j|t).$$

Another interpretation is in terms of variances (see Light and Margolin, 1971). In a node t, assign all class j objects the value 1, and all other objects the value 0. Then the sample variance of these values is $p(j|t)(1 - p(j|t))$. If this is repeated for all J classes and the variances summed, the result is

$$\sum_j p(j|t)(1 - p(j|t)) = 1 - \sum_j p^2(j|t).$$

Finally, note that the Gini index considered as a function $\phi(p_1, \ldots, p_J)$ of the $p_1, \ldots, p_J$ is a quadratic polynomial with nonnegative coefficients. Hence, it is concave in the sense that for $r + s = 1$, $r \geq 0$, $s \geq 0$,

$$\phi(rp_1 + sp_1', rp_2 + sp_2', \ldots, rp_J + sp_J') \geq r\phi(p_1, \ldots, p_J) + s\phi(p_1', \ldots, p_J').$$

This ensures (see the appendix) that for any split s,

$$\Delta i(s, t) \geq 0.$$

Actually, it is strictly concave, so that $\Delta i(s, t) = 0$ only if $p(j|t_R) = p(j|t_L) = p(j|t)$, $j = 1, \ldots, J$.

The Gini index is simple and quickly computed. It can also incorporate symmetric variable misclassification costs in a natural way (see Section 4.4.2).

4.3.2 The Twoing Criterion

The second approach to the multiclass problem adopts a different strategy. Denote the class of classes by C, i.e.,

$$C = \{1, \ldots, J\}.$$

At each node, separate the classes into two superclasses,

$$C_1 = \{j_1, \ldots, j_n\}, \; C_2 = C - C_1.$$

Call all objects whose class is in C_1 class 1 objects, and put all objects in C_2 into class 2.

For any given split s of the node, compute $\Delta i(s, t)$ *as though it were a two-class problem.* Actually $\Delta i(s, t)$ depends on the selection of C_1, so the notation

$$\Delta i(s, t, C_1)$$

is used. Now find the split $s^*(C_1)$ which maximizes $\Delta i(s, t, C_1)$. Then, finally, find the superclass C_1^* which maximizes

$$\Delta i(s^*(C_1), t, C_1).$$

The split used on the node is $s^*(C_1^*)$.

The idea is then, at every node, to select that conglomeration of classes into two superclasses so that considered as a two-class problem, the greatest decrease in node impurity is realized.

This approach to the problem has one significant advantage: *It gives "strategic" splits* and informs the user of class similarities. At each node, it sorts the classes into those two groups which in some sense are most dissimilar and outputs to the user the optimal grouping C_1^*, C_2^* as well as the best split s^*.

The word *strategic* is used in the sense that near the top of the tree, this criterion attempts to group together large numbers of classes that are similar in some characteristic. Near the bottom of the tree it attempts to isolate single classes. To illustrate, suppose that in a four-class problem, originally classes 1 and 2 were grouped together and split off from classes 3 and 4, resulting in a node with membership

Class:	1	2	3	4
No. cases:	50	50	3	1

Then on the next split of this node, the largest potential for decrease in impurity would be in separating class 1 from class 2.

Spoken word recognition is an example of a problem in which twoing might function effectively. Given, say, 100 words (classes), the first split might separate monosyllabic words from multisyllabic words. Future splits might isolate those word groups having other characteristics in common.

As a more concrete example, Figure 4.5 shows the first few splits in the digit recognition example. The 10 numbers within

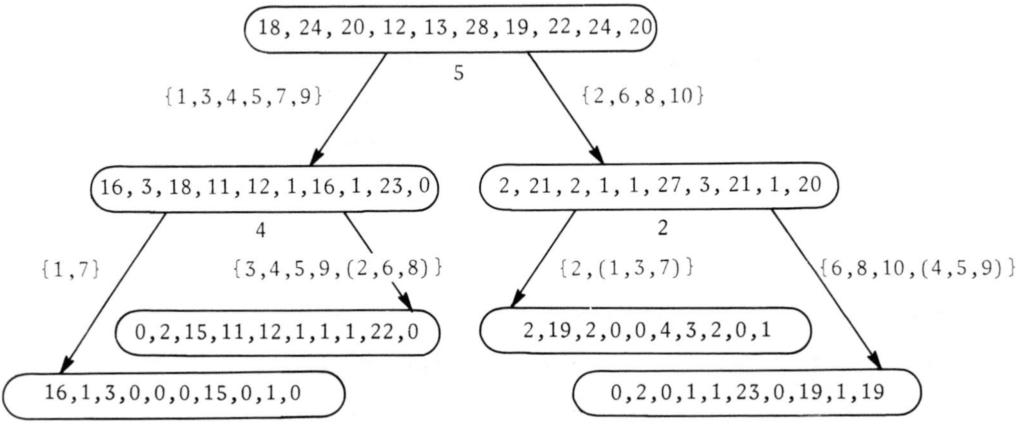

FIGURE 4.5

each node are the class memberships in the node. In each split the numbers in brackets by the split arrows are the superclasses C_1^*, C_2^*, for the split. In parentheses in the brackets are the classes whose populations are already so small in the parent node that their effect in the split is negligible. Zero populations have been ignored.

Recall that the lights are numbered as

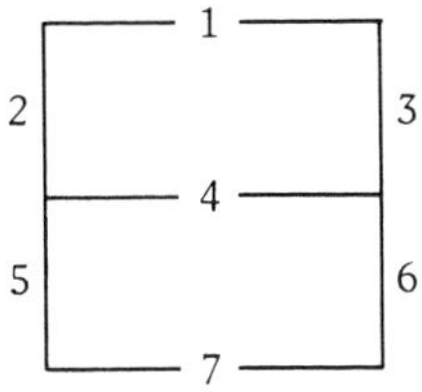

The first split, on the fifth light, groups together classes 1, 3, 4, 5, 7, 9 and 2, 6, 8, 10. Clearly, the fifth light should be off for 1, 3, 4, 5, 7, 9 and on for the remaining digits. The next split on the left is on light 4 and separates classes 1, 7 from classes 3, 4, 5, 9. On the right, the split on light 2 separates class 2 from 6, 8, 10.

Although twoing seems most desirable with a large number of classes, it is in such situations that it has an apparent disadvantage in computational efficiency. For example, with J classes, there are 2^{J-1} distinct divisions of C into two superclasses. For $J = 10$, $2^{J-1} \simeq 1000$. However, the following result shows, rather surprisingly, that twoing can be reduced to an overall criterion, running at about the same efficiency as the Gini criterion.

THEOREM 4.10. *Under the two-class criterion* $p(1|t)p(2|t)$, *for a given split* s, *a superclass* $C_1(s)$ *that maximizes*

$$\Delta i(s, t, C_1)$$

is

$$C_1(s) = \{j: p(j|t_L) \geq p(j|t_R)\}$$

and

$$\max_{C_1} \Delta i(s, t, C_1) = \frac{p_L p_R}{4}\left[\sum_j |p(j|t_L) - p(j|t_R)|\right]^2.$$

COROLLARY 4.11. *For any node* t *and split* s *of* t *into* t_L *and* t_R, *define the twoing criterion function* $\Phi(s, t)$ *by*

$$\Phi(s, t) = \frac{p_L p_R}{4}\left[\sum_j |p(j|t_L) - p(j|t_R)|\right]^2 .$$

Then the best twoing split $s^*(C_1^*)$ *is given by the split* s^* *which maximizes* $\Phi(s, t)$ *and* C_1^* *is given by*

$$C_1^* = \{j: p(j|t_L^*) \geq p(j|t_R^*)\},$$

where t_L^*, t_R^* *are the nodes given by the split* s^*.

The proof of Theorem 4.10 is in the appendix to this chapter together with a proof that $\Phi(s, t) \geq 0$ for all splits s with equality only if $p(j|t_R) = p(j|t_L) = p(j|t)$, $j = 1, \ldots, J$. These results tend to support the choice of $p(1|t)p(2|t)$ as a preferred two-class measure of node impurity, since we have been unable to extend Theorem 4.10 to other functions in F.

The idea of twoing can also be applied to situations in which the classes in C, though categorical, are also naturally ordered (see Hills, 1967; McCullagh, 1980; and Anderson and Philips, 1981). For instance, in a study of the symptoms of back pain after treatment, the classes were defined as {worse, same, slight improvement, moderate improvement, marked improvement, complete relief}.

In such applications, it is natural to consider the *ordered twoing* criterion given by

$$\Phi(s, t) = \max_{C_1} \Delta i(s, t, C_1),$$

where C_1, C_2 are partitions of $C = \{1, \ldots, J\}$ into two superclasses restricted by the condition that they be of the form

$$C_1 = \{1, \ldots, j_1\},\ C_2 = \{j_1 + 1, \ldots, J\},\ j_1 = 1, \ldots, J - 1.$$

The twoing splitting process does not fit into the general framework discussed earlier. It does not operate on an overall measure of node impurity $i(t)$ to achieve a maximum value of $\Delta i(s, t)$. As a result, there is no corresponding measure of tree impurity $I(T)$. This is not a disadvantage. A splitting criterion should be judged primarily in terms of how well it performs in tree construction.

4.3.3 Choice of Criterion: An Example

Both the Gini and twoing criteria have been implemented in CART. Each method has its own advantages. In either case, we have not succeeded in finding an extension of Theorem 4.5 on handling categorical data. If the Gini index is used, the number of categories in any variable should be kept moderate to prevent exhaustive subset searches. If twoing is used and there are a small number of classes, then for each fixed superclass selection, Theorem 4.5 can be used on the categorical variables, and then a direct search can be made for the best superclass.

Choice of a criterion depends on the problem and on what information is desired. The final classification rule generated seems to be quite insensitive to the choice. To illustrate, both Gini and twoing were used on a replicate data set in the digit recognition problem, with final tree selection using the 1 SE rule.

In both cases, trees with 10 terminal nodes were selected. Both trees have the same test sample accuracy (.33). Figure 4.6 shows the two trees. The numbers underneath the nodes are the coordinates split on. The numbers in the terminal nodes indicate node class assignment.

The two trees are very similar. At the node indicated by the arrow, the Gini criterion prefers a split on coordinate 6, while twoing selects the second coordinate. For the Gini criterion, the split on the second coordinate was the second best split, and for twoing, the split on the sixth coordinate was the second best.

The class membership of this node is

Class no.:	1	2	3	4	5	6	7	8	9	10
N_j:	0	19	15	0	4	1	3	14	20	9.

It mainly consists of classes 2, 3, 8, 9, 10, whose errorless configurations are 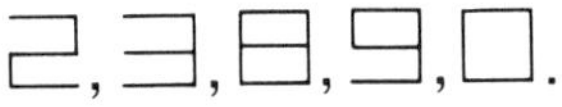 In terms of coordinates, these configurations are characterized by

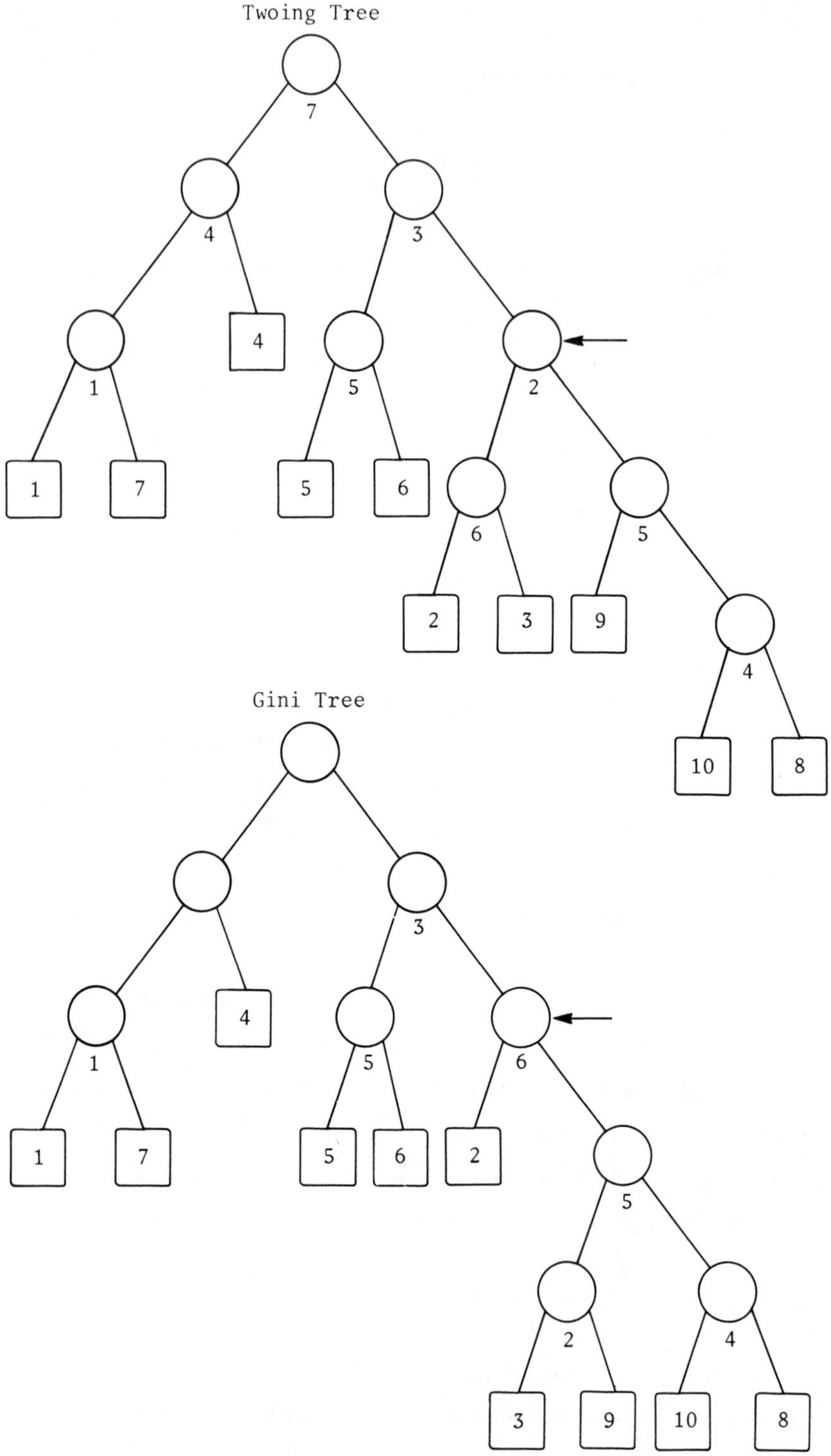

FIGURE 4.6

Class 2: $x_2 = 0$, $x_6 = 0$
3: $x_2 = 0$, $x_5 = 0$
8: no zeros
9: $x_5 = 0$
10: $x_4 = 0$

The split on $x_6 = 0$ preferred by Gini separates out class 2 and sends it left. The twoing split on $x_2 = 0$ groups together classes 2 and 3 and sends them left with 9 and 10 going right.

Different data sets were generated from the digit recognition model by changing the random number generator seed. Trees were grown on the data sets using both the Gini and twoing criteria. The preceding example was chosen as best illustrating the difference. In general, the first few splits of the two trees are the same, and the two trees selected have comparable accuracy. Where they differ, Gini tends to favor a split into one small, pure node and a large, impure node. Twoing favors splits that tend to equalize populations in the two descendant nodes.

In the waveform recognition example, the trees constructed by Gini and twoing and pruned back by the 1 SE rule had (to our surprise) identical splits. When a new seed was used to generate another waveform data set, the two trees differed slightly. Both had nine terminal nodes. The Gini tree had slightly better test sample accuracy (.28) than the twoing tree (.30). The right branch leading from the root node had identical splits in both trees. Where they differed, in the left branch, the same phenomenon was observed as in the digit data. The twoing splits tended to produce descendant nodes of more equal size than the Gini splits.

There are usually only slight differences between the Gini and twoing trees. In balance, comparing the two on many data sets, where they differ, the Gini splits generally appear to be better. In fact, one can give examples of two candidate splits of a node, one of which is clearly superior to another in terms of producing pure descendant nodes, in which twoing (but not Gini) selects the poorer split. For these reasons, we usually prefer the use of the Gini criterion.

4.4 PRIORS AND VARIABLE MISCLASSIFICATION COSTS

The parameters that can be set in tree structured classification include the priors $\{\pi(j)\}$ and variable misclassification costs $\{C(i|j)\}$. These are interrelated to the extent discussed in Section 4.4.3.

4.4.1 Choice of Priors

The priors are a useful set of parameters, and intelligent selection and adjustment of them can assist in constructing a desirable classification tree.

In some studies, the data set may be very unbalanced between classes. For example, in the mass spectra data base nonchlorine compounds outnumbered chlorine compounds by 10 to 1. If the priors are taken proportional to the occurrence of compounds in the data base, then we start with a misclassification rate of 10 percent: Everything is classified as not containing chlorine. Growing a classification tree using such priors decreases the misclassification rate to about 5 percent. But the result is that nonchlorines have a 3 percent misclassification rate, while chlorines have a 30 percent misclassification rate.

The mechanism producing this disparity is that if equal numbers of chlorines and nonchlorines are misclassified, the effect on the chlorine classification rate will be much larger than on the nonchlorine rate.

The priors can be used to adjust the individual class misclassification rates in any desired direction. For example, taking equal priors tends to equalize the misclassification rates. In the chlorine example, equal priors resulted in a 9 percent misclassification rate for chlorine and a 7 percent rate for nonchlorines. Putting a larger prior on a class will tend to decrease its misclassification rate, and vice versa.

If the initial choice of priors gives questionable results, we suggest the growing of some exploratory trees using different priors as outlined in Chapter 5.

4.4.2 Variable Misclassification Costs via Gini

In Section 4.3 the assumption was made that the cost of misclassifying a class j case as a class i case was equal to 1 for all $i \neq j$. In general, if variable misclassification costs $\{C(i|j)\}$ are specified, then the question arises of how to incorporate these costs into the splitting rule. For the Gini index there is a simple extension. Again, consider the suboptimal classification rule which, at a node t, assigns an unknown object into class j with estimated probability $p(j|t)$. Note that the estimated expected cost using this rule is

$$\sum_{j,i} C(i|j)p(i|t)p(j|t). \tag{4.12}$$

This expression is used as the Gini measure of node impurity $i(t)$ for variable misclassification costs.

In the two-class problem, (4.12) reduces to

$$(C(2|1) + C(1|2))p(1|t)p(2|t),$$

giving the same splitting criterion (essentially) as in the unit cost case. This points up a difficulty, noted by Bridges (1980), in the way in which the Gini index deals with variable costs. The coefficient of $p(i|t)p(j|t)$ in (4.12) is $C(i|j) + C(j|i)$. The index therefore depends only on the symmetrized cost matrix and does not appropriately adjust to highly nonsymmetric costs.

Another, more theoretical, problem is that $i(t)$ defined by (4.12) is not necessarily a concave function of the $\{p(j|t)\}$, and so $\Delta i(s, t)$ could conceivably be negative for some or all splits in S.

4.4.3 Variable Misclassification Costs via Altered Priors

Suppose that in a two-class problem with equal priors, it is twice as expensive to misclassify a class 2 case as it is to misclassify a class 1 case; that is, $C(1|2) = 2$, $C(2|1) = 1$. As compared to the equal cost situation, we want a tree that misclassifies fewer class 2 cases.

Another way to look at it is that every case in class 2 misclassified counts double, so the situation is similar to that if the prior on class 2 is taken twice as large as the prior on class 1.

Pursuing this idea, let $Q(i|j)$ be the proportion of the class j cases in $\mathcal{L}$ classified as class i by a tree T. Then the resubstitution estimate for the expected misclassification cost is

$$R(T) = \sum_{i,j} C(i|j)Q(i|j)\pi(j).$$

Let $\{\pi'(j)\}$ and $\{C'(i|j)\}$ be altered forms of $\{\pi(j)\}$ and $\{C(i|j)\}$ such that

$$C'(i|j)\pi'(j) = C(i|j)\pi(j), \quad i,\ j \in C. \tag{4.13}$$

Then $R(T)$ remains the same when computed using $\{\pi'(j)\}$ and $\{C'(i|j)\}$.

Take $\{C'(i|j)$ to be the unit cost matrix and suppose that altered priors $\{\pi'(j)\}$ can be found satisfying (4.13). Then the cost structure of T is equivalent, in the above sense, to a unit cost problem with the $\{\pi(j)\}$ replaced by the $\{\pi'(j)\}$.

If the costs are such that for each class j, there is a constant misclassification cost $C(j)$ regardless of how it is misclassified, that is, if

$$C(i|j) = C(j), \quad i \neq j,$$

then $C'(i|j)$ can be taken as unit costs with the altered priors

$$\pi'(j) = C(j)\pi(j)/\textstyle\sum_j C(j)\pi(j). \tag{4.14}$$

This suggests that a natural way to deal with a problem having a constant cost structure $C(j)$ for the jth class is to redefine priors by (4.14) and proceed as though it were a unit cost problem.

In general, the $\{\pi'(j)\}$ should be chosen so that the $\{C'(i|j)\}$ are as close as possible to unit cost. This has been implemented in CART by defining the $\{\pi'(j)\}$ through (4.14) using

$$C(j) = \sum_i C(i|j). \tag{4.15}$$

4.5 TWO EXAMPLES

In the waveform recognition problem, recall that the classes are superpositions of two waveforms as sketched in Figure 4.7. Classes

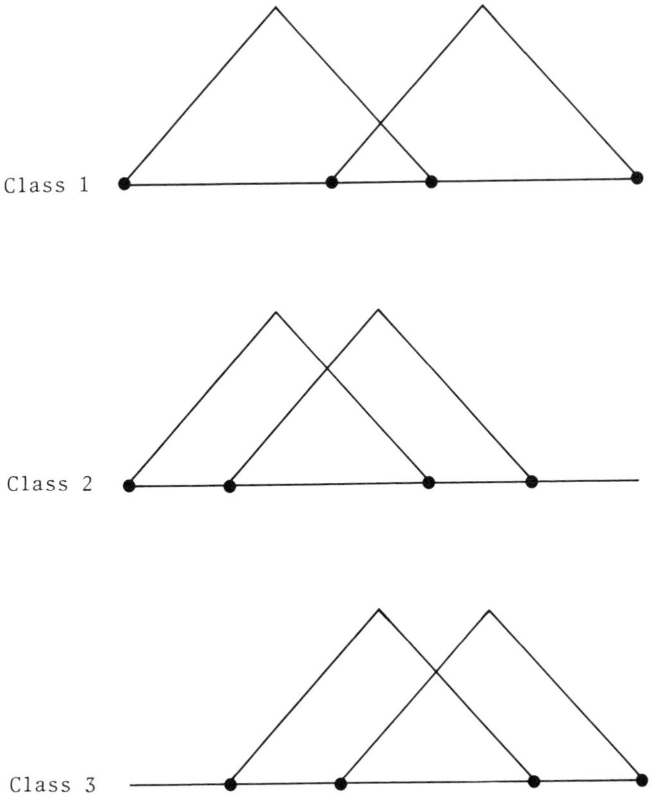

FIGURE 4.7

2 and 3 are mirror images of each other. Since class 1 is the odd class, we decided to track the results of varying misclassification costs by making it more costly to misclassify class 1 as a class 2 or 3. The misclassification cost matrix $C(i|j)$ given in (4.16) was used.

$$C(i|j) \quad j \begin{array}{c} \\ 1 \\ 2 \\ 3 \end{array} \begin{array}{c} i \\ \begin{array}{ccc} 1 & 2 & 3 \end{array} \\ \begin{pmatrix} 0 & 5 & 5 \\ 1 & 0 & 1 \\ 1 & 1 & 0 \end{pmatrix} \end{array} \tag{4.16}$$

In this case, there are constant class misclassification costs; $C(1) = 5$, $C(2) = 1$, $C(3) = 1$, and altering the priors is the preferred procedure. The example was run using altered priors and the unit cost Gini criterion and rerun using the original priors and the Gini criterion incorporating the symmetrized cost matrix.

In the second example, the cost matrix was taken as the symmetrized version of (4.16):

$$C(i|j) \quad j \begin{array}{c} \\ 1 \\ 2 \\ 3 \end{array} \begin{array}{c} i \\ \begin{array}{ccc} 1 & 2 & 3 \end{array} \\ \begin{pmatrix} 0 & 3 & 3 \\ 3 & 0 & 1 \\ 3 & 1 & 0 \end{pmatrix} \end{array}.$$

In this example, use of the Gini criterion with varying costs seems preferred. This is contrasted with running the example using the Gini criterion with unit costs and altered priors defined by (4.14) and (4.15).

The results of the first example are summarized in Table 4.1 and Figure 4.8. Except for the last split in the Gini case, which only skims seven cases off to the left, the trees are very similar,

TABLE 4.1 Costs

	R^{cv}	R^{ts}
Altered priors	.53 ± .05	.46
Symmetric Gini	.64 ± .07	.46

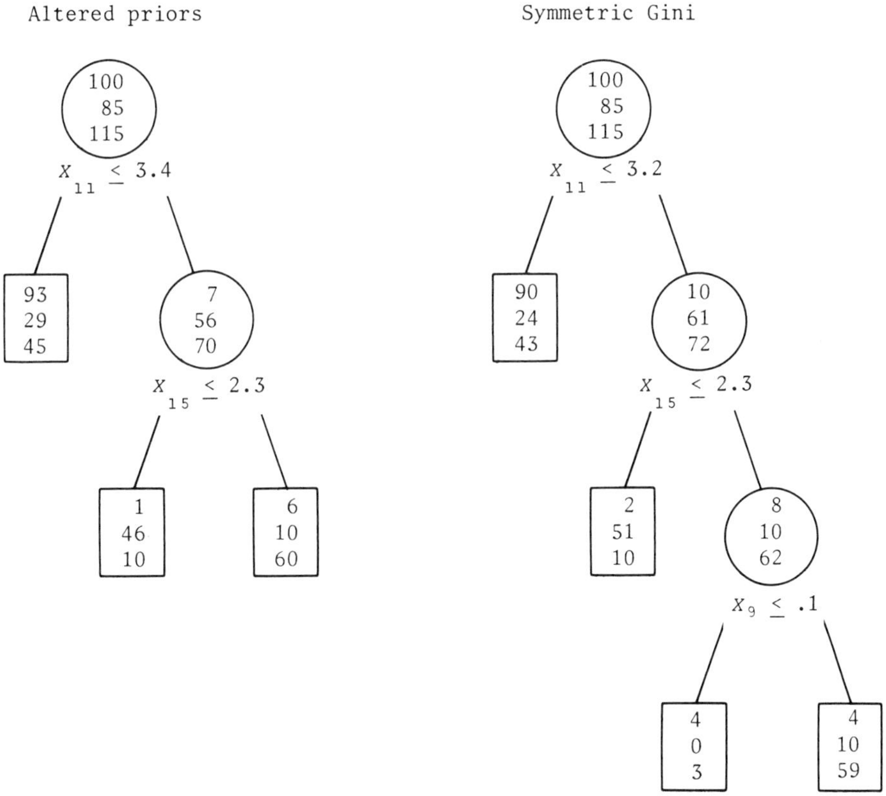

FIGURE 4.8 Tree diagrams.

and their test set costs are the same. The Gini R^{CV} is high, but we suspected that this was a random fluctuation. To check, we replicated with a different seed. In the replication, the Gini R^{CV} and R^{ts} differed by less than 1 SE.

The size is the major difference between these trees and the tree grown using unit costs. These trees are smaller with only two to three splits. The reason is this: Because it costs five times as much to misclassify a class 1 case, the main emphasis in the tree construction will be to guard class 1 cases from misclassification. This is mainly done in the first split with all but a few of the class 1s going into the left node.

The next split on the right separates classes 2 and 3, but not very thoroughly. However, since the cost of misclassifying $2 \leftrightarrow 3$ or $2 \rightarrow 1$, $3 \rightarrow 1$, is low, the pruning upward deletes the splits that further separate out classes 2 and 3. The result is that for the altered priors tree, the misclassification rate on the test sample for class 1 is .09. For classes 2 and 3 it is .45 and .46. Similar estimates hold for the Gini tree.

Using the tree grown on unit costs and the original priors, but computing costs for the test sample using the variable cost matrix results in the estimate .90. Therefore, in this example, costs can be halved by incorporating the variable costs into the tree construction and pruning.

In general, if one or a few classes have high misclassification costs in comparison to the other classes, then the tree construction will tend to ignore both the separation between the lower cost classes and their separation out at nodes already identified as being higher cost class nodes.

The costs and tree diagrams in the second example are given in Table 4.2 and Figure 4.9.

In the cost structure of this example, the mistake $2 \leftrightarrow 3$ is only 1/3 as costly as the mistakes $1 \leftrightarrow 2$ and $1 \leftrightarrow 3$. The priors are altered in the proportions 3:2:2. This does not produce much

TABLE 4.2 Costs

	R^{cv}	R^{ts}
Altered priors	.81 ± .07	.81
Symmetric Gini	.71 ± .07	.75

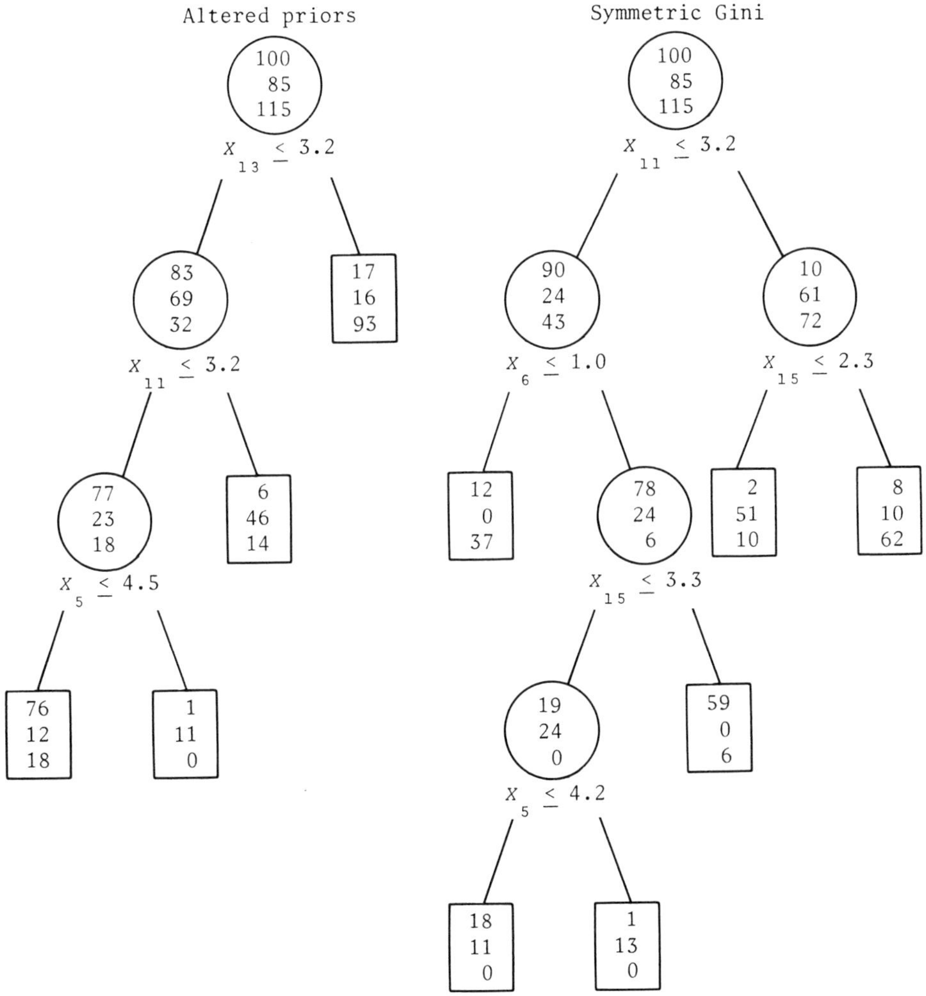

FIGURE 4.9 Tree diagrams.

difference in the priors, and the two trees are decidedly different. For instance, in the symmetric Gini, the first split selected is in the midregion, where class 1 can most easily be separated from 2 and 3.

There are some interesting facets in these examples. First, since the Gini criterion symmetrizes the loss matrix, the same criterion was used to grow the symmetric Gini trees in both examples. The only difference is that in the first example the pruning up uses the nonsymmetric losses. In the second example, the pruning uses the symmetric losses. The difference in the tree structures is substantial.

Second, take the tree grown on the waveform data using unit costs and the original priors (Figure 2.17). Use the 5000-case test sample together with the symmetric variable cost matrix to estimate the cost of this tree. The answer is .72, less than either of the two trees illustrated in Figure 4.9. Yet this tree was grown without regard for the variable misclassification costs.

The major reason for this apparently odd result is fairly simple. In the first example, class 1 could be accurately classified as long as the misclassification of classes 2 and 3 as class 1s could be more or less ignored. However, using univariate splits, there is a lower limit to the *mutual* separation between class 1 and classes 2 and 3. Regardless of how high $1 \leftrightarrow \{2, 3\}$ is weighted, matters will not improve much.

Third, in both examples, the higher losses inceases the SE's of the cross-validation estimates. Using the 1 SE rule allows a considerable increase in the cost estimate. A smaller increase may be desirable. For instance, in the altered prior tree (second example), the minimum cross-validated cost tree had a test sample cost of .71, as against the .81 cost of the tree selected by the 1 SE rule.

4.6 CLASS PROBABILITY TREES VIA GINI

4.6.1 Background and Framework

In some problems, given a measurement vector $\mathbf{x}$, what is wanted is an estimate of the probability that the case is in class j, $j = 1, \ldots, J$.

For instance, in a medical diagnosis situation, suppose the patient may have one of three diseases $D1$, $D2$, $D3$. Instead of classifying the patient as having one of the three, and ruling out the other two, it may be preferable to estimate the relative probabilities of his having $D1$, $D2$, or $D3$.

More precisely, in terms of the probability model given in Section 1.4, suppose the data are drawn from the probability distribution

$$P(A, j) = P(\mathbf{X} \in A, Y = j).$$

Then we want to construct estimates for the probabilities

$$P(j|\mathbf{x}) = P(Y = j|\mathbf{X} = \mathbf{x}), \; j = 1, \ldots, J.$$

In other words, given that we observe $\mathbf{x}$, estimate the probability that the case is in class j, $j = 1, \ldots, J$.

For this type of problem, instead of constructing classification rules, we want to construct rules of the type

$$\mathbf{d}(\mathbf{x}) = (d(1|\mathbf{x}), \ldots, d(J|\mathbf{x}))$$

with $d(j|\mathbf{x}) \geq 0$, $j = 1, \ldots, J$, and

$$\textstyle\sum_j d(j|\mathbf{x}) = 1, \text{ all } \mathbf{x}.$$

Such rules will be called class probability estimators.

Obviously, the best estimator for this problem, which we will call the Bayes estimator and denote by $\mathbf{d}_B$, is

$$\mathbf{d}_B(\mathbf{x}) = (P(1|\mathbf{x}), \ldots, P(J|\mathbf{x})).$$

A critical question is how to measure the accuracy of an arbitrary class probability estimator. Let E denote the expectation of a random variable. We adopt the following.

DEFINITION 4.17. *The accuracy of a class probability estimator* $\mathbf{d}$ *is defined by the value*

$$E[\sum_j (P(j|\mathbf{X}) - d(j|\mathbf{X}))^2].$$

However, this criterion poses an awkward problem, since its value depends on the unknown $P(j|\mathbf{x})$ that we are trying to estimate. Fortunately, the problem can be put into a different setting that resolves the difficulty. Let $\mathbf{X}$, Y on $X \times C$ have the distribution $P(A, j)$ and define new variables z_j, $j = 1, \ldots, J$, by

$$z_j = \begin{cases} 1 \text{ if } Y = j \\ 0 \text{ otherwise.} \end{cases}$$

Then

$$\begin{aligned} E(z_j|\mathbf{X} = \mathbf{x}) &= P(Y = j|\mathbf{X} = \mathbf{x}) \\ &= P(j|\mathbf{x}). \end{aligned}$$

Let $\mathbf{d}(\mathbf{x}) = (d(1|\mathbf{x}), \ldots, d(J|\mathbf{x}))$ be any class probability estimator.

DEFINITION 4.18. *The mean square error (MSE)* $R^*(\mathbf{d})$ *of* $\mathbf{d}$ *is defined as*

$$E[\textstyle\sum_j (z_j - d(j|\mathbf{X}))^2].$$

Thus, the MSE of $\mathbf{d}$ is simply the sum of its mean square errors as a predictor of the variables z_j, $j = 1, \ldots, J$.

The key identity is

PROPOSITION 4.19. *For any class probability estimator* $\mathbf{d}$,

$$R^*(\mathbf{d}) - R^*(\mathbf{d}_B) = E[\sum_j (P(j|\mathbf{X}) - d(j|\mathbf{X}))^2]. \tag{4.20}$$

The proof is a standard and simple exercise in conditional probabilities, which we omit. There are two interesting pieces of information in Proposition 4.19. The first is that among all class probability estimators, $\mathbf{d}_B$ has minimum MSE.

The second, and more important, is that *the accuracy of* $\mathbf{d}$ as defined in Definition 4.17, *differs from* $R^*(\mathbf{d})$ *only by the constant term* $R^*(\mathbf{d}_B)$. Therefore, *to compare the accuracy of two estimators* $\mathbf{d}_1$ *and* $\mathbf{d}_2$, *we can compare the values of* $R^*(\mathbf{d}_1)$ *and* $R^*(\mathbf{d}_2)$.

The significant advantage gained here is that $R^*(\mathbf{d})$ can be estimated from data, while accuracies cannot. We focus, then, on the problem of using trees to produce class probability estimates with minimal values of R^*.

4.6.2 Growing and Pruning Class Probability Trees

Assume that a tree T has been grown on a learning sample $(\mathbf{x}_n, j_n)$, $n = 1, \ldots, N$, using an unspecified splitting rule and has the set of terminal nodes $\tilde{T}$.

Associated with each terminal node t are the resubstitution estimates $p(j|t)$, $j = 1, \ldots, J$, for the conditional probability of being in class j given node t.

The natural way to use T as a class probability estimator is by defining: If $\mathbf{x} \in t$, then

$$\mathbf{d}(\mathbf{x}) = \big(p(1|t), \ldots, p(J|t)\big).$$

Stretching notation, either $\mathbf{d}$ or T will be used to denote this estimator, depending on which is more appropriate.

For each case $(\mathbf{x}_n, j_n)$ in the learning sample, define J values $\{z_{n,i}\}$ by

$$z_{n,i} = \begin{cases} 1 \text{ if } j_n = i \\ 0 \text{ otherwise.} \end{cases}$$

Then the resubstitution estimate $R(T)$ of $R^*(T)$ can be formed by this reasoning: For all $(\mathbf{x}_n, j_n)$ with $\mathbf{x}_n \in t$, $j_n = j$,

$$\sum_i (z_{n,i} - d(i|\mathbf{x}_n))^2 = (1 - p(j|t))^2 + \sum_{i \neq j} p^2(i|t)$$
$$= 1 - 2p(j|t) + S$$

where

$$S = \sum_i p^2(i|t).$$

Then put

$$R(\mathbf{d}) = \sum_{t \in \tilde{T}} \sum_j (1 - 2p(j|t) + S)p(j, t)$$
$$= \sum_{t \in \tilde{T}} \sum_j (1 - 2p(j|t) + S)p(j|t)p(t). \tag{4.21}$$

Evaluating the sum over j in the last expression gives

$$R(\mathbf{d}) = \sum_{t \in \tilde{T}} (1 - S)p(t). \tag{4.22}$$

The surprising thing is that

$$1 - S = 1 - \sum_j p^2(j|t)$$

is exactly the Gini diversity index (4.9). So growing a tree by using the Gini splitting rule continually minimizes the resubstitution estimate $R(T)$ for the MSE. In consequence, use of the Gini splitting rule is adopted as the best strategy for growing a class probability tree.

The major difference between classification trees grown using the Gini rule and class probability trees is in the pruning and selection process. Classification trees are pruned using the criterion $R(T) + \alpha|\tilde{T}|$, where

$$R(T) = \sum_{t \in \tilde{T}} r(t)p(t) \tag{4.23}$$

and $r(t)$ is the within-node misclassification cost.

Class probability trees are pruning upward using $R(T) + \alpha|\tilde{T}|$ *but with $r(t)$ the within node Gini diversity index.*

Take T_{max} to be grown as before, and prune upward, getting the sequence $T_1 > T_2 > \cdots > \{t_1\}$. To get test sample estimates

$R^{ts}(T)$ of $R^*(T)$ for T any of the T_k, run all the $N_j^{(2)}$ class j cases in the test sample down the tree T. Define

$$R_j^{ts}(T) = \frac{1}{N_j^{(2)}} \sum_{n,i} (z_{n,i} - d(i|\mathbf{x}_n))^2,$$

where the sum is over the $N_j^{(2)}$ test sample cases. Then we put

$$R^{ts}(T) = \sum_j R_j^{ts}(T)\pi(j). \tag{4.24}$$

If the priors are data estimated, the test sample estimates of them are used in (4.24).

If $T_1, \ldots, T_V$ are the V cross-validation trees associated with T, let $\mathbf{d}^{(v)}$, $v = 1, \ldots, V$, denote the corresponding class probability estimators. Define

$$R_j^{cv}(T) = \frac{1}{N_j} \sum_v \sum_{n,i} (z_{n,i} - d^{(v)}(i|\mathbf{x}_n))^2, \tag{4.25}$$

where the inner sum is over all class j cases in the vth test sample $\mathcal{L}_v$. Now put

$$R^{cv}(T) = \sum_j R_j^{cv}(T)\pi(j). \tag{4.26}$$

If the priors are data estimated, the entire learning sample estimates are used to estimate the $\pi(j)$ in (4.26).

Standard errors for the R^{ts} and R^{cv} estimates are derived in Chapter 11.

4.6.3 Examples and Comments

Class probability trees were constructed for both the digit and waveform data using the 1 SE rule. The results are summarized in Table 4.3.

TABLE 4.3

	$\|\tilde{T}\|$	R^{cv}	R^{ts}
Digit	11	.472 ± .045	.537 ± .009
Waveform	7	.489 ± .037	.449 ± .008

We calculated the value of the Gini index for the tree grown on the digit data using twoing to split and the misclassification rate to prune. The result, using the test sample data, was .553. This was also done for the waveform tree grown using the Gini index and pruned using the misclassification rate. The result, on test sample data, was .459.

The improvement, as measured by the R^{ts} values, is less than spectacular. Still, this procedure has not been extensively tested. There may be situations where the improvement is more significant.

APPENDIX

Proof of a Generalized Proposition 4.4

We give a proof of the following result, which includes Proposition 4.4 as a special case.

Proposition A.1 *Let* $\phi(p_1, \ldots, p_J)$ *be a strictly concave function on* $0 \le p_j \le 1$, $j = 1, \ldots, J$, $\sum_j p_j = 1$. *Then for*

$$i(t) = \phi(p(1|t), \ldots, p(J|t))$$

and any split $\mathcal{s}$,

$$\Delta i(\mathcal{s}, t) \ge 0$$

with equality if, and only if, $p(j|t_L) = p(j|t_R) = p(j|t)$, $j = 1, \ldots, J$.

By the strict concavity of ϕ,

$$\begin{aligned} i(t_L)p_L + i(t_R)p_R &= \phi(p(1|t_L), \ldots, p(J|t_L))p_L \\ &\quad + \phi(p(1|t_R), \ldots, p(J|t_R))p_R \\ &\leq \phi(p_L p(1|t_L) + p_R p(1|t_R), \ldots, p_L p(J|t_L) \\ &\quad + p_R p(J|t_R)) \end{aligned}$$

with equality holding if, and only if, $p(j|t_L) = p(j|t_R)$, $j = 1, \ldots, J$. Now

$$\begin{aligned} p_L p(j|t_L) + p_R p(j|t_R) &= [p(j, t_L) + p(j, t_R)]/p(t) \\ &= p(j|t). \end{aligned} \tag{A.2}$$

This implies that

$$i(t_L)p_L + i(t_R)p_R \leq i(t)$$

with equality if, and only if, $p(j|t_L) = p(j|t_R)$, $j = 1, \ldots, J$. If this latter holds, using (A.2) again gives $p(j|t_L) = p(j|t_R) = p(j|t)$, $j = 1, \ldots, J$.

Proof of Theorem (4.10) and Its Corollaries

Recall that for a given split s and a division of the classes $\{1, \ldots, J\}$ into the two superclasses C_1 and C_2, $\Delta i(s, t, C_1)$ denotes the decrease in node impurity computed as in a two-class problem using the $p(1|t)p(2|t)$ criterion. The problem is to find, for a given s, a superclass $C_1(s)$ so that

$$\Delta i(s, t, C_1(s)) = \max_{C_1} \Delta i(s, t, C_1).$$

For any node t, define

$$p(C_1|t) = \sum_{j \in C_1} p(j|t), \quad p(C_2|t) = \sum_{j \in C_2} p(j|t).$$

The following identity is proved later:

$$\Delta i(s, t, C_1) = p_L p_R [p(C_1|t_L) - p(C_1|t_R)]^2. \tag{A.3}$$

Now, define $z_j = p(j|t_L) - p(j|t_R)$, so $\sum_j z_j = 0$. For any real z, let z^+ and z^- be its positive and negative parts: $z^+ = z$ and $z^- = 0$ if $z \geq 0$; $z^+ = 0$ and $z^- = -z$ if $z \leq 0$. Then $z = z^+ - z^-$, and $|z| = z^+ + z^-$.

From (A.3), since p_L, p_R depend only on s and not on C_1, then $C_1(s)$ either maximizes $S = \sum_{j \in C_1} z_j$ or minimizes it. The maximum value of S is achieved by taking $C_1 = \{j;\ z_j \geqslant 0\}$ and equals $\sum_j z_j^+$. The minimum value is achieved by $C_1 = \{j;\ z_j < 0\}$ and equals $-\sum_j z_j^-$.

Noticing that

$$\sum_j z_j^+ - \sum_j z_j^- = \sum_j z_j = 0,$$

we see that the maximum value of S equals the absolute value of the minimum value and that both equal

$$\Big(\sum_j z_j^+ + \sum_j z_j^-\Big)/2 = \Big(\sum_j |z_j|\Big)/2.$$

Then, using (A.3),

$$\max_{C_1} \Delta i(s, t, C_1) = p_L p_R \Big[\sum_j |p(j|t_L) - p(j|t_R)|\Big]^2/4,$$

and a maximizing superclass is $C_1(s) = \{j;\ p(j|t_L) \geqslant p(j|t_R)\}$. This proves Theorem 4.10 once (A.3) is derived.

To get (A.3), write

$$\Delta i(s, t) = p(1|t)p(2|t) - p_L p(1|t_L)p(2|t_L) - p_R p(1|t_R)p(2|t_R).$$

Using $p(2|\cdot) = 1 - p(1|\cdot)$ for t, t_L, t_R gives

$$\Delta i(s, t) = p(1|t) - p(1|t_L)p_L - p(1|t_R)p_R \qquad \text{(A.4)}$$
$$+ p^2(1|t_L)p_L + p^2(1|t_R)p_R - p^2(1|t).$$

Replacing $p(1|t)$ in (A.4) by $p_L p(1|t_L) + p_R p(1|t_R)$ leads to

$$\begin{aligned} \Delta i(s,\ t) &= p^2(1|t_L)p_L + p^2(1|t_R)p_L - (p_L p(1|t_L) + p_R p(1|t_R))^2 \\ &= p_L p_R p^2(1|t_L) + p_L p_R p^2(1|t_R) - 2p_L p_R p(1|t_L)p(1|t_R) \\ &= p_L p_R (p(1|t_L) - p(1|t_R))^2. \end{aligned}$$

This last is equivalent to (A.3).

5

STRENGTHENING AND INTERPRETING

5.1 INTRODUCTION

The methodological development produced some features that were added to the basic tree structure to make it more flexible, powerful, and efficient.

The tree growing procedure described in the previous chapters uses splits on one variable at a time. Some problems have a structure that suggests treatment through combinations of variables. Three methods for using combinations are given in Section 5.2.

Section 5.3 deals with predictive association between splits and the definition of surrogate splits. This is a useful device which is analogous to correlation in linear models. It is used to handle data with missing variables and give a ranking of variable importance.

Although cross-validation gives accurate estimates of overall tree cost, it is not capable of improving the resubstitution estimates of the individual terminal node costs. Two heuristic methods for improved within-node cost estimates are discussed in Section 5.4.

In Section 5.5 the important issues are interpreting and exploring the data structure through trees. The instability of the tree topology is illustrated. Two avenues for tree interpretation are discussed: first, a close examination of the tree output; second, the growing of exploratory trees, both before and after the main tree procedure. A method is given for rapid computation of exploratory trees.

The question of computational efficiency is covered in Section 5.6. Some benchmarks are given for tree construction time. With large data sets, a method of subsampling can be used which significantly decreases the time requirement while having only a minor effect on the tree structure.

Finally, Section 5.7 gives a comparison of tree structured classification with nearest neighbor and discriminant function methods as applied to the digit and wave recognition examples.

The appendix gives a description of the search algorithm for finding best linear combination splits.

5.2 VARIABLE COMBINATIONS

5.2.1 Introduction

In Chapter 2 we noted that at times the data structure may be such that it makes more sense to split on combinations of variables than on the individual original variables.

We have found three useful combination procedures. The first is a search for a best linear combination split; the second uses Boolean combinations; and the third is through the addition of features—ad hoc combinations of variables suggested by examination of the data.

5.2.2 Linear Combinations

In some data, the classes are naturally separated by hyperplanes not perpendicular to the coordinate axes. These problems are difficult for the unmodified tree structured procedure and result in large trees as the algorithm attempts to approximate the hyperplanes by multidimensional rectangular regions. To cope with such situations, the basic structure has been enhanced to allow a search for best splits over linear combinations of variables.

The linear combination algorithm works as follows. Suppose there are M_1 ordered variables (categorical variables are excluded). If there are missing data, only those cases complete in the ordered variables are used. At every node t, take a set of coefficients $\mathbf{a} = (a_1, \ldots, a_{M_1})$ such that $\|\mathbf{a}\|^2 = \sum_m a_m^2 = 1$, and search for the best split of the form

$$\sum_m a_m x_m \leq c \tag{5.1}$$

as c ranges over all possible values. Denote this split by $s^*(\mathbf{a})$ and the corresponding decrease in impurity by $\Delta i(s^*(\mathbf{a}), t)$. The best set of coefficients $\mathbf{a}^*$ is that $\mathbf{a}$ which maximizes $\Delta i(s^*(\mathbf{a}), t)$. That is,

$$\Delta i(s^*(\mathbf{a}^*), t) = \max_{\mathbf{a}} \Delta i(s^*(\mathbf{a}), t).$$

This produces a linear split of the form

$$\sum_m a_m^* x_m \leq c^*.$$

Although the concept is easily stated, the implementation in terms of an effective search algorithm for maximizing $\Delta i(s^*(\mathbf{a}), t)$ over the large set of possible values of $\mathbf{a}$ is complex. The details of the algorithm we use are given in the appendix to this chapter. It is not guaranteed to produce $\mathbf{a}^*$. Like other search algorithms, it may get trapped in local maxima.

For high-dimensional data this gives a complicated tree structure. At each node one has to interpret a split based on a linear combination of all ordered measurement variables. But some of the variables in the combination may contribute very little to the effectiveness of the split. To simplify the structure, we weed out these variables through a backward deletion process.

For m ranging from 1 to M_1, vary the threshold constant c and find the best split of the form

$$\sum_{m' \neq m} a^*_{m'} x_{m'} \leq c_m.$$

That is, we find the best split using the linear combination with coefficients **a*** but deleting x_m and optimizing on the threshold c. Denote the decrease in impurity using this split by Δ_m and

$$\Delta^* = \Delta i(s^*(a^*), t).$$

The most important single variable to the split $s^*(a^*)$ is the one whose deletion causes the greatest deterioration in performance. More specifically, it is that variable for which Δ_m is a minimum. Similarly, the least important single variable is the one for which Δ_m is a maximum.

Measure the deterioration due to deleting the most important variable by $\Delta^* - \min_m \Delta_m$, and the deterioration due to deleting the least important variable by $\Delta^* - \max_m \Delta_m$. Set a constant β, usually .2 or .1, and if

$$\Delta^* - \max_m \Delta_m < \beta(\Delta^* - \min_m \Delta_m),$$

then delete the least important variable.

Now repeat this procedure on the remaining undeleted variables. Keep iterating until no more deletion occurs. Denote the indices of the undeleted variables by $\{m_1\}$. Now the search algorithm is used again to find the best split of the form

$$\sum_{m \in \{m_1\}} a^{**}_m x_m \leq c^{**}$$

under the restriction

$$\sum_{m \in \{m_1\}} (a_m^{**})^2 = 1.$$

Compared to stepwise methods in regression or discrimination, where variables are retained even if they give only slight increases in the stepping criterion, the tree deletion procedure is a much more stringent weeding out of variables. This is justified on two grounds. First, because of the tree structure, the algorithm always has another crack at the data. Second, the inclusion of too many variables in any single split gives a confusing tree structure.

The introduction of linear combinations allows the tree-growing procedure to discover and use any linear structure in the data. But the results are no longer invariant under monotone transformations of individual variables. However, because a direct optimization search is used, the effect of outliers or long-tailed distributions will be small compared with their effect on discriminant methods.

A second difficulty is the loss in interpretability of the results. One of the advantages of the unmodified tree process is the ready and simple understanding conveyed by univariate splits on the original variables. But if the data do possess a strong linear structure, then the simplicity of the unmodified tree structure can be quite deceptive. From the many splits generated to approximate the linear separability, it will be difficult to tell, for instance, that a split of the form $x_1 + x_2 \leq 3$ gives an optimal separation.

To illustrate, the linear combination procedure was run on the waveform recognition example with $\beta = .2$. The tree selected (1 SE rule) had a cross-validated error rate of $.23 \pm .02$, with a test sample rate (5000 samples) of .20.

In contrast to the tree illustrated in Figure 2.17, the linear combination tree has only three terminal nodes. It is diagrammed in Figure 5.1. The linear splits used are given beneath the nodes, and the learning sample class populations are given inside the terminal nodes.

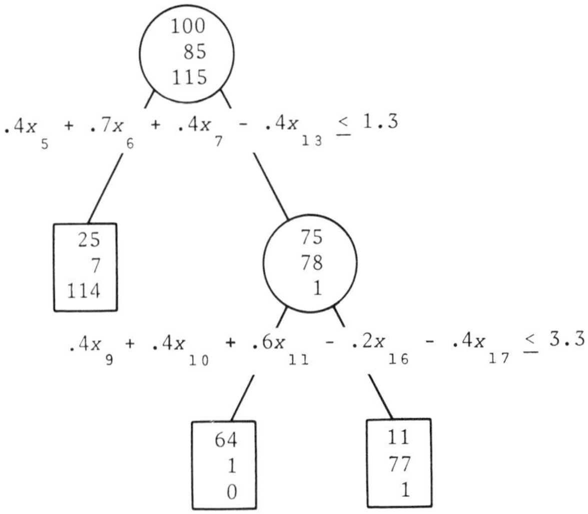

FIGURE 5.1

The first linear combination separates off class 3 as being low in coordinates 5, 6, 7 and high on the thirteenth coordinate. Then it separates class 1 from class 2 as being low in coordinates 9, 10, 11 and high in the sixteenth and seventeenth.

The test set error rate is improved by .08 through the use of linear combinations, and the structure of the tree is considerably simplified.

The linear combination splits are *added* to the univariate splits. So even with this option, it may happen that the split selected at a node is univariate.

For other methods of constructing piecewise linear classifiers, see Meisel (1972), Friedman (1977), and Sklansky (1980).

5.2.3 *Boolean Combinations*

Another class of problems which are difficult for the basic tree procedure is characterized by high dimensionality and a Boolean structure. A common class of such problems occurs in medical diagnostic data sets consisting of a large number of binary variables indicating the presence or absence of certain symptoms and other variables which are medical test results. Another example is in the classification of chemical compounds through the peaks in their mass spectra.

Both doctors and mass spectrographers look for certain combinations of variables as being significant, that is, the presence of symptom 3 together with a positive result on test 5, or the presence of a peak at location 30 together with another peak at location 44. The general kind of thing that is being looked for is of the form: Does the case have property (D_1 and D_2) or (D_3 and D_5), etc.? We refer to such an expression as a Boolean combination.

As in the linear case, the basic tree structure may eventually uncover the structure of the data, but only at the cost of many splits and a resulting disguise of the structure. The method outlined below has been designed to deal more effectively with a Boolean structure.

Since the class of all Boolean combinations of splits is extremely large and can lead to a confusing tree, the class of Boolean combinations considered is restricted to splits of the form

$$\{\text{Is } x_{m_1} \in B_1 \text{ and } x_{m_2} \in B_2 \text{ and } \ldots \text{ and } x_{m_h} \in B_h?\}. \tag{5.2}$$

This includes combinations of the form: Does the patient have symptom 3 and read positive on test 5? Or, does the spectrum have a peak at 30 with intensity greater than e_1 and a peak at 44 with intensity greater than e_2? When the complementary node is considered, it also includes splits of the form

$$\{\text{Is } x_{m_1} \in B_1 \text{ or } x_{m_2} \in B_2 \text{ or } \dots \text{ or } x_{m_h} \in B_h?\}. \tag{5.3}$$

The class (5.2) of Boolean splits is denoted as

$$s_{m_1} \cap s_{m_2} \cap \dots \cap s_{m_n}$$

and interpreted as the set of all cases sent to t_L by *every* split in the set $\{s_{m_1}, \dots, s_{m_n}\}$. Denote the decrease in impurity of the node t by the split as

$$\Delta i(s_{m_1} \cap s_{m_2} \cap \dots \cap s_{m_n}, t). \tag{5.4}$$

Theoretically, the optimal procedure is to maximize (5.4) over all splits on variables $x_{m_1}, \dots, x_{m_n}$ and then to maximize over all subsets $\{m_1, \dots, m_n\} \subset \{1, \dots, M\}$. At present, we do not know of a feasible way to implement this direct maximization procedure. Instead, a stepwise method is used.

If a split s on an ordered variable x is of the form $\{\text{Is } x \leq c?\}$, let $\bar{s}$ be the split $\{\text{Is } x > c?\}$. If s is a split on a categorical variable x of the form $\{\text{Is } x \in \{b_1, \dots, b_h\}?\}$, denote by $\bar{s}$ the split $\{\text{Is } x \notin \{b_1, \dots, b_h\}?\}$.

DEFINITION 5.5. *If s is any split of the form* $\{\text{Is } x \in B?\}$, then *the complementary split $\bar{s}$ to s is defined as* $\{\text{Is } x \in B^C?\}$.

Let s_m^* be the best split on the variable x_m and take S^* to be the set of splits

$$S^* = \{s_1^*, \bar{s}_1^*, s_2^*, \bar{s}_2^*, \dots, s_M^*, \bar{s}_M^*\}.$$

Then the stepwise process goes this way.

1. If $s_{m_1}^*$ is the best split in S^*, find the $s \in S^*$ that maximizes

$$\max(\Delta i(s_{m_1}^* \cap s, t), \Delta i(\bar{s}_{m_1}^* \cap s, t)).$$

If that maximum value is $\Delta i(s_{m_1}^* \cap s^*, t)$, denote $s_1^* \cap s_2^*$

$= s^*_{m_1} \cap s^*$. If the maximum value is $\Delta i(\bar{s}^*_{m_1} \cap s^*, t)$, denote $s^*_1 \cap s^*_2 = \bar{s}^*_{m_1} \cap s^*$.

2. Find the s in S^* that maximizes $\Delta i(s^*_1 \cap s^*_2 \cap s, t)$. If the maximum is achieved at $s = s^*$, denote $s^*_1 \cap s^*_2 \cap s^*_3 = s^*_1 \cap s^*_2 \cap s^*$. Continue adding splits to this intersection until step 3 is satisfied.
3. Fix $\beta > 0$; if at any stage in step 2

$$\Delta i(s^*_1 \cap \cdots \cap s^*_{n+1}, t) \leq (1 + \beta)\Delta i(s^*_1 \cap \cdots \cap s^*_n),$$

then stop and use the split $s^*_1 \cap \cdots \cap s^*_n$.

If data are missing, the algorithm operates on all data which are complete for the splits being considered.

The Boolean combination method preserves the invariance of the tree structure under monotone univariate transformations and maintains the simplicity of interpretation. It has been successfully used in a number of applications. An example is given in Chapter 7.

5.2.4 Using Features

The term *feature* is used in the pattern recognition literature to denote a variable that is manufactured as a real-valued function of the variables that were originally measured.

Features are generally constructed for one of two reasons:

1. The original data are high dimensional with little usable information in any one of the individual coordinates. An attempt is made to "concentrate" the information by replacing a large number of the original variables by a smaller number of features.
2. On examination, the structure of the data appears to have certain properties that can be more sharply seen through the values of appropriate features.

A powerful aspect of the tree structured approach is that it permits the introduction of a large number of candidate features and then selects the best among them to split the data on. An example of this is in the reduction of the high-dimensional chemical spectra problem described in Chapter 7.

We use the waveform recognition data to illustrate the construction and use of features. Our claim here is that even if we did not know the mechanism that generated the waveform data, examination of the data would show that an individual waveform tended to be consistently high in some areas and low in others and, furthermore, that these regions of highs and lows were the characteristics that discriminated between classes.

Following this intuitive appraisal, 55 new features were constructed. These were averages, $\bar{x}_{m_1,m_2}$, over the variables from m_1 to m_2 for m_1, m_2 odd, that is

$$\bar{x}_{m_1,m_2} = \frac{1}{m_2 - m_1 + 1} \sum_{m=m_1}^{m_2} x_m, \quad m_2 > m_1.$$

The tree procedure was then run using the original 21 variables and the 55 added features, for a total of 76 variables.

The tree selected by the 1 SE rule had a cross-validated error of .20 and a test sample (5000 samples) error rate of .20. The tree had only four terminal nodes (see Figure 5.2).

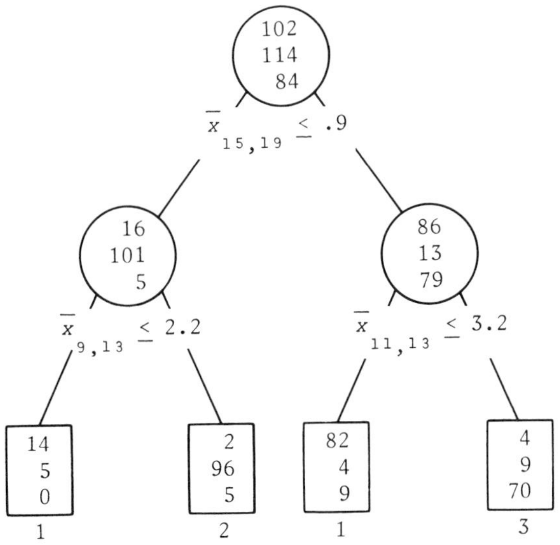

FIGURE 5.2

The structure in this tree is more discernible than in the unvariate split tree. The first split separates out class 2 by its low average over x_{15} to x_{19}. Then, on the right, classes 1 and 3 are separated by their average values over x_{11} to x_{13}. Then the left node, mostly class 2 but with 16 class 1 cases, is further purified by splitting on the average over x_9 to x_{13}.

Construction of features is a powerful tool for increasing both accuracy and understanding of structure, particularly in high dimensional problems. The selection of features is an art guided by the analyst's intuition and preliminary exploration of the data. However, since the tree structure allows an almost unlimited introduction of features, inventiveness is encouraged.

On the other hand, cross-validation and pruning prevent the new features from overfitting the data and keep the tree honest.

5.3 SURROGATE SPLITS AND THEIR USES

5.3.1 Definitions

At any given node t, let s^* be the best split of the node into t_L and t_R. In the basic tree structure, s^* is the best univariate split. In general, s^* can be the best linear or Boolean combination split.

Assume standard structure. Take any variable x_m. Let S_m be the set of all splits on x_m, and $\bar{S}_m$ the set of splits complementary to S_m. For any split $s_m \in S_m \cup \bar{S}_m$ of the node t into t_L' and t_R', let $N_j(LL)$ be the number of cases in t that both s^* and s_m send left, that is, that go into $t_L \cap t_L'$. By our usual procedure, we estimate the probability that a case falls into $t_L \cap t_L'$ as

$$p(t_L \cap t_L') = \sum_j \pi(j)N_j(LL)/N_j.$$

Then we define the estimated probability $p_{LL}(s^*, s_m)$ that *both* s^* and s_m send a case in t left as

$$p_{LL}(s^*, s_m) = p(t_L \cap t'_L)/p(t).$$

Similarly, define $p_{RR}(s^*, s_m)$.

Suppose we tried to predict the results of the split s^* knowing only the results of the split s_m. If s_m sends a case to t'_L, predict that s^* sends it to t_L, and for cases sent to t'_R, predict that they are sent to t_R. Then the probability that s_m predicts s^* correctly is estimated by

$$p(s^*, s_m) = p_{LL}(s^*, s_m) + p_{RR}(s^*, s_m). \tag{5.6}$$

DEFINITION 5.7. *A split* $\tilde{s}_m \in S_m \cup \bar{S}_m$ *is called a surrogate split on* x_m *for* s^* *if*

$$p(s^*, \tilde{s}_m) = \max_{s_m} p(s^*, s_m),$$

where the maximum is over $S_m \cup \bar{S}_m$.

Usually, the surrogate split $\tilde{s}_m$ is unique and can be interpreted as the split on x_m that most accurately predicts the action of s^*.

To define a measure of how good a predictor $\tilde{s}_m$ is of s^* requires some more consideration. Suppose that s^* sends the cases in t left with relative probability p_L and right with relative probability $p_R (p_L + p_R = 1)$. If a new case falls into t, then a natural way to predict whether it will send it right or left is the rule: Predict t_L if $p_L = \max(p_L, p_R)$; otherwise t_R.

Given that a case falls into t, this rule has error probability $\min(p_L, p_R)$. It may be that using $\tilde{s}_m$ gives a prediction rule with error probability just a bit smaller than the $\max(p_L, p_R)$ rule. Then little is gained by using $\tilde{s}_m$ to predict s^*. In consequence, to get a measure of how good a predictor $\tilde{s}_m$ is, its error probability should be compared with $\min(p_L, p_R)$.

DEFINITION 5.8. *Define the predictive measure of association* $\lambda(s^*|\tilde{s}_m)$ *between* s^* *and* $\tilde{s}_m$ *as*

$$\lambda(s^*|\tilde{s}_m) = \frac{\min(p_L, p_R) - (1 - p(s^*, \tilde{s}_m))}{\min(p_L, p_R)}.$$

This measure is the relative reduction in error gotten by using $\tilde{s}_m$ to predict s^* as compared with the $\max(p_L, p_R)$ prediction. If $\lambda(s^*|\tilde{s}_m) \leq 0$, $\tilde{s}_m$ is no help in predicting s^* and is discarded as a surrogate split.

The surrogate splits and their measures of predictive association have three major uses: handling missing data, variable importance ranking, and detection of masking. The last use will be discussed in Section 5.5.3.

5.3.2 Missing Data

The missing data algorithm is designed to accomplish two purposes simultaneously: first, to make maximum use of the data cases, complete or not, in the tree construction; second, to construct a tree that will classify any case dropped into it, even if the case has some variable values missing. This differs from the usual missing value procedures in regression or classification, where the covariance matrix is filled in and then used to produce a single prediction equation defined only on complete cases.

The missing value algorithm works as follows. Suppose that the best split s^* on a node is being found. If there are missing values, the best split s_m^* on x_m is computed using all cases containing a value of x_m and then s^* selected as that split s_m^* which maximizes $\Delta i(s_m^*, t)$. In linear combinations, the best split is computed using all cases complete in the ordered variables. For a Boolean split, all cases complete in the variables appearing in the Boolean expression are used.

If a case has missing values so that s^* is not defined for that case, proceed as follows. Among all nonmissing variables in the case, find that one, say, x_m, with $\tilde{s}_m$ having the highest measure of predictive association with s^*. Then split the case using $\tilde{s}_m$.

This procedure is analogous to replacing a missing value in a linear model by regressing on the nonmissing value most highly correlated with it. However, it is more robust. For example, in linear regression, when missing values are filled in by regressing on nonmissing values, the results can sometimes be catastrophic. The regression coefficients are computed by inverting the covariance matrix and are therefore sensitive to the smaller eigenvalues. The bias introduced by filling in the missing values can sometimes result in nearly singular covariance matrices and generally produces poor estimates unless the proportion of missing data is small. Even the sophisticated EM algorithm which produces maximum likelihood estimates under the assumption of joint normality tends to break down when the missing value pattern is not random or the predictor variables have asymmetric distributions.

In the tree structured algorithm, the cases with missing values in the selected splitting variable do not determine which way the other cases will go. The worst case situation is that the highest available $\lambda(s^*|\tilde{s}_m)$ is small. Then the case will be assigned to t_L or t_R with error probability approximately $\min(p_L, p_R)$. Frequently, in a high-dimensional problem, the proportion of missing values in most of the variables is small. Thus, the number of cases affected at any given node will generally be small. Furthermore, since the splitting continues, there is always the potential that the cases which were sent the wrong way by the missing value algorithm will still be classified correctly.

See Friedman (1977) for an earlier approach to the missing data problem in classification trees.

5.3.3 Examples with Missing Data

To see how (and how much) missing data affects tree construction, the 7-variable digit recognition data and the 21-variable waveform recognition data had 5 percent, 10 percent, and 25 percent of the data deleted completely at random. On the average, the number of variables deleted per case and the percent of complete cases are given in Tables 5.1 and 5.2.

TABLE 5.1 Digit Recognition Data

% Data Missing	No. Variables Deleted/Case	% Complete Cases
5%	.35	70
10%	.70	48
25%	1.75	13

TABLE 5.2 Waveform Recognition Data

% Data Missing	No. Variables Deleted/Case	% Complete Cases
5%	1.05	34
10%	2.10	11
25%	5.25	0.2

When we look at the effects of missing data, the most reasonable assumption is that future data will contain the same proportion and kind of missing values as the present data. This is implicitly assumed in the construction of the cross-validation estimates, since the test samples contain about the same proportion of missing values as the learning samples.

Therefore, in the first experiment, trees were grown on the 5 percent, 10 percent, and 25 percent deleted data sets and checked against 5000-case test samples with the same percentage of deleted variables. The results are given in Tables 5.3 and 5.4 for the 1 SE trees.

TABLE 5.3 Digit Recognition Data

% Data Missing	$\|\widetilde{T}\|$	R^{CV}	R^{ts}
0	10	.30 ± .03	.30
5	10	.34 ± .03	.32
10	12	.39 ± .03	.35
25	10	.48 ± .04	.44

TABLE 5.4 Waveform Recognition Data

% Data Missing	$\|\tilde{T}\|$	R^{cv}	R^{ts}
0	11	.29 ± .03	.28
5	12	.33 ± .03	.28
10	8	.34 ± .03	.32
25	5	.35 ± .03	.32

The effect of missing data is apparent in the digit recognition problem, with the error rate rising to 44 percent. In the waveform data, for each class, the correlations between variables are generally high. So it is not surprising that surrogate splits can take the place of the optimal split without much degradation. In the digit data, for each class the variables are independent and the effectiveness of the surrogate splits is based on the redundancy in the noiseless representation of the digits.

The loss of accuracy due to missing values is caused by an interaction of two factors: first, the tree constructed using incomplete data is less accurate; second, it is more difficult to classify future cases having missing values.

To estimate these effects separately in the digit recognition data, the performance of the 1 SE trees constructed using 5 percent, 10 percent, 25 percent deleted data was checked using a 5000-case test sample of complete data vectors. In a second experiment, a tree was constructed using the complete data. Then 5 percent, 10 percent, and 25 percent of the data were deleted at random for the 5000-case test sets and these were run through the tree. The results are given in Table 5.5.

In this example, the major effect is in the difficulty in classifying new data vectors having missing values. The loss in accuracy in classifying complete data vectors due to the use of incomplete data in tree construction is minimal. In fact, for 25 percent missing in construction, the tree with 14 terminal nodes

TABLE 5.5

First Experiment		Second Experiment	
% Data Missing (Learning Sample)	R^{ts}	% Data Missing (Test Sample)	R^{ts}
0	.30	0	.30
5	.30	5	.32
10	.30	10	.35
25	.30	25	.43

immediately preceding the 1 SE tree in the $\{T_k\}$ sequence had a test sample error rate of .28.

The difficulty in classifying measurement vectors with missing variables might have been anticipated. In the model generating the digit data, the overall correlations between variables are in the .1 to .2 range. If a variable is missing, there are no good surrogate splits on it.

In contrast, when the same two experiments were run on the waveform data, the R^{ts} values were all in the .29 to .31 range. With the highly correlated waveform variables, good surrogate splits were available, and the tree had no difficulty in classifying cases even with 25 percent of the variables missing.

Caution should be used in generalizing from these examples. If the missingness does not occur at random, the effects can be larger than just noted. For example, if, whenever a variable is missing, the variables containing the best surrogate splits also tend to be missing, then the effect will be magnified. Chapter 8, on regression, has an example which shows how significant missing data can be.

5.3.4 Variable Ranking

A question that has been frequent among tree users is: Which variables are the most important? The critical issue is how to rank those variables that, while not giving the best split of a node, may give the second or third best. For instance, a variable x_1 may never occur in any split in the final tree structure. Yet, if a masking variable x_2 is removed and another tree grown, x_1 may occur prominently in the splits, and the resulting tree may be almost as accurate as the original. In such a situation, we would require that the variable ranking method detect "the importance" of x_1.

The most satisfactory answer found to date is based on the surrogate splits $\tilde{s}_m$. Let T be the optimal subtree selected by the cross-validation or test sample procedure. If the Gini splitting rule has been used, then at each node $t \in T$, compute $\Delta I(\tilde{s}_m, t)$. If twoing is used, define $\Delta I(\tilde{s}_m, t) = \max_{C_1} \Delta I(\tilde{s}_m, t, C_1)$.

DEFINITION 5.9. *The measure of importance of variable x_m is defined as*

$$M(x_m) = \sum_{t \in T} \Delta I(\tilde{s}_m, t).$$

(If there is more than one surrogate split on x_m at a node, use the one having larger ΔI.)

The concept behind the variable ranking is this: If the best split of node t in the univariate case is on x_{m_1}, and if x_{m_2} is being masked at t, that is, if x_{m_2} can generate a split similar to $s^*_{m_1}$ but not quite as good, then at t, $\Delta I(\tilde{s}_{m_2}, t)$ will be nearly as large as $\Delta I(s^*_{m_1}, t)$.

However, the idea is somewhat more subtle than it appears. This is seen by contrasting it with another approach that was eventually discarded. In the latter, at each t, $\Delta I(s^*_m, t)$ was computed, and the measure of importance was defined by

$$\sum_{t \in T} \Delta I(s^*_m, t).$$

That is, the measure of importance was the sum over all nodes of the decrease in impurity produced by the best split on x_m at each node.

This is unsatisfactory in the following sense. Suppose that at node t, the split $s^*_{m_1}$ has low association with s^*_m but that $\Delta I(s^*_{m_1}, t)$ ranks high among the values $\Delta I(s^*_m, t)$, $m = 1, \ldots, M$. Because of the low association, when t is split into t_L and t_R by s^*, it may happen that the optimal splits on x_{m_1} in either t_L or t_R (or both) are close to $s^*_{m_1}$ and have comparatively large ΔI values. Then, essentially the same split has contributed to the variable importance of x_{m_1}, not only at t but also at t_L and/or t_R. Further, it can keep contributing at the nodes below t_L and t_R until the split is used or its splitting power dissipated. Thus, the importance given x_{m_1} will be misleadingly high.

Definition 5.9 does not have any obvious drawbacks, and in simulation examples where it is possible to unambiguously define variable importance from the structure of the example, it has given results in general agreement with expectations.

Since only the relative magnitudes of the $M(x_m)$ are interesting, the actual measures of importance we use are the normalized quantities $100M(x_m)/\max_m M(x_m)$. The most important variable then has measure 100, and the others are in the range 0 to 100.

For example, in the digit recognition data, the variable importances computed are given in Figure 5.3. Adding 17 zero-one noise variables to the data gives the measures of importance shown in Figure 5.4.

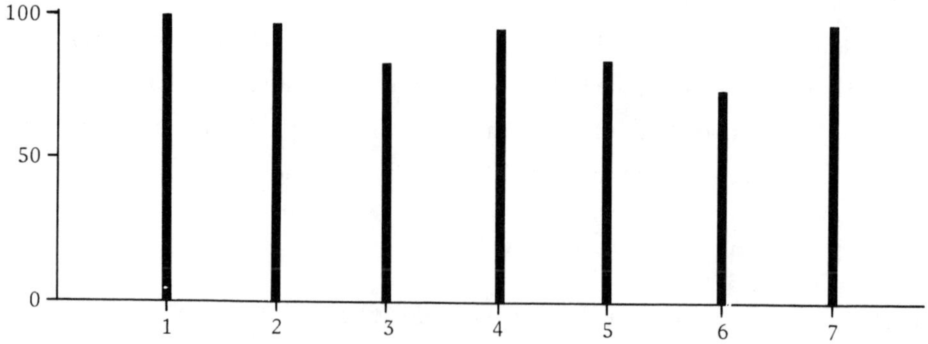

FIGURE 5.3

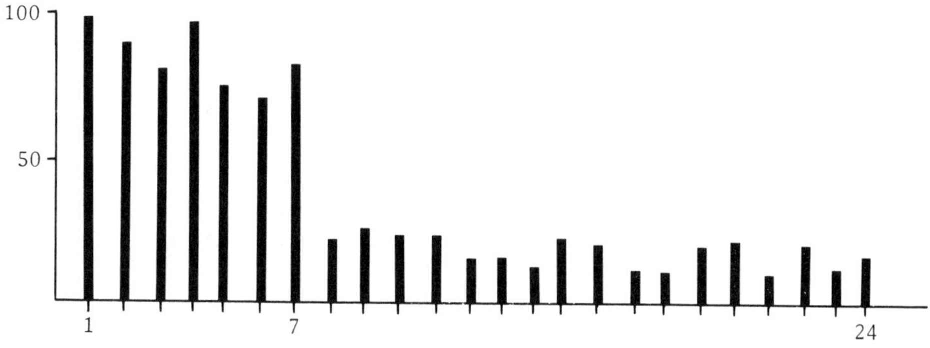

FIGURE 5.4

In the waveform data, the measures of importance are as shown in Figure 5.5. The only variables that are split in the tree selected are 6, 7, 10, 11, 15, 16, 17. The importance measure indicates that the other variables, with the exception of the first and last few, also carry significant splitting power but are being masked.

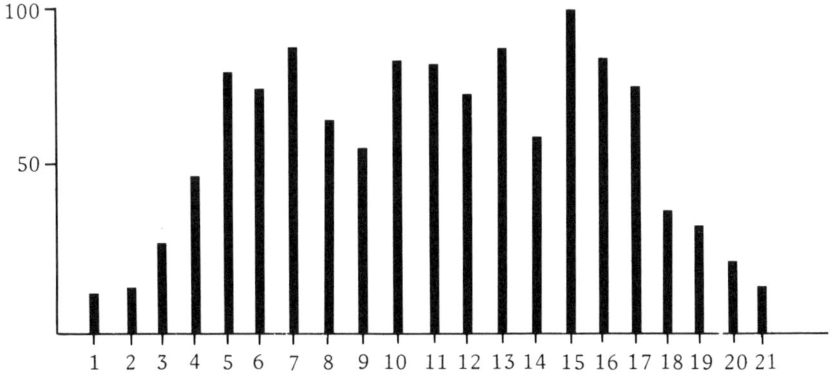

FIGURE 5.5

When this same example is run with 19 noise variables added having a $N(0, 1)$ distribution, the importances for the first 21 variables are as shown in Figure 5.6. The 19 noise variable importances are shown in Figure 5.7.

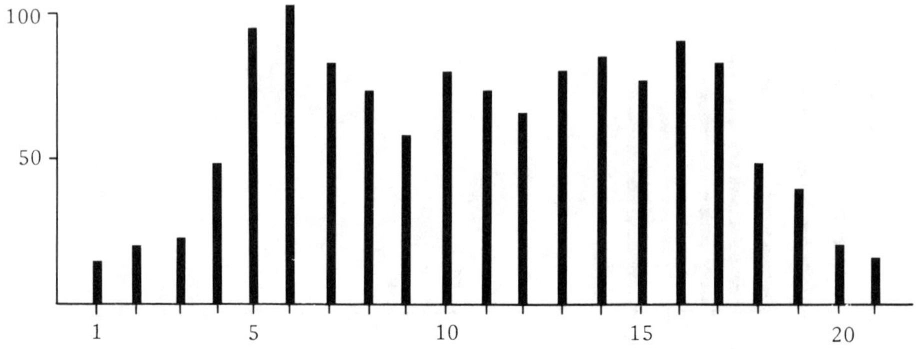

FIGURE 5.6

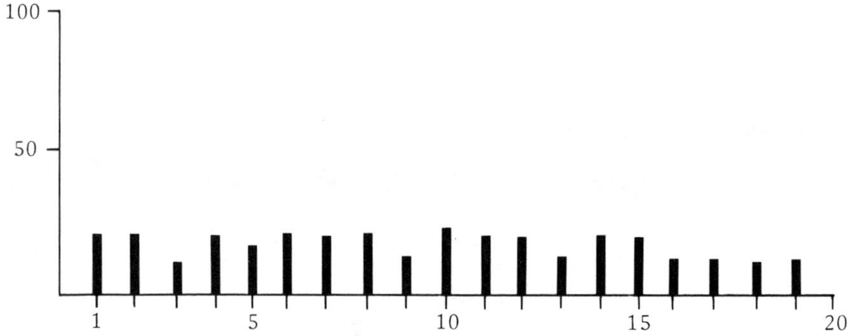

FIGURE 5.7

Caution should be used in interpreting these variable importances. Importance can be defined in many different ways. We make no claim that ours is intrinsically best. Furthermore, they can be appreciably altered by random fluctuations in the data (see Section 5.5.2).

See Darlington (1968) for a discussion of measures of variable importance in the context of multiple linear regression.

5.4 ESTIMATING WITHIN-NODE COST

In many applications, the user not only wants a classification for any future case, but also some information about how sure he should be that the case is classified correctly. The true value of the probability of misclassification given that the case falls into terminal node t provides this information. The value $r(t)$ is the resubstitution estimate for this probability in the unit cost case. With variable costs, $r(t)$ is an estimate for the expected misclassification cost given that the case falls into t.

Unfortunately, the node estimates can be even more misleading than the tree resubstitution estimate $R(T)$. This is particularly so if the terminal node t is the result of a number of splits and has a relatively small population. At each split the tree algorithm tries to slice the data so as to minimize the impurity of the descendants. The terminal nodes look much purer in terms of the learning sample than on subsequent test samples. For instance, Table 5.6 gives the resubstitution and test sample estimates for the 11 terminal nodes of the waveform tree drawn in Figure 2.17.

TABLE 5.6

Terminal Node	Population	$r(t)$	$r^{ts}(t)$
1	27	.26	.49
2	24	.17	.25
3	15	.33	.32
4	96	.17	.23
5	8	.12	.46
6	7	.00	.43
7	41	.02	.20
8	5	.00	.41
9	13	.31	.32
10	9	.00	.36
11	55	.09	.25

This table shows clearly how unreliable the estimates $r(t)$ are. Unfortunately, we have not found a completely satisfactory way of getting better estimates. The cross-validation trees are usually different. There is no way to set up a correspondence between the nodes of the cross-validation trees and the main tree.

We have developed an ad hoc method which gives estimates $\hat{r}(t)$ that in every example examined are a significant improvement over $r(t)$. The intuitive idea is that the bias in $r(t)$ is roughly inversely proportional to the size of t as measured by $p(t)$. Therefore, set

$$\hat{r}(t) = r(t) + \frac{\varepsilon}{p(t) + \lambda}, \tag{5.10}$$

where ε and λ are constants to be determined as follows. The resubstitution estimates satisfy

$$R(T) = \sum_{t \in \tilde{T}} r(t)p(t).$$

We first of all insist that the altered estimates add up to the cross-validated estimate $R^{CV}(T)$ instead of to $R(T)$. That is,

$$R^{CV}(T) = \sum_{t \in \tilde{T}} \hat{r}(t)p(t) \tag{5.11}$$

must be satisfied.

Now consider a tree T' grown so large that each terminal node contains only one case. Then, assuming priors estimated from the data, $p(t) = 1/N$ for every $t \in \tilde{T}'$, and $\varepsilon/(p(t) + \lambda) \simeq \varepsilon/\lambda$. In the two-class problem discussed in the appendix to Chapter 3, a heuristic argument was given to show that $R(T') \leq 2R_B$, where R_B is the Bayes rate. Assuming the inequality is not too far from equality gives

$$2R_B \simeq \varepsilon/\lambda. \tag{5.12}$$

The discussion in the appendix to Chapter 3 is limited to the two-class problem, but it can be generalized to the multiclass problem with symmetric misclassification costs. Thus, (5.12) is more generally applicable.

Now we further assume that R_B can be adequately approximated by $\min_k R^{CV}(T_k)$. This leads to the equation

$$\varepsilon = 2\lambda \cdot \min_k R^{CV}(T_k). \tag{5.13}$$

Solving (5.12) and (5.13) together gives

$$\lambda \sum_{\tilde{T}} \frac{p(t)}{p(\lambda) + \lambda} = \frac{R^{CV}(T) - R(T)}{2 \min_k R^{CV}(T_k)} . \tag{5.14}$$

If $R^{CV}(T) \leq R(T)$, the original resubstitution estimates $r(t)$ are used. Otherwise, Equation (5.14) has a unique solution for $\lambda > 0$, which can be easily computed by a simple search algorithm.

Table 5.7 gives $\hat{r}(t)$ for the waveform tree of Figure 2.17.

TABLE 5.7

Terminal Node	$r(t)$	$\hat{r}(t)$	$r^{ts}(t)$
1	.26	.44	.49
2	.17	.36	.25
3	.33	.58	.32
4	.17	.23	.23
5	.12	.47	.46
6	.00	.36	.43
7	.02	.15	.20
8	.00	.40	.41
9	.31	.58	.32
10	.00	.33	.36
11	.09	.19	.25

To compare overall error, note that

$$\begin{aligned} \sum_{\tilde{T}} |r^{ts}(t) - r(t)| &= 2.26 \\ \sum_{\tilde{T}} |r^{ts}(t) - \hat{r}(t)| &= .89. \end{aligned} \tag{5.15}$$

The $\hat{r}(t)$ estimates do much better than $r(t)$. Where they have large errors, it is on the conservative side. Measured overall by (5.15), their error rate is only 39 percent as large as that of $r(t)$.

These results are typical of the 10 examples we have examined, including both real and simulated data. As compared with test sample results, $\hat{r}(t)$ has overall error consistently less than half that of $r(t)$.

Another method for estimating within-node class probabilities and costs has been tested only on the simulated waveform data. However, the results were promising enough to warrant a brief description.

Assume that all measurement variable $x_1, \ldots, x_M$ are ordered. In the learning sample $\mathcal{L}$, let $s_1, \ldots, s_M$ be the sample standard deviations of the variables $x_1, \ldots, x_M$. Take $h > 0$ to be a small fixed real number. Then construct a "noisy copy" of the learning sample as follows: For each n, $n = 1, \ldots, N$, let

$$\mathbf{x}'_n = \mathbf{x}_n + h\mathbf{z}_n \quad \text{and} \quad j'_n = j_n,$$

where

$$\mathbf{z}_n = (s_1 w_1^{(n)}, \ldots, s_M w_M^{(n)})$$

and the $w_1^{(n)}, \ldots, w_M^{(n)}$ are drawn from a $N(0, 1)$ distribution, independently for different m and n.

Repeat this procedure using more independent $N(0, 1)$ variables until there are enough noisy copies to total about 5000 cases. Run these cases through the tree T selected and compute their misclassification cost $R^{nc}(T, h)$. Gradually increase h until a value h_0 is found such that

$$R^{nc}(T, h_0) = R^{cv}(T).$$

Now use these 5000 cases with $h = h_0$ to estimate within-node class probabilities and costs.

Looked at another way, this method is similar to the modified bootstrap procedure. That is, the learning sample is being used to construct Gaussian Kernel estimates for the densities $f_j(\mathbf{x})$, using the sharpness parameter h. Then a pseudo-test sample of about size 5000 is gotten by randomly sampling from these estimated densities. Finally, h is adjusted so that the misclassification cost using the pseudo-test sample matches the cross-validated cost.

This method, applied to the waveform data, produced estimates of node class probabilities that were in surprisingly good agreement with the results gotten from the 5000-case test sample. Its use cannot be recommended until it is more extensively tested. Furthermore, it is difficult to see how it can be extended to include categorical measurement variables.

5.5 INTERPRETATION AND EXPLORATION

5.5.1 Introduction

Data analysts continually refer to the *structure* of the data and the search for the structure of the data. The concept is that the data are roughly generated as structure (or signal) plus noise. Then the analyst's function is to find the underlying mechanism that is generating the data, that is, to separate the signal from the noise and to characterize the signal.

We have found this a hazardous and chancy business. Even though the tree structured approach generally gives better insights into the structure than most competing methods, extensive exploration and careful interpretation are necessary to arrive at sound conclusions. (See Einhorn, 1972, and Doyle, 1973, for similar remarks regarding AID.)

Even with caution, data analysis is a mixture of art and science, involving considerable subjective judgment. Two reasonable analysts, given the same data, may arrive at different conclusions. Yet both, in an appropriate sense, may be right.

Three subsections cover some of the difficulties and tools to deal with them. The first illustrates some of the problems. The second briefly covers use of the basic output in interpretation. The third discusses the growing of exploratory trees.

5.5.2 Instability of Tree Structures

One problem in uncovering the structure is that the relation between class membership and the measured variables is made complex by the associations between the measured variables. For instance, the information about class membership in one group of measured variables may be partially duplicated by the information given in a completely different group. A variable or group of variables may appear most important, but still contain only slightly more information than another group of variables.

A symptom of this in the tree structure is that at any given node, there may be a number of splits on different variables, all of which give almost the same decrease in impurity. Since data are noisy, the choice between competing splits is almost random. However, choosing an alternative split that is almost as good will lead to a different evolution of the tree from that node downward.

To illustrate, four sets of digit and recognition data were generated in exactly the same way as described previously, but each with a different random number seed. For side-by-side comparison, the trees grown on the original data and one of the trees grown on the new data set are shown in Figures 5.8 and 5.9. In Figure 5.8 the numbers under a split, for example, the number 5 indicates the split, Is $x_5 = 0$?. The numbers in the terminal nodes are class identifications.

The two trees are very different. In this example the difference reflects the fact that the information in seven digits is somewhat redundant, and there are a number of different classification rules all achieving nearly the same accuracy. Then, depending on chance fluctuations, one or the other of the rules will be selected.

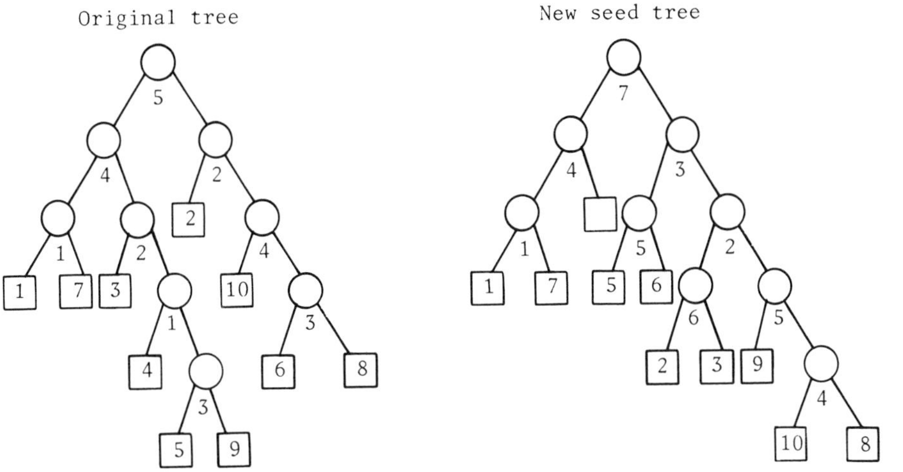

FIGURE 5.8 Digit recognition tree.

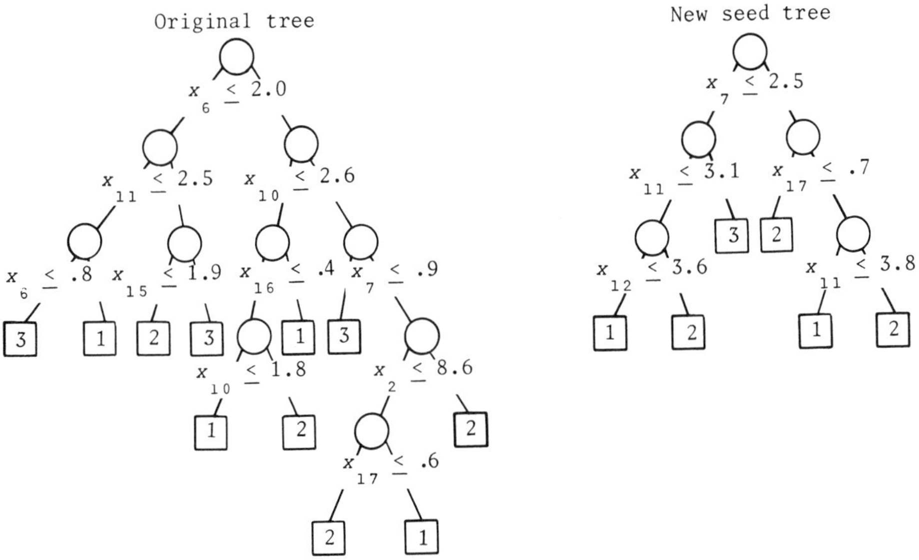

FIGURE 5.9 Waveform recognition data.

In contrast to the digit problem, although the two trees in Figure 5.9 are much different, the decision rules are similar. For instance, in the first tree, most of class 1 is in the second and seventh terminal nodes from the left. The second terminal node is characterized by moderately low x_6 and low x_{11}. The seventh terminal node is characterized by higher x_6, low x_{10}, and x_{16} not too small. In the second tree, the leftmost class 1 terminal node is formed by the combination of low x_7, low x_{11} and x_{12}. Since x_6 and x_7 are highly correlated, as are x_{11} and x_{12}, these two nodes are similarly formed. The paths to the seventh terminal node in the first tree and the other class 1 node in the second tree have a similar resemblance.

Along with the potential for instability in the tree topology, there are some facets that are fairly stable. Overall accuracy is relatively constant. The results for the five data sets are given in Tables 5.8 and 5.9.

TABLE 5.8 Digit Recognition Data

Data Set	$\|\tilde{T}\|$	R^{cv}	R^{ts}
Original	10	.30	.30
2nd	10	.31	.30
3rd	26	.34	.31
4th	10	.36	.30
5th	10	.33	.31

TABLE 5.9 Waveform Recognition Data

Data Set	$\|\tilde{T}\|$	R^{cv}	R^{ts}
Original	11	.29	.29
2nd	6	.25	.29
3rd	8	.29	.33
4th	10	.32	.29
5th	5	.32	.34

The variable importance values are more affected by random fluctuations. Figures 5.10 and 5.11 give the maximum and minimum values over the five replicate data sets for each of the two examples.

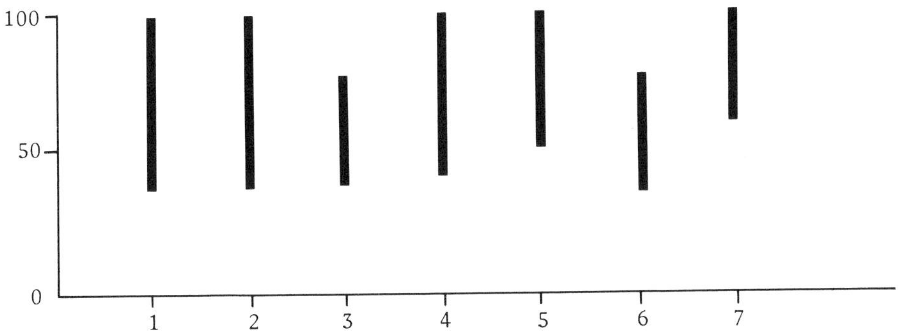

FIGURE 5.10 Digit recognition data: Variable importance.

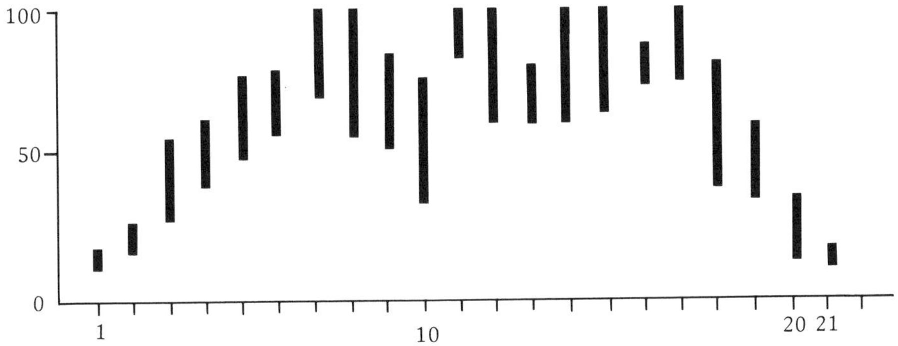

FIGURE 5.11 Waveform recognition data: Variable importance.

In practice, tree instability is not nearly as pronounced as in these two simulated examples. With real data, at least for nodes near the top of the tree, the best split is usually significantly better than the second best. Adding small amounts of noise to the data does not appreciably change the tree structure, except, perhaps, for a few of the lower nodes.

5.5.3 *Using Output*

The crux of the matter is that while a diagram of a tree grown on a data set gives an easily interpretable picture of a structure for the data, it may not be a very complete picture.

A number of diagnostics can be gotten from the tree output to check for masking and alternative structure. First, the tree diagrams for the cross-validation trees can be examined to see how closely they correspond to the "master" tree and what alternative structures appear. Second, possible alternative paths and masking at each node may be detected by looking at the best surrogate splits and their association with the optimal split. Also, looking at the decreases in impurity given by the best splits on the individual variables and noting splits which closely compete with the optimal split can give clues in a similar direction. Third, information about overall masking may be gotten from the variable importance values.

5.5.4 Growing Exploratory Trees

Growing a large tree T_{max} and pruning it up using a test sample or cross-validation can be computationally expensive. The analyst may want to grow some exploratory trees both before and after the primary tree growing.

Some questions that might warrant preliminary study are, Should variable combinations be used? and What selection of priors gives the best performance? A major question following a primary tree growing procedure is: What is the effect of deleting certain variables?

An inexpensive exploratory capability is contained in CART as an optional feature. The user specifies a value of the complexity parameter α. Then the program produces the optimal tree $T(\alpha)$. There are two parts that make this construction fast. The first is that growing a very large tree T_{max} and pruning it upward is not necessary. A condition can be given for growing a *sufficient tree* $T_{suff}(\alpha)$ which is guaranteed to contain $T(\alpha)$ but is usually only slightly larger.

Another large improvement in efficiency can be made by using subsampling (see Section 5.6.1). This procedure determines the splits in each node by using a randomly selected fraction of the data. Comparative computer timings using sufficient trees with and without subsampling will be discussed in Section 5.6.2.

The sufficient tree $T_{\text{suff}}(\alpha)$ is given by

DEFINITION 5.16. *For any node t, let* $t_1, t_2, t_3, \ldots, t_h, t$ *be the sequence of nodes leading from* t_1 *to t. Define*

$$S(t) = \min_{1 \le i \le h} (R(t_i) - \alpha(h - i + 1)).$$

If $S(t) > 0$, *split the node t using the optimal split* s^*. *If* $S(t) \le 0$, *declare t a terminal node of* $T_{\text{suff}}(\alpha)$.

(Even if $S(t) > 0$, t may be declared terminal for the same reasons a node is terminal in $T_{\max}$, i.e., $N(t) \le N_{\min}$.)

PROPOSITION 5.17. $T(\alpha) \subset T_{\text{suff}}(\alpha)$.

The proof is not difficult and is given in Chapter 10 (see Theorem 10.32).

As we pass down through the tree, $S(t)$ is defined recursively by

$$\begin{aligned} S(t_1) &= R(t_1) - \alpha \\ S(t_L) &= \min(S(t) - \alpha,\ R(t_L) - \alpha) \\ S(t_R) &= \min(S(t) - \alpha,\ R(t_R) - \alpha). \end{aligned} \tag{5.18}$$

Then the stopping rule for $T_{\text{suff}}(\alpha)$ can be checked at each node t without searching up through the path from t_1 to t.

The minimal cost-complexity tree $T(\alpha)$ is gotten from $T_{\text{suff}}(\alpha)$ by pruning upward until a weakest link $\bar{t}_k$ is found that satisfies $g_k(\bar{t}) \ge \alpha$. Then $T(\alpha)$ equals T_k (see Section 3.3).

A substantial part of the computation in the primary tree growing procedure is put into the lower branches of the large initial tree $T_{\max}$. For instance, typically the number of nodes in $T_{\max}$ is four to five times the number in the 1 SE tree. Use of the sufficient tree thus produces a significant reduction in running time.

A useful and straightforward application of exploratory trees is in tracking the effects of variable deletion once a primary tree has been grown. The primary tree output gives the value α_k corresponding to the selected tree. Then deleting one or more variables and growing the tree $T(\alpha_k)$ on the remaining variables give an indication of the effects of the deletion. The bias $R^{CV} - R$ is known for the original tree, so that reasonable estimates for the true cost of the deleted variable trees can be gotten by assuming the same bias.

The preprimary tree growing exploration is more problematic. There are two difficulties. One is how to choose α. The second is that too much initial exploration and data dredging may result in implicitly biasing the primary tree growing process. This is particularly true if the entire data set is used in the initial exploratory phase.

There are some general guidelines for selecting α. For $T(\alpha) \neq \{t_1\}$,

$$R_\alpha(T(\alpha)) \leq R_\alpha(\{t_1\})$$

or

$$R(T(\alpha)) + \alpha|\widetilde{T}(\alpha)| \leq R(t_1) + \alpha.$$

Put $\alpha = R(t_1)/H$. Then the inequality becomes

$$HR(T(\alpha)) \leq R(\{t_1\})(H + 1 - |\widetilde{T}(\alpha)|),$$

giving the loose upper bound

$$|\widetilde{T}(\alpha)| \leq H + 1.$$

A rough rule of thumb is that using $\alpha = R(t_1)/2H$ will produce a tree with about H terminal nodes.

Since the prior exploratory phase is aimed at broad features, such as the usefulness of variable combinations and rough adjustment of priors, we suggest that α not be taken too small. If the learning sample is large, to speed up computations, we advise the use of the subsampling feature discussed in Section 5.6.1.

5.6 COMPUTATIONAL EFFICIENCY

Since tree construction involves a large number of sorts and optimizations, computational efficiency is important. The first subsection discusses the use of subsampling to increase efficiency. The second subsection gives some benchmark timing runs on a number of problems with various options.

5.6.1 Subsampling

The idea is this: In a class J problem, an upper sample size limit N_0 is set. As the tree is grown, starting from the root node on down, subsampling is done at every node until the total node population falls below N_0.

If the population in node t is $N_1(t)$, ..., $N_j(t)$, with

$$\sum_j N_j(t) > N_0,$$

select a random subsample of size N_1', ..., N_J' having the following properties (assume N_0 is divisible by J):

$$\begin{aligned} &\text{(i)}\quad N_j' \le N_j(t),\ j = 1, \ldots, J, \\ &\text{(ii)}\quad \sum_j N_j' = N_0, \\ &\text{(iii)}\quad \sum_j \left(N_j' - \frac{N_0}{J}\right)^2 = \text{minimum}. \end{aligned} \qquad (5.19)$$

That is, subsample to get total sample size N_0 and such that the individual class sample sizes are as nearly equal as possible. If for every j, $N_j(t) \ge N_0/J$, then the optimal subsample is clearly $N_j' = N_0/J$.

The difficulty arises when some of the $N_j(t)$ are less than N_0/J. Order the classes so that $N_1(t) \le N_2(t) \le \cdots \le N_j(t)$. Then

PROPOSITION 5.20. *The* N_j', $j = 1, \ldots, J$, *satisfying (5.10) are defined recursively by: If* N_1', ..., N_j' *are the first* j *best choices, then*

$$N'_{j+1} = \min\left(N_{j+1}(t),\ \frac{N_0 - N'_1 - N'_2 - \cdots - N'_j}{J - j}\right).$$

The argument is this: The problem is equivalent to minimizing $\sum_j (N'_j)^2$ under the constraints $0 \leq N'_j \leq N_j(t)$ and $\sum_j N'_j = N_0$. If $N_1(t) \geq N_0/J$, the solution is to take $N'_j = N_0/J$, $j = 1, \ldots, J$. Otherwise, take $N'_1 = N_1(t)$. The problem is now reduced to the $J - 1$ class problem of minimizing $\sum_{j \geq 2} (N'_j)^2$ under the modified constraint $\sum_{j \geq 2} N'_j = N_0 - N'_i$. Repeating the argument for $J - 1$ instead of J gives

$$N'_j = \min\left(N_2(t),\ \frac{N_0 - N'_1}{J - 1}\right)$$

Now repeat until $j = J$.

For example, in a four-class problem with $N_0 = 200$, if the node population was

30 40 90 150,

then the subsample is

30 40 65 65.

The best split s^* is found *using only the subsample*. But once the best split is found, then *the entire original population in the node is split using* s^*. Subsampling affects only those upper nodes with total population greater than N_0. As the tree splitting continues and smaller nodes are formed with population less than N_0, all of the sample is used to determine the best split.

The subsampling procedures used, say, in linear regression with a large data set usually subsample once and do all the regression on the subsample. The rest of the information is lost forever. In the tree structure, the subsampling affects only the first few splits in the tree. But in these initial nodes, there are usually a few adjacent splits on one variable that are markedly superior to all other splits. This clear-cut information is reflected in the subsample. In the splits further down, the full sample size is available.

Some care must be taken to weight the subsample appropriately. The class j proportions in the subsample are not the same as in the original sample in the node. Recall that

$$p(j|t) = \pi(j)(N_j(t)/N_j)/\sum_j \pi(j)N_j(t)/N_j.$$

A weighting w_j of each case in the jth subsample class is necessary to adjust to these original node proportions. This is done using the weights

$$w_j = \pi(j)N_j(t)/N_jN'_j. \tag{5.21}$$

Then defining, for the subsample,

$$p'(j|t) = w_jN'_j/\sum_j w_jN'_J$$

gives $p'(j|t) = p(j|t)$. If $N'_{j,L}$, $N'_{j,R}$ of the N'_j cases in the subsample go to t_L, t_R, respectively, then define

$$\begin{aligned} &\text{(i)}\ \ p'(j|t_L) = w_jN'_{j,L}/\sum_j w_mN'_{j,L}, \\ &\text{(ii)}\ \ p'(j|t_R) = w_jN'_{j,R}/\sum_j w_jN'_{j,R}. \end{aligned} \tag{5.22}$$

Denote the denominators in (5.22) as $p'(t_L)$, $p'(t_R)$ and use these definitions to determine the best split on the subsample.

Unequal priors lead to the question of whether the subsample should be taken, as far as possible, with equal sample sizes from each class or divided in another proportion. A large-sample theoretical result indicated that the optimal proportions vary between being equal and being proportional to $\sqrt{\pi(j)}$. Since this result, at worst, gives sample proportions less disparate than the priors, the lure of simplicity prevailed, and the subsampling rule implemented is that outlined above.

Some approximate calculations can be made on the effects of subsampling. The computational burden at each node (assuming M

ordered and no categorical variables) consists of a sort on each variable and then an evaluation of $\Phi(s, t)$ for each split on each variable. If the population size in a node is $N(t)$, then the quick sort used in CART requires $C_1 N(t) \log N(t)$ operations on each variable. The Φ computations require $N(t)$ evaluations on each variable.

If the tree is grown to a uniform depth D, that is, to 2^D terminal nodes, then the sort time is proportional to

$$N(D + 1)(\log_2 N - \frac{D}{2})$$

and the evaluation time to $(D + 1)N$.

If subsampling is used, with maximum node size N_0, so that the tree is grown to depth $D_0 = \log_2(N/N_0)$ before subsampling stops, then the sort time is proportional to

$$N \log_2 N_0 + N(D + 1 - D_0)(\log_2 N - \frac{D}{2} - \frac{D_0}{2}).$$

The ratio of sort time using subsampling to total sort time is approximately

$$1 - \frac{D_0}{D + 1}. \tag{5.23}$$

A similar argument for the Φ evaluations gives the ratio

$$1 - \frac{D_0 - 1}{D + 1}. \tag{5.24}$$

The reduction is most significant when D_0 is nearly equal to D.

In growing the large preliminary tree T_{max}, take $N_{min} = \frac{\sqrt{N}}{2}$. Using $N_0 = 256$, the ratios (5.23) and (5.24) are as follows:

	(5.23)	(5.24)
$N = 2000$	.6	.7
$N = 16{,}000$	.3	.4

There is substantial savings only for very large data sets. In growing exploratory trees, the required depth D is smaller, so the computational savings using subsampling are more significant.

The most useful application of subsampling has been in those problems where the data base is too large to be held in fast memory. One pass through the disk file for a node puts the subsample in core; the split is then found, and another pass through the disk puts the rest of the data down the split.

5.6.2 How Fast is CART?

The current CART program is running on two machines that are extremes in computing power. One is a powerful IBM 3081. The other is a small VAX 11/750. Timing runs were made on both machines. The examples used were the digit recognition data, the digit recognition data with 17 noise variables added, the waveform recognition data, and the waveform data with 19 noise variables added.

The CPU times listed in Table 5.10 are based on tenfold cross-validation. They include the entire construction process, including the cross-validation and pruning necessary to produce the final tree.

TABLE 5.10 CPU Seconds

	Digit (7 Variables)	Digit and Noise (24 Variables)	Waveform (21 Variables)	Waveform and Noise (40 Variables)
IBM	3	9	25	47
VAX 11/750	64	228	510	1020

In another timing experiment, exploratory trees were grown using the method outlined in Section 5.5.4 with α equal to the value corresponding to the tree selected by cross-validation and the 1 SE rule. The time results are given in Table 5.11.

TABLE 5.11 CPU Seconds

	Digit (7 Variables)	Digit and Noise (24 Variables)	Waveform (21 Variables)	Waveform and Noise (40 Variables)
IBM	.2	.7	2.5	4.1
VAX 11/750	6.	21	43	90

5.7 COMPARISON OF ACCURACY WITH OTHER METHODS

Two other classification methods were compared with the tree structured results on the simulated data sets. The first was nearest neighbor classification. The second was the stepwise linear discriminant program in BMDP.

In nearest neighbor classification, the learning sample is used as a collection of "templates." A new observation is classified by assigning to it the class of its nearest neighbor in the learning sample.

Four experiments were carried out using the following learning samples:

1. Digit recognition
2. Digit recognition plus 17 noise variables
3. Waveform
4. Waveform plus 19 noise variables

In (1) and (2) the distance between two sequences of zeros and ones was defined as the number of places in which they are different. In case of ties, nearest neighbors were selected at random. In samples (3) and (4), the distance was the euclidean distance between the observation vectors.

A test set of 5000 was used to estimate the misclassification rates. The results are given in Table 5.12.

TABLE 5.12 Misclassification Rates

Data Set	Nearest Neighbor	Tree
(1) Digit	.29	.30
(2) Digit and noise	.59	.30
(3) Waveform	.22	.28
(4) Waveform and noise	.62	.28

The accuracy of nearest neighbor is significantly better than the tree algorithm only in the waveform data. As one might expect, the addition of noise variables seriously degrades the performance of nearest neighbor classifiers. Still, if noise variables are weeded out and a sensible metric used, nearest neighbor methods can be reasonably accurate. Our objections to these methods, stated in Chapter 1, are based on other characteristics.

Our resident BMDP expert, Alan Hopkins, obligingly ran the BMDP stepwise linear discriminant program on both the 7-variable digit data and the 21-variable waveform data. Standard default values for F-to-enter and remove were used. Test samples of size 5000 gave accurate misclassification rate estimates.

In the digit data, all seven variables were entered. The test sample misclassification rate was .25. This is surprisingly low considering the nonnormal nature of the data.

The classifier has the following form. For a new measurement vector x, evaluate each of the following linear combinations:

Class 1: $-12.8 - .8x_1 + 3.0x_2 + 8.0x_3 - 1.8x_4 - .6x_5 + 12.3x_6 + 1.3x_7$

Class 2: $-24.3 + 10.8x_1 + .8x_2 + 6.4x_3 + 9.2x_4 - 13.2x_5 - .4x_6 + 10.8x_7$

Class 3: $-23.8 + 10.4x_1 + 1.4x_2 + 8.3x_3 + 8.1x_4 + 1.2x_5 + 9.1x_6 + 10.5x_7$

Class 4: $-21.8 - 1.3x_1 + 11.5x_2 + 7.5x_3 + 11.3x_4 - .8x_5 + 10.3x_6 + 2.7x_7$

Class 5: $-24.0 + 8.5x_1 + 9.0x_2 + .8x_3 + 11.8x_4 + .9x_5 + 9.1x_6 + 9.5x_7$

Class 6: $-31.9 + 11.2x_1 + 9.1x_2 + .9x_3 + 11.3x_4 + 13.7x_5 + 7.7x_6 + 10.7x_7$

Class 7: $-14.0 + 10.6x_1 + 1.2x_2 - 6.8x_3 - 1.3x_4 + 1.8x_5 + 9.1x_6 + 1.7x_7$

Class 8: $-31.7 + 9.0x_1 + 9.5x_2 + 7.0x_3 + 9.8x_4 + 12.7x_5 + 9.0x_6 + 9.9x_7$

Class 9: $-28.2 + 8.9x_1 + 9.2x_2 + 8.2x_3 + 10.3x_4 - .4x_5 + 10.0x_6 + 11.0x_7$

Class 10: $-28.1 + 10.0x_1 + 10.0x_2 + 7.2x_3 + .1x_4 + 13.0x_5 + 9.6x_6 + 8.4x_7$

Then classify $\mathbf{x}$ as that class corresponding to the largest value of the 10 linear combinations.

Although the accuracy of this rule is better than that of the tree classifier, its form is complex and sheds little light on the structure of the data. However, it turns out that the Bayes rule for this problem does have the preceding form with slightly different coefficients from those shown here. This may explain why the discriminant rate of .253 is about 1 SE (5000 test sample) below the Bayes rate of .26. This value is not typical of other replicate data sets. A more typical value of .31 resulted on the replicate set used in Section 4.3.3. The tree rate for these data is .33.

With the waveform data, seven variables were entered: x_5, x_{10}, x_{13}, x_{14}, x_{15}, x_{19}, x_{20}. The test sample misclassification estimate was .26. This is slightly better than the standard tree classifier (.28), but not as good as the tree grown using linear combinations (.20).

In both examples (using the replicate digit data), the linear discriminant resubstitution error rate was misleadingly low, .24 in the digit problem and .20 in the waveform data. The BMDP program gives jackknifed (N-fold cross-validation) estimates of the rates as .26 and .23.

The two simulated examples used in the last few chapters were not designed to make tree structured classification look good as compared to other methods. In fact, nearest neighbor and linear discriminant methods offer competitive accuracies.

It is not hard to construct examples in which one method will do poorly while another does well. The comparative accuracy of different methods depends on the data. Still, in all applications to real data that we know of where various methods have been compared, the accuracy of tree structured classifiers has generally been either best or close to best.

APPENDIX

The Linear Combination Search Algorithm

Suppose that at a given node t, a goodness of split criterion $\Phi(s, t)$ is defined. Hereafter, when we say "best split," we mean with respect to the $\Phi(s, t)$ criterion.

At the node there are $N = N(t)$ measurement vectors $\mathbf{x}_1, \ldots, \mathbf{x}_N$ with M_1 ordered coordinates. Let the ordered coordinates of the nth case be $x_{m,n}$, $m = 1, \ldots, M_1$. At the first step of the algorithm, the N values $x_{m,n}$, $n = 1, \ldots, N$, are centered at their median and divided by their interquartile range for each m, $m = 1, \ldots, M_1$.

Only these normalized values are used in the algorithm, and, to avoid a plethora of notation, we denote them also by $x_{m,n}$.

The algorithm cycles through the variables $x_1, \ldots, x_{M_1}$, at each step doing a search for an improved linear combination split. At the beginning of the Lth cycle, let the current linear combination split be $v \leq c$, where $v = \sum_{m=1}^{M_1} \beta_m x_m$.

For fixed γ, consider the problem of finding the best split of the form

$$v - \delta(x_1 + \gamma) \leq c, \tag{A.1}$$

where δ ranges over all possible values. This is done by rewriting (A.1) as

$$\delta(x_1 + \gamma) \geq v - c$$

so that

$$\delta \geq \frac{v - c}{x_1 + \gamma}, \quad x_1 + \gamma \geq 0$$

$$\delta \leq \frac{v - c}{x_1 + \gamma}, \quad x_1 + \gamma < 0.$$

Now the procedure is similar to the univariate case. Let

$$u_n = \frac{v_n - c}{x_{1,n} + \gamma}.$$

Order the values of u_n. For all u_n such that $x_{1,n} + \gamma \geq 0$, find the best split of the form $u \leq \delta_1$ over all values of δ_1. For all u_n such that $x_{1,n} + \gamma < 0$, find the best split of the form $u \geq \delta_2$. Take the best of these two splits and let δ be the corresponding threshold.

Actually, there is a neater way of finding the best δ that combines the two searches over u_n into a single search. But its description is more complicated and is therefore omitted.

This procedure is carried out for three values of γ: $\gamma = 0$, $\gamma = -.25$, $\gamma = .25$. For each value of γ, the best split of the form (A.1) is found. Then these three splits are compared and δ, γ corresponding to the best of the three are used to update v as follows. Let

$$v' = \sum_{m=1}^{M_1} \beta'_m x_m,$$

where

$$\beta'_1 = \beta_1 - \delta, \; \beta'_m = \beta_m, \; m > 1$$

and

$$c' = c + \delta\gamma.$$

Now this is repeated using variable x_2; that is, a search is made for the best split of the form

$$v' + \delta'(x_2 + \gamma') \leq c', \; \gamma' = 0, -.25, .25.$$

Then variable x_3 is entered, and so on, until variable x_{M_1} is used. At this point the updated split is of the form $v_1 \leq c_1$. Then the cycle is completed by finding the best split of the form $v_1 \leq c_1'$ as c_1' ranges over all values. This split forms the initial split for the next cycle. The initial split for the first cycle is the best univariate split.

Let s_L be the linear combination split produced at the end of the Lth cycle. The cycling terminates if $|\Phi(s_L, t) - \Phi(s_{L-1}, t)| \leq \varepsilon$, where ε is a small threshold. After the final linear combination split is determined, it is converted into the split $\sum_1^{M_1} a_m x_m \leq c$ on the original nonnormalized variables.

This algorithm is the result of considerable experimentation. Two previous versions were modified after we observed that they tended to get trapped at local maxima. There is still room for improvement in two directions. First, it is still not clear that the present version consistently gets close to the global maximum. Second, the algorithm is CPU expensive. On the waveform example, tree building took six times as long as using univariate splits only. We plan to explore methods for increasing efficiency.

The linear combination algorithm should not be used on nodes whose population is small. For instance, with 21 ordered variables, finding the best linear combination split in a node with a population not much larger than 21 will produce gross overfitting of the data. For this reason, the program contains a user-specified minimum node population N_c. If $N(t) \leq N_c$, the linear combination option at t is disabled and only univariate splits are attempted.

6

MEDICAL DIAGNOSIS AND PROGNOSIS

Tree structured classification and regression techniques have been applied by the authors in a number of fields. Our belief is that no amount of testing on simulated data sets can take the place of seeing how methods perform when faced with the complexities of actual data.

With standard data structures, the most interesting applications have been to medical data. Some of these examples are covered in this chapter. The next chapter discusses the mass spectra projects involving nonstandard data structures.

In Section 6.1 we describe 30-day prognoses for patients who were known to have suffered heart attacks. Section 6.2 contains a report on CART's performance in diagnosing heart attacks. The subjects were other patients who entered hospital emergency rooms with chief complaints of acute chest pain. Section 6.3 gives applications of CART to the diagnosis of cancer. The descriptive variables of that study are measures of immunosuppression. In Section 6.4 the classification of age by measurements of gait is used to illustrate the application of CART to the detection of outliers. A brief final section contains references to related work on computer-aided diagnosis. In each of these sections, the medical terminology needed is explained briefly but in sufficient detail to render the material accessible to all who understand the previous chapters.

The chapter contains some, but rather few, explicit comparisons with parametric competitors. We justify this choice with the presumption that our readers are more familiar with them than with CART; in any case, this is a book about CART. In practice, more often than not CART has outperformed its parametric competitors on problems of classification in medicine. Usually, the reductions in misclassification cost have been less than 15 percent of those of the "best" parametric procedures. Notwithstanding substantial contributions to nonparametric statistics in recent years and the clear mathematical and simulation arguments for their use, it is our experience that parametric procedures (such as the Fisher linear discriminant and logistic regression) hold up rather well in medical applications. Thus, for CART to do as well as it does is an accomplishment. It will be emphasized that the ease with which the tree can be interpreted and applied renders it an important alternative to other procedures even when its performance is roughly comparable to theirs.

6.1 PROGNOSIS AFTER HEART ATTACK

As the ability of modern medicine to provide costly but potentially beneficial treatment grows, the ability to identify patients at high risk of morbidity or mortality correspondingly grows in importance. One area in which the development of interventions is most active is heart disease; in particular, an epoch of special attention is the time following acute myocardial infarction (heart attack). Substantial energy, time, and funds have been spent, for example, on assessing the value of (beta-adrenergic blocking) drugs and other interventions in limiting the size of infarction. (See summaries of work on the timolol trial by the Norweigian Multicenter Study Group, 1981, and also the report of the Beta-

Blocker Heart Attack Trial Study Group, 1981.) This section is a report on one attempt to identify patients who are at risk of dying within 30 days from among those who have suffered heart attacks and who have survived at least 24 hours past admission to the University of California, San Diego Medical Center. The data are part of a larger study conducted by the Specialized Center for Research on Ischemic Heart Disease at the University of California, San Diego (see Henning et al., 1976, and Gilpin et al., 1983).

The diagnosis of heart attack, a sometimes difficult problem, involved the following criteria: (1) history of characteristic chest pain; (2) indicative electrocardiograms; and (3) characteristic elevations of enzymes that tend to be released by damaged heart muscle.

Present discussion focuses on 215 patients, of whom 37 died not more than 30 days following heart attack, and 178 of whom did not. The "survivors" are called class 1 and the "early deaths" class 2. On the basis of substantial experience, the physicians involved in the project gave prior probabilities $\pi(1) = .8$ and $\pi(2) = .2$. There seemed to be little variability across physicians or statisticians to this assessment. Note that the empirical fraction 37/215 matches the prior probability quite well. The reader may be concerned that there are only 37 class 2 patients—or perhaps that there are only 215 patients in all. It is unfortunately the case that in many medical studies, particularly (prospective) studies of ongoing events, the patient group whose health is most threatened is represented by rather small numbers of subjects.

The assessment of costs of misclassification led to a disagreement regarding the relative cost of misclassifying an early death to the cost of misclassifying a survivor, that is, regarding $C(1|2)/C(2|1)$. Preliminary analyses were made solely by statisticians, whose views were those of potential patients in possible need of substantial care. Thus, in preliminary analyses the ratio was taken to be 12. Later consultation with the physicians led to a reduction of the ratio for the final analyses to 8/3. A moment's reflection makes clear the discrepancy. The most expensive and detailed care should be reserved for patients truly threatened with loss of life. It is to the detriment of these patients that valuable staff and other resources be spent on patients at relatively low risk of dying. A result of the above considerations is that a terminal node of any CART-designed tree is a class 1 node if the number of survivors at the node exceeds 3.2 times the number of early deaths, that is, if $37\pi(1)C(2|1)N_1(t) > 178\pi(2)C(1|2)N_2(t)$.

From among about 100 variables screened, 19 were selected for inclusion in CART. Thirteen of the variables were chosen because when class 1 and class 2 patients were compared, lowest attained significance levels resulted from two-sample *t* tests on the differences (continuous variables) or from chi square tests for independence (dichotomous variables); 6 of the variables included were chosen because related studies by others suggested that they have predictive power for the question at hand. Given the implicit concern CART has with relationships among variables, it may have been preferable to use a variable selection scheme (possibly CART itself) that looked directly at "interactions" and not merely "main effects." Twelve of the 19 included variables are dichotomous; the remaining are continuous and nonnegative. Note that for all 215 patients there were complete data on all 19 included variables. All included variables are *noninvasive*; that is, they can be measured without use of a catheter. (A catheter is a small hollow tube that is passed within the arteries or veins to reach the heart. It is used to measure pressures inside the heart and to inject liquids used for studies by X ray.) There is substantial evidence that classification more successful than what is being reported here can be made with the help of invasive measurements. However, the process of catheterization presents some risk and also possible discomfort to the patient.

Figure 6.1 shows the tree CART produced for the problem; the Gini splitting criterion was used with a minimum node content of 5 observations, and tenfold cross-validation was employed. The displayed tree is that with the smallest cross-validated risk.

TABLE 6.1 Summary of CART's Resubstitution Performance

Classified \ True	Class 1 "Survivors"	Class 2 "Early Deaths"	Total
Class 1 "Survivors"	158	9	167
Class 2 "Early Deaths"	20	28	48
Total	178	37	215

If priors and costs are normalized so that $\pi(1)C(2|1) = .6$ and $\pi(2)C(1|2) = .4$, then it follows from the data in Table 6.1 that the resubstitution overall misclassification cost is $(.6)\left(\frac{20}{178}\right) + (.4)\left(\frac{9}{37}\right) \doteq .17$. Since the no-data optimal rule has misclassification cost .4, about 59 percent of the overall cost of misclassification appears to be saved by our data and CART. (A no-data optimal rule assigns any observation to a class j that minimizes $\sum_i \pi(i)C(j|i)$. In case of ties, the convention used here is to choose the lowest value of j.) Of course, this may be somewhat optimistic, and therefore results of the tenfold cross-validation are of interest. The minimum cross-validated misclassification cost is $.19\,[\doteq (.6)(.14) + (.4)(.28)]$, as is evident from Table 6.2, which gives results for resubstitution and cross-validation for the trees that arose by optimal pruning. The previously estimated 59 percent reduction of overall misclassification cost is reduced by cross-validation to 53 percent.

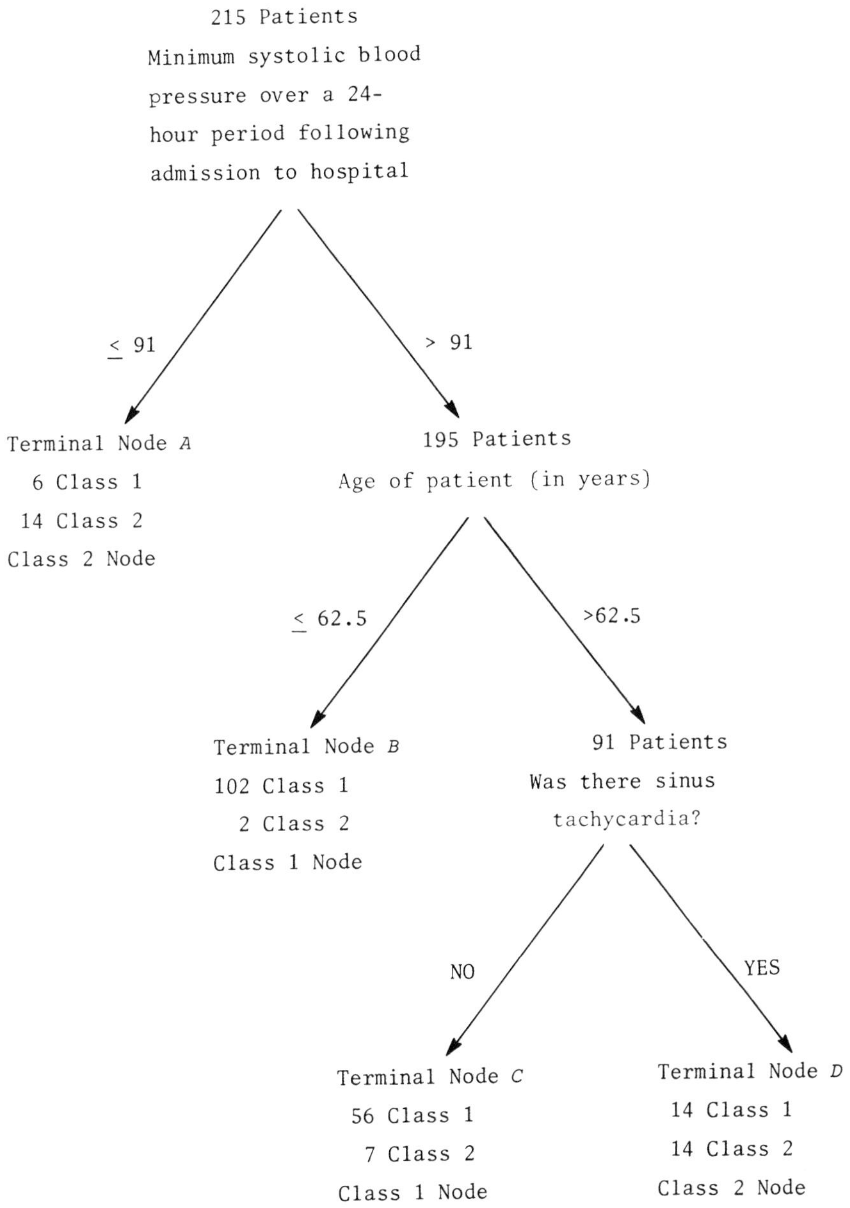

FIGURE 6.1 Note that the systolic blood pressure is the maximum blood pressure that occurs with each heart cycle during contraction of the left-sided pumping chamber; this pressure is measured with a blood pressure cuff and stethoscope. By definition, there was sinus tachycardia present if the sinus node heart rate ever exceeded 100 beats per minute during the first 24 hours following admission to the hospital; the sinus node is the normal electrical pacemaker of the heart and is located in the right atrium.

TABLE 6.2

Number of Terminal Nodes	Resubstitution Misclassification Cost	Cross-Validated Misclassification Cost
13	.10	.24
11	.11	.24
10	.11	.22
8	.12	.22
7	.13	.22
6	.14	.23
4	.17	.19
2	.27	.27
1	.4	

Since N_2 is only 37, the alert reader may ask whether cross-validation might have been more accurate if it had been done by randomly dividing each of the two classes into groups of about 10 percent each and randomly combining. Indeed, at least one of the 10 percent subgroups in the cross-validation we did has no early deaths at all. Our current thinking is that stratification can be used to produce more accurate cross-validation estimates of the risks of tree structured classifiers. (See Section 8.7.2 for a similar conclusion in the regression context.)

Perhaps the most striking aspect of the tree presented in Figure 6.1 is its simplicity. Not only is it simple in appearance, but also the variables required to use it are easy to measure. Note, however, that two of the three splits require information not necessarily available upon admission to the hospital, and some physicians may find the possible 24-hour period past admission an unreasonably long time to wait for the results of CART's classification.

Several procedures were compared with CART in their abilities to classify. The BMDP stepwise discriminant program (see Jennrich and Sampson, 1981) began with the same 19 variables as did CART. Its solution retained 12 variables, and resubstitution overall classification cost was .20, which tenfold cross-validation adjusted to $.21 \doteq (.6)(.13) + (.4)(.33)$. Also, logistic regression (DuMouchel, 1981) was employed. Ten of the original 19 variables were used. (The decision on which to include was based on considerations not reported here.) But logistic regression was given additional help; namely, three two-factor interactions suggested by various CART runs were included in the analysis. (Morgan and Sonquist, 1963, emphasized that tree structured rules for classification and regression can be used as detectors of interactions.) In view of which variables appear at adjacent nodes, the tree presented in Figure 6.1 suggests two of those three interactions: age by minimum systolic blood pressure and age by heart rate. Upon resubstitution, logistic regression produced the same class contributions to overall misclassification cost as did the stepwise discriminant (.10 from class 1 and .35 from class 2). However, the two procedures did not classify each patient identically. A tenfold cross-validation adjusted the logistic regression resubstitution figure (.20) to $.21 \doteq (.6)(.12) + (.4)(.36)$. In other areas of the study from which this section was taken, logistic regression with main effects and interactions suggested by CART performed well. We mention in passing that attempts to employ nearest neighbor classification rules were unsuccessful because the procedures performed quite poorly and were computationally expensive as well.

Particularly in a medical context, it is important to emphasize that even though the linear classification rule given by the stepwise discriminant and the quadratic rule given by logistic regression are easy to apply, neither is as easy as the simple displayed tree.

In summary, for the problem of distinguishing early deaths and survivors in the context presented, with complete data from 215 patients, it seems possible to reduce the misclassification cost to about 50 percent its value for the no-data optimal rule. And no procedure we tried performed quite as well as the rule given by CART.

6.2 DIAGNOSING HEART ATTACKS

This section, like the previous section, is about heart attacks, but its concern is with diagnosis rather than prognosis. The specific issue is this. A patient enters the emergency room of a hospital with the complaint of acute chest pain. Is the patient suffering (or has he or she just suffered) an acute myocardial infarction? In certain respects the problem of diagnosis may be less difficult than that of making 30-day prognoses, since changes in treatment protocols from one hospital to another and population bases over time probably have less impact on successful diagnoses of heart attacks than on prognoses.

The work reported in this section is part of a study done by Lee Goldman and others based on data gathered at the Yale-New Haven Hospital and at the Brigham and Women's Hospital, Boston (see Goldman et al., 1982). Goldman is with the Department of Medicine, Brigham and Women's Hospital and Harvard Medical School.

The way heart attacks were diagnosed in this study involved criteria (2) and (3) of Section 6.1: indicative electrocardiograms and characteristic elevations of levels of enzymes that tend to be released by damaged heart muscle. The measurements of the enzymes can take time, which is especially precious when, on the one hand, a patient just presented may be undergoing a heart attack, and, on the other, coronary care units for treating patients who have undergone heart attacks are heavily used and expensive. Thus, CART was employed with noninvasive, relatively easily gathered data and the hope of providing quick and accurate diagnoses.

The initial phase of this study involved every patient at least 30 years of age who came to the Yale-New Haven Hospital emergency room between August and November, 1977, with the chief complaint of chest pain that did not come from known muscle damage or pneumonia and that might have been caused by a heart attack. Of the 500 such patients, 482 had data sufficiently complete to be included in the learning sample. The cited reference contains an explanation of why neither the exclusion of 18 patients from the learning sample nor the absence of some data for those patients retained introduced any appreciable biases into the analyses. Of the 482 members of the training sample, 422, the class 1 patients, did not suffer heart attacks, although some were diagnosed as suffering from acute (ischemic) heart disease. In all reported growing of trees the empirical prior probabilities $\pi(1) = .88$ and $\pi(2) = .12$ were used, as was the Gini splitting criterion.

About 100 noninvasive variables were culled from data forms completed by emergency room interns or residents at times when the patients' post-emergency room courses and enzyme levels were unknown. Univariate criteria like those described in Section 6.1 determined the 40 variables retained for use in CART. These retained variables were largely dichotomous or categorical.

The work which led to the tree of Figure 6.2 was done before cross-validation was a standard component of CART. Instead, the process of finding optimally pruned trees employed the bootstrap, which is described in Section 11.7. The pruned tree with smallest bootstrapped misclassification cost was trimmed further in a subjective fashion. That trimming and the choice of misclassification costs involve the notions of sensitivity and specificity.

The *sensitivity* of a diagnostic test (sometimes called the *true positive rate*) is the percentage of correctly diagnosed patients from among those who have suffered heart attacks. The

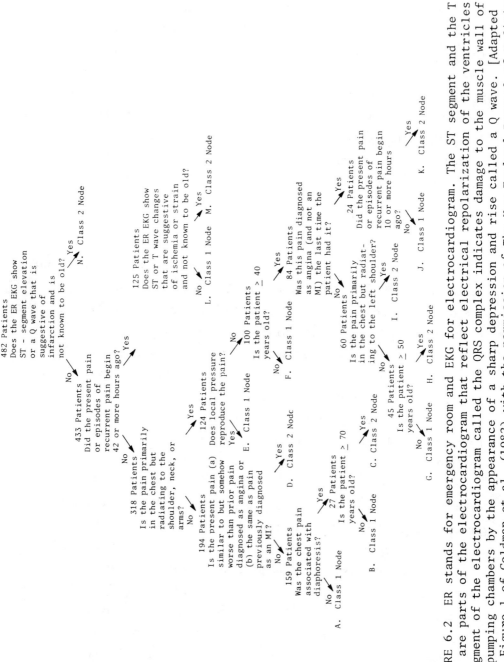

FIGURE 6.2 ER stands for emergency room and EKG for electrocardiogram. The ST segment and the T wave are parts of the electrocardiogram that reflect electrical repolarization of the ventricles. A segment of the electrocardiogram called the QRS complex indicates damage to the muscle wall of the pumping chambers by the appearance of a sharp depression and rise called a Q wave. [Adapted from Figure 1 of Goldman et al. (1982) with the permission of *The New England Journal of Medicine*.]

specificity (or *true negative rate*) is the percentage of correctly diagnosed patients from among those who have not suffered heart attacks.

The only values of $C(1|2)$ and $C(2|1)$ ever considered satisfy $C(1|2) > C(2|1)$, for it seems more important to diagnose correctly a patient who has suffered a heart attack than one who has not. Even so, if the ultimate rule for classification is to be of real use, it should strive to minimize the *false positive rate* (the complement of the specificity). With these considerations in mind, it was determined (by Goldman) that $C(1|2)/C(2|1)$ should be approximately as small as possible subject to the constraint that the resulting optimal tree and classification rule have 100 percent resubstitution sensitivity. These considerations yielded $C(1|2)/C(2|1)$ slightly in excess of 14. If costs are normalized so that $\pi(1)C(1|2) + \pi(2)C(1|2) = 1$, then $\pi(1)C(2|1) = .33$ and $\pi(2)C(1|2) = .67$. Thus, the no-data optimal rule has misclassification cost .33.

Trimming of the CART- and bootstrap-generated tree mentioned previously was motivated as follows. The cited cost considerations dictated that a terminal node t is a class 1 node if the class 1 patients at t are 15 or more times as numerous as the class 2 patients, that is, if $N_1(t) \geq 15N_2(t)$. If $N_1(t) + N_2(t) < 15$ and $N_2(t) = 0$, it is perhaps not clear whether t ought to be a class 1 or class 2 node. Thus, in general, if any terminal node t satisfied $N_1(t) + N_2(t) < 15$ and $N_2(t) = 0$, the split that led to t was deleted, even if CART and the bootstrap dictated it should be retained. (Node G is the exception.) All told, the subjective trimming process eliminated two splits. The compositions of the terminal nodes will be given in the discussion of the validation set.

Seven or eight years ago, when various ideas surrounding cross-validation techniques and CART were not nearly so well developed as they are now, various trees were grown, and these trees invariably were trimmed in a subjective fashion by investigators who relied on their knowledge of subject matter and context. Not surprisingly, knowledgeable investigators often can trim large trees to ones that then turn out to be useful.

Table 6.3 gives CART's resubstitution performance. The resubstitution overall misclassification cost is .07. Since the no-data optimal rule has misclassification cost .33, about 80 percent of the cost of misclassification appears to be saved by the data and CART as amended.

TABLE 6.3 Resubstitution Performance of the Tree of Figure 6.2 on the Yale-New Haven Hospital Data

Classified \ True	Class 1 Not Heart Attack	Class 2 Heart Attack	Total
Class 1 Not Heart Attack	339	0	339
Class 2 Heart Attack	83	60	143
Total	422	60	482

In Figure 6.2, the node that is three left daughter nodes below the root node differs from the other nodes in that its question has two parts: a "yes" to either sends a candidate patient to terminal node D. This combination of questions was imposed by Goldman for reasons based on his knowledge of cardiology: the first question was suggested by CART; the second makes clinical sense in view of the ancestor nodes. Both the imposed split and the subjective trimming that have been discussed remind us that CART is a valuable companion to, but not a replacement for, substantive knowledge. (Of course, the same can be said for the best of other statistical procedures.)

One especially interesting aspect of the study reported is that the tree was used to classify patients at another hospital.

TABLE 6.4

Node	Class	Yale-New Haven Hospital Training Sample	Brigham and Women's Hospital Validation Sample	Misclassification Cost per Terminal Node for Validation Sample
A	1	132	82	.024
	2	0	2	
B	1	20	19	0
	2	0	0	
C[a]	1	5	2	.002
	2	2	1	
D[a]	1	31	33	.037
	2	4	4	
E	1	24	12	0
	2	0	0	
F	1	16	12	0
	2	0	0	
G	1	13	6	.012
	2	0	1	
H[a]	1	23	6	.007
	2	9	4	
I[a]	1	7	14	.016
	2	8	1	
J	1	19	17	.012
	2	0	1	
K[a]	1	3	8	.009
	2	2	1	
L	1	115	63	.012
	2	0	1	
M[a]	1	9	12	.013
	2	1	4	
N[a]	1	5	16	.018
	2	34	35	

[a]Class 2 node; all other nodes are class 1 nodes.

Source: Adapted from Table 3 of Goldman et al. (1982) with the permission of *The New England Journal of Medicine*.

We now turn to the larger of the two studies at the Brigham and Women's Hospital. This validation study involved patients who presented to the emergency room between October 1980 and August 1981 under the same conditions as had the Yale-New Haven patients. The minimum age was reduced from 30 to 25. Also, those patients who were not admitted to the hospital were precluded from the study if they did not consent to return to the hospital in 48 to 72 hours for certain noninvasive follow-up tests. About 80 percent of the nonadmitted patients agreed to return, and 85 percent of those actually returned for the tests. The other 15 percent were known to be alive and, indeed, well at least one week after leaving the emergency room. In all, 357 patients were included in the validation sample, 302 class 1 and 55 class 2.

Table 6.4 summarizes the class memberships for both training and validation samples for each terminal node of the tree. From the data of Table 6.5 it follows that the given classification rule had sensitivity 91 percent and specificity 70 percent. Also, the validation sample overall misclassification cost is .16, about 48 percent of its no-data value.

TABLE 6.5 Performance of the Classification Rule on the Brigham and Women's Hospital Validation Sample of Patients

Classified \ True	Class 1 Not Heart Attack	Class 2 Heart Attack	Total
Class 1 Not Heart Attack	211	5	216
Class 2 Heart Attack	91	50	141
Total	302	55	357

The most interesting and important competitors of the classification rule we have been discussing are the decisions of the emergency room physicians. The cited reference has a discussion of many aspects of those decisions; however, the simplest way of quantifying them is by identifying the decision to admit a patient to the hospital coronary care unit with a diagnosis of heart attack. Physicians' classifications are slightly less accurate than the tree, with sensitivity 91 percent and specificity 67 percent. Note that the 50 (of 55) correctly diagnosed class 2 patients are not identical to the 50 class 2 patients correctly diagnosed by the amended CART-generated classification rule. If the two classification rules are integrated so that a patient is classified to class 2 if the tree indicates class 2 *and* the emergency room physician admits the patient to the coronary care unit (or, in fact, to any hospital bed), then the resulting sensitivity drops to 88 percent, but the specificity rises to 77 percent. The overall cost of misclassification is thereby reduced to .15, which is 46 percent of its no-data value.

To summarize, an amended CART-generated classification rule, or better, that rule supplemented by the judgments of emergency room physicians, can reduce the cost of misclassification to less than 50 percent of its no-data value for an extremely important and sometimes very difficult problem in clinical medicine. A reader might inquire whether other technologies do as well or better. Analyses that utilize logistic regression, for example, and which are not reported in detail here, show that it does nearly but not quite as well as the rule presented. And even though the tree in this section is not so simple as the one in Section 6.1, it is still much easier to use accurately (by emergency room personnel) than is a linear function of, say, nine variables.

6.3 IMMUNOSUPPRESSION AND THE DIAGNOSIS OF CANCER

A depressed immune system is often a concomitant of cancer, or at least of advanced cancer. Thus, if it were possible by laboratory test to quantify the level of immunocompetence of an individual, it might be possible to determine if he or she has cancer. (Someone whose immune system is functioning satisfactorily is said to be immunocompetent.) This section is a report on the use of 21 continuous laboratory variables related to immunocompetence, age, sex, and smoking habits in an attempt to distinguish patients with cancer from controls. The study was undertaken by Robert Dillman and James Koziol of the University of California, San Diego Cancer Center (see Dillman and Koziol, 1983, and Dillman et al., 1983).

The class 2 subjects in this section are 64 patients with advanced refractory cancer, that is, patients whose diseases were resistant to cure. The histologic types of their diseases were diverse, at least 13 in all. The class 1 subjects, the controls, were 64 healthy individuals. There were some missing values for 58 of the 128 individuals whose data are the subject of this section.

It will be clear shortly that CART does very well in providing simple rules for distinguishing between class 1 and class 2 individuals. Moreover, the study on which this section is based is an important contribution. However, it might be subject to criticism applicable to many investigations of its type; namely, the controls involved were healthy and not afflicted by inflammatory diseases other than cancer. The separation of patients with cancer from other ill patients by procedures such as CART would be an especially difficult and important task.

In the problem of this section, unlike those of Sections 6.1 and 6.2, issues of variable selection did not arise. Those measurements available and given any a priori chance of being helpful to classification were used in CART. We took $\pi(1) = .5 = \pi(2)$ and $C(2|1) = C(1|2)$, which for convenience was taken to be 1; thus, $\pi(1)C(2|1) = \pi(2)C(1|2) = .5$. Any classification rule that ignores the covariates is a no-data optimal rule and has misclassification cost .5. In the trees we displayed, a terminal node t is a class 1 node if $N_1(t) > N_2(t)$, a class 2 node if $N_2(t) > N_1(t)$. If one thinks of the trees as being used in diagnosis, then typically the resulting prior probabilities and losses will be quite different from the choices we made. However, it may still be true that, at least approximately, $\pi(1)C(2|1) = \pi(2)C(1|2)$.

Both trees were grown using the Gini splitting criterion, and the pruned tree was chosen to minimize the tenfold cross-validated misclassification cost. The first was grown with the 24 cited variables, the second with 9 preselected variables, those initially thought to be most important in distinguishing cancer patients from controls (see Figure 6.3).

Notice that the splits in the first tree all appeared in the second, but that one additional split was made in the second tree. Also, terminal node A in the second tree has one fewer class 2 observation than the corresponding node in the first. The missing observation seems to have appeared in terminal node D in the second tree. In fact, the first surrogate split at the root node of the first tree is on a variable that was not included on the list of variables for the second tree. When CART was run on a collection of variables that included the preselected nine and the missing surrogate variable, that one observation again appeared at node A, but the tree was otherwise unchanged.

CART's resubstitution cost of misclassification was .094 for both trees, which is an apparent 81 percent reduction in the corresponding no-data value. That the sensitivities and specificities differ is clear from Tables 6.6 and 6.7.

Although the first tree appears to be more sensitive but less specific than the second, cross-validation indicates that just the opposite is true, for the cross-validated misclassification probabilities are as shown in Table 6.8. As with resubstitution, upon cross-validation the trees had identical overall costs of misclassification. Each cost was .22, which amounts to a saving of roughly 57 percent of the no-data misclassification cost. The cited

Tree grown with 21 continuous variables that measure immunosuppression, sex, age, and smoker (or not)

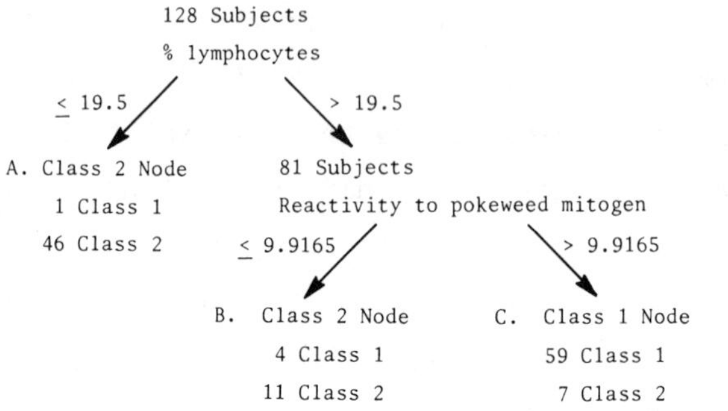

Tree grown with 9 variables thought to measure immunosuppression

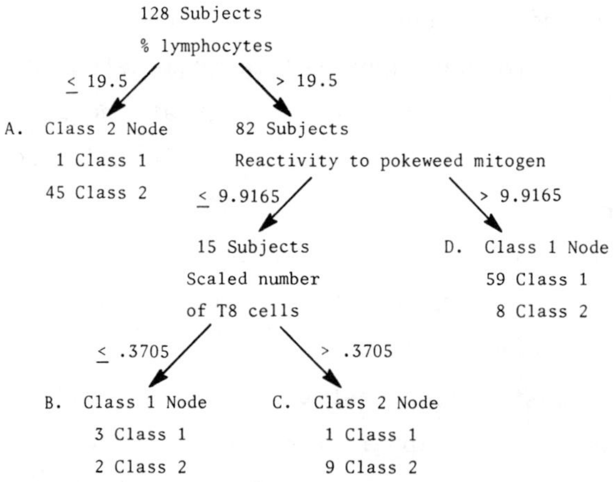

FIGURE 6.3 There are many classifications of white blood cells. For example, they are dichotomized as to B or T cells and as lymphocytes or not. Lymphocytes are the main components of the body's immune system. By percent lymphocytes is meant the number of lymphocytes as a percentage of the overall white blood count. T8 cells are a type of T cell that may suppress an individual's immune reaction to a challenge. One challenge that is widely used in studies of immunosuppression is that presented by pokeweed mitogen. It is well understood from other sources that immunosuppressed individuals tend to have low lymphocyte counts, low reactivity to pokeweed mitogen, and large numbers of T8 cells relative to other lymphocytes. [Adapted from Figure 3 of Dillman and Koziol (1983) with the permission of *Cancer Research*.]

TABLE 6.6 Resubstitution: 24-Variable Tree

Classified \ True	Class 1 Controls	Class 2 Cancer	Total
Class 1 Controls	59	7	66
Class 1 Cancer	5	57	62
Total	64	64	128

TABLE 6.7 Resubstitution: 9-Variable Tree

Classified \ True	Class 1 Controls	Class 2 Cancer	Total
Class 1 Controls	62	10	72
Class 2 Cancer	2	54	56
Total	64	64	128

TABLE 6.8

	Class 1	Class 2
24-variable tree	.13	.30
9-variable tree	.18	.25

papers of Dillman and Koziol suggest that in detecting cancer, neither linear discrimination nor logistic regression is more sensitive than CART and that they are both less specific.

From the studies summarized in this section, we see that only the percent lymphocytes, reactivity to pokeweed mitogen, and number of T8 cells are used as primary splitting variables. These are relatively old tests for immunosuppression. By and large, the newer and much more expensive tests were not helpful as supplemental information, although two of the six variables that scored highest in the measure of importance arose from the new tests. A final interesting fact that surfaced from the analyses under discussion is this: the ratio of T4 cells to T8 cells, which was once thought to be a powerful diagnostic tool for the problem at hand, turns out to be one of the least important variables (and never appeared as even a surrogate splitting variable).

6.4 GAIT ANALYSIS AND THE DETECTION OF OUTLIERS

This section is different in emphasis from the earlier sections in this chapter. The context in which the data were gathered may be less familiar to readers than are heart attacks and cancer, though this subject matter, too, has important clinical applications. In addition, we concentrate on the detection of outliers as opposed to classification.

The data arose from a study of the development of gait, that is, walking, in 424 normal children aged 1 to 7. All studies were performed at the Motion Analysis Laboratory at Children's Hospital and Health Center, San Diego. The Laboratory is directed by David H. Sutherland, Chief of Orthopaedic Surgery at Children's Hospital and a member of the Department of Surgery at the University of California, San Diego.

Gait is useful as a barometer of (the top down) neurologic development of normal children—and also of children suffering from cerebral palsy or muscular dystrophy. A supposedly normal child whose walk is markedly different from that of others his or her age may have neurologic abnormalities. On the other hand, a child known to be neurologically impaired might have that impairment quantified in part by gait measurements. So, it should be of interest to see how well the age of a normal child can be deduced from gait measurements (which do not a priori dictate that child's age or size) and thereby to learn something of the relationship between chronologic age and neurologic development. The prediction of age from gait measurements is the classification problem of the present section. Readers desiring more information on gait analysis are referred to work by Sutherland and colleagues (Sutherland et al., 1980; Sutherland et al., 1981; and Sutherland and Olshen, 1984).

Children were studied at ages 1 through 4 in increments of six months, and also at ages 5, 6, and 7. (In all but the length of step and walking velocity, the gait of a 7-year-old resembles that of an adult so closely as to render gait studies of maturation uninteresting in normal children over 7 years.) The classes are numbered 1 through 10 in order of increasing age. Each child was studied within 30 days of the date required for his or her class. While some children were studied at two or even three ages, most were studied just once. Although there were not exactly equal numbers of children in each group, the numbers were close enough to equality that we took $\pi(i) \equiv .1$, $i = 1, \ldots, 10$. Also, we took $C(i|j) = |i - j|^{1/2}$ so that each no-data optimal rule assigns every observation to class 5 or class 6 independent of the covariates; and each such rule has expected cost of misclassification approximately 1.45. The exact choice of C and possible competitors are discussed later in this section.

During a gait study a child is attired in minimal clothing and has sticks placed perpendicular to the sacrum (at the lower end of the spine), facing forward midway between the hips, and facing forward on each tibia (shinbone). Also, four markers are placed on the lower extremities. The child walks at a speed comfortable to him or her down the level walkway. Data are gathered on the motions of the joints in three planes, the walking velocity, cadence, time spent on each leg alone, and the ratio of the width of the pelvis to the distance between the ankles while walking. In addition, both the sex and the dominant side, if it is discernible, are included as variables. Many measurements were gathered with high-speed cameras and later digitized from film. (We mention in passing the paucity of work by statisticians on digitizing processes.) Figure 6.4 depicts the layout of the laboratory.

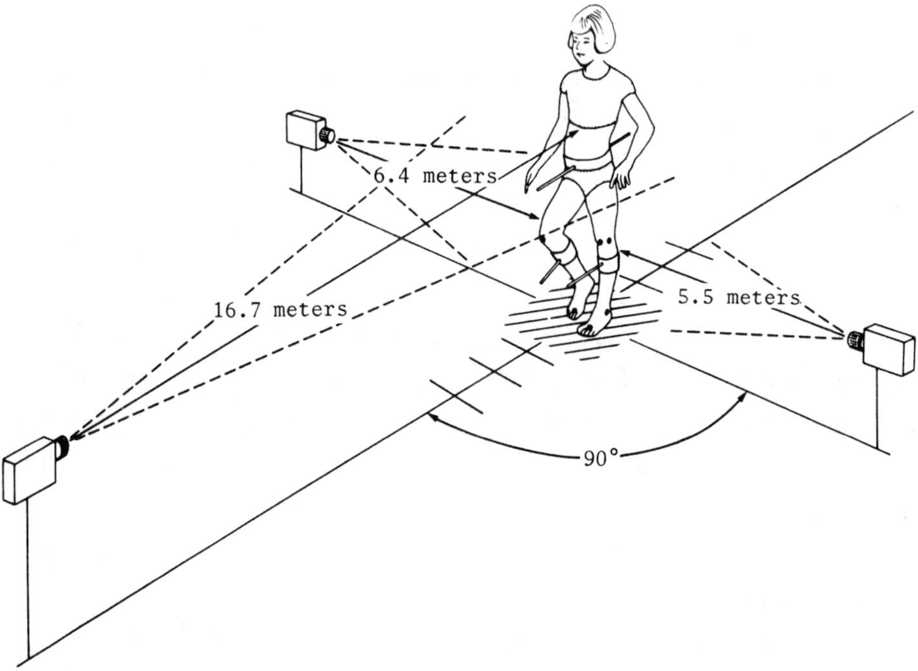

FIGURE 6.4 Camera layout. [Reprinted from Figure 2 of Sutherland et al. (1980) with the permission of *The Journal of Bone and Joint Surgery*.]

Each child is observed for at least three passes down the walkway. A pass includes about six steps, that is, three full cycles. Data on seven important variables from the three passes are averaged. The cycle for which the seven variables most closely resemble their averages is chosen as the representative cycle for that child. Because joint motions are approximately periodic, an overall mean and the first six sine and cosine Fourier coefficients are used as summary statistics of the motions and as variables in CART. In all, there are well over 100 variables that might have been used for classification. Various CART runs on subsets of variables available at this writing, univariate statistics like those of Sections 6.1 and 6.2, and Sutherland's judgment led to a much reduced list of 18 variables, the independent variables of the analysis reported here. Trees were generated with the twoing splitting criterion. Tenfold cross-validation showed that the tree with smallest cross-validated misclassification cost has 19 terminal nodes and that its estimated misclassification cost is .84. The tree gives patent evidence of the outliers that are the focal point of this section. On subjective grounds, an optimal tree with 16 terminal nodes was chosen for presentation here. To describe the tree in detail would take us too far afield, so instead in Table 6.9 we report the contents and class assignments of its 16 terminal nodes.

CART's resubstitution performance is summarized in Table 6.10. Notice the three circled observations: a 5-year-old (J. W.) classified as a $1\frac{1}{2}$-year-old, another 5-year-old (J. C.) classified as a 2-year-old, and a 7-year-old (L. H.) classified as a 2-year-old.

Since the three children whose observations are circled were classified as being only 2 years old and at least 3 years younger than their true ages, their films and written records were reviewed to examine qualitative aspects of their walks and to verify values of their motion variables. Furthermore, their data from subsequent years, when available, were analyzed with a view toward detecting possible developmental abnormalities. No such abnormalities were found, but still, three interesting facts emerged.

J. W. was and remains a normal child. For him the original measurements were all correct. However, possibly because of his fear of the laboratory, his walk was deliberately such that the heel for each step followed directly on the toe of the previous step; this is called "tandem walking." What resulted was a large

TABLE 6.9

Terminal Node	Assigned Age	True Ages Age 1.0	Age 1.5	Age 2.0	Age 2.5	Age 3.0	Age 3.5	Age 4.0	Age 5.0	Age 6.0	Age 7.0	Total
A	1.0	33	1	0	0	0	0	0	0	0	0	34
B	1.5	12	24	9	4	0	0	0	1	0	0	50
C	2.0	2	7	13	3	0	0	1	0	0	0	26
D	3.0	0	0	3	9	18	6	10	3	3	1	53
E	2.0	1	5	13	10	9	1	1	1	0	1	42
F	6.0	0	0	0	0	0	0	0	0	15	8	23
G	7.0	0	0	0	0	1	1	2	4	6	16	30
H	3.5	0	0	0	4	8	17	3	4	1	0	37
I	4.0	0	0	0	0	4	3	8	5	2	0	22
J	2.5	0	0	3	5	4	0	2	0	0	0	14
K	1.5	2	3	2	0	0	0	0	0	0	0	7
L	4.0	0	0	1	0	0	4	5	2	0	0	12
M	7.0	0	0	0	0	0	1	2	5	5	19	32
N	5.0	0	0	0	1	0	1	0	4	0	0	6
O	5.0	0	0	0	0	1	1	2	7	1	0	12
P	6.0	0	0	0	0	1	5	2	4	11	1	24
	Total	50	40	44	36	46	40	38	40	44	46	424

TABLE 6.10

Classified \ True	Age 1.0	Age 1.5	Age 2.0	Age 2.5	Age 3.0	Age 3.5	Age 4.0	Age 5.0	Age 6.0	Age 7.0	Total
Age 1.0	33	1	0	0	0	0	0	0	0	0	34
Age 1.5	14	27	11	4	0	0	0	(1)	0	0	57
Age 2.0	3	12	26	13	9	1	2	(1)	0	(1)	68
Age 2.5	0	0	3	5	4	0	2	0	0	0	14
Age 3.0	0	0	3	9	18	6	10	3	3	1	53
Age 3.5	0	0	0	4	8	17	3	4	1	0	37
Age 4.0	0	0	1	0	4	7	13	7	2	0	34
Age 5.0	0	0	0	1	1	2	2	11	1	0	18
Age 6.0	0	0	0	0	1	5	2	4	26	9	47
Age 7.0	0	0	0	0	1	2	4	9	11	35	62
Total	50	40	44	36	46	40	38	40	44	46	424

external rotation of the hip, as is characteristic of the $1\frac{1}{2}$-year-olds to which he was classified.

In the class of L. H., all measurements used in CART except walking velocity agreed with those of the second analysis of the data. Her walking velocity was stated in laboratory records to be 81, which is correct if velocity is measured in meters per minute. For all other children, velocity was measured in centimeters per second; L. H.'s 81 thus should have been 136. With the correction, but the same classification scheme as before, L. H. moves from terminal node E to terminal node P. She is thereby reclassified from age 2 to age 6, only 1 year below her true age.

Like L. H., J. C. was originally a member of terminal node E. While her other measurements were unchanged after reanalysis, her walking velocity was found to have been coded originally as 58 rather than the true value 109. Once the mistake is corrected, J. C. moves to terminal node O, and she is classified as the 5-year-old she was at the time of her study.

The 7-year-old in terminal node D might be thought of as a candidate outlier, though we have no evidence that such is the case. Of course, the membership of that terminal node is quite diverse. Also, from the paper of Sutherland et al. (1980), it seems that for a 7-year-old to be classified by gait as a 3-year-old is not as surprising as for a 5-year-old (such as J. W. and J. C.) to be classified as a 2-year-old.

The reader should note the ease with which outliers were detected and also note that although nothing of medical significance was learned from the detection, two mistakes were found and corrected and a third set of data were deleted from further analyses.

CART has contributed to important understanding of the gait data beyond the detection of outliers. For example, in the forthcoming monograph of Sutherland and Olshen (1984), it is argued from several points of view that dominant side bears no discernible relationship to the development of mature gait. This somewhat surprising result was suggested by preliminary CART runs in the process of variable selection, for dominance seldom appeared as much as a surrogate splitting variable. Its coordinate importance was always low. Because plausibility arguments suggested that dominance should play some part in our classification problem, it was retained in the final list of 18 variables, despite the lack of any quantitative evidence for the retention. In the CART run reported here, dominance appeared once as a fourth surrogate splitting variable, and its coordinate importance was about 60 percent that of the second least important variable.

We close this section with brief comments on the choice of the cost function C. The three mentioned outliers show up clearly when trees are grown with the default zero-one cost function $C(i|j) = \delta_{ij}$, and it is evident why this might occur. With that choice CART tries to have each terminal node consist of children of one age; failing that, a 6-year-old looks no worse at a 2-year-old node than does a $2\frac{1}{2}$-year-old. In view of the natural ordering of the classes, the zero-one cost function does not seem sensible. The first alternative we tried has $C(i|j) = |i - j|$. This led to larger trees than did the previous function, and furthermore, those trees obscured two of the three outliers. Again, explanation is simple. With cost six times as large for classifying a 6-year-old as a 2-year-old than for so classifying a $2\frac{1}{2}$-year-old, CART will work hard to ensure that the variabilities of ages within terminal nodes are small. The final choice $C(i|j) = |i - j|^{1/2}$ seems to incorporate the best features of its two cited competitors and also to be rather simple, as they are. Thus, it was the choice for presentation here.

6.5 RELATED WORK ON COMPUTER-AIDED DIAGNOSIS

The paper by Goldman et al. (1982) contains a nice survey of previous contributions to uses of statistical techniques in the diagnosis of myocardial infarction. The contribution of Pozen, D'Agostino, and Mitchell (1980), who utilized logistic regression, is particularly noteworthy. Goldman observes that the approach to the problem through CART was more accurate than previous attempts.

The use of statistical approaches to combining cancer markers and thereby aiding the automatic diagnosis of cancer is quite new. The cited contributions of Dillman and Koziol and their colleagues (1983) are at the forefront of the field. Further work is being carried out under the auspices of the National Cancer Institute's Biological Response Modifier Program, wherein a number of agents are being evaluated for their collective impact on the immune system in a variety of individuals with and without cancer. CART will be employed in that study.

One obstacle to the successful use of CART or any other technique that utilizes an ongoing data base in medical prognosis is that data bases change over time. Patient mixes and hospital protocols change enough to affect both unconditional distributions of class membership and conditional distributions of predictors given class membership. Feinstein (1967) has been skeptical of some statistical approaches to diagnosis and prognosis for this reason. His skepticism has been advanced by practitioners of artificial intelligence (see Szolovits, 1982).

In the recent paper by Duda and Shortliffe (1983), explicit criticism is made of "decision-tree or statistically based programs." Their approach to medical diagnosis involves modeling so-called expert opinion and claims to offer explanations of diagnoses. The approach through CART to medical diagnosis and prognosis attempts to supplement trees constructed from data by expert opinion, but not to do away altogether with data as a principal tool. Goldman et al. (1982) indicate that sound statistical methodology applied to good data sometimes provides more accurate medical diagnoses than do experts.

7

MASS SPECTRA CLASSIFICATION

7.1 INTRODUCTION

The preceding chapters have discussed and used the standard structure built into CART. The data structure was assumed to consist of a fixed number of variables and the set of splits S is built into the program.

However, the general tree methodology is a flexible tool for dealing with general data structures. It may be applied to a general measurement space X without assumptions regarding its structure. The critical element, which must be defined by the analyst, is a set of binary questions or, equivalently, a set S of splits s of X into two disjoint subsets. A measure of node impurity is then selected, and at each node a search is made over S for that split s^* which most reduces the impurity.

The burden is on the analyst in constructing a set of questions which can effectively extract information from the data. However, the tree structure permits considerable "overkill." A large number of splits can be searched over, allowing the inclusion of many questions that may or may not turn out to be informative.

The program then selects the best of these questions. For example, in the ship classification problem a set of 5000 splits was used. In the chemical spectra problems, about 2000 questions were used.

Even if the data base contains a fixed number of variables, it may not be desirable to use the standardized set as splits. For example, in word recognition, a common procedure is to use band pass filters to decompose the amplitude into amplitudes over M frequency ranges. A single word is then represented by M waveforms. Even if these waveforms are discretized over time into N time samples, there are MN coordinates, where MN is generally large.

Using univariate splits on these coordinates is not a promising approach. In speech recognition, as in other pattern recognition problems, the common approach is to define "features" consisting of combinations of variables suggested either by the physical nature of the problem or by preliminary exploration of the data. To make the problem amenable to available techniques, usually a fixed number of features are extracted and then some standard classification technique such as discriminant analysis or kth nearest neighbor is employed.

In contrast, a tree approach permits the use of a large and variable number of features. This is particularly useful in problems where it is not clear which features and what threshold values are informative.

Given only a set S of splits, there are no longer any entities such as variables, and the variable oriented features of CART are not applicable. What remains applicable is discussed in Section 7.2. The general structure and question formulation are illustrated in Section 7.3 through a description of the mass spectra tree grown to recognize the presence of the element bromine in compounds.

7.2 GENERALIZED TREE CONSTRUCTION

The only parts of CART which depend on a fixed cases by variables data matrix are the measure of predictive association between variables and the linear combination of variables. Boolean combinations of splits may be used in the general case. The missing value algorithm and variable ranking are no longer applicable.

All other parts of the tree methodology are applicable, including the pruning algorithm, the test sample and cross-validation estimates, and the standard error estimates.

We have not set up a program which accepts a general data structure and permits the user specification of the set of splits $\mathcal{S}$. Each nonstandard tree has been programmed from scratch incorporating pruning, estimation, etc., as subroutines. Building a general program is feasible and may be undertaken if need develops.

7.3 THE BROMINE TREE: A NONSTANDARD EXAMPLE

7.3.1 Background

In 1977-1978 and again in 1981, the EPA funded projects for the construction of classification trees to recognize the presence of certain elements in compounds through the examination of their mass spectra. In the initial project, a tree algorithm for recognition of chlorine was constructed (Breiman, 1978). During the second project (Breiman, 1981), trees were constructed to detect the presence of bromine, an aromatic phenyl ring, and nitrogen. In this chapter we describe the construction of the bromine tree.

In brief, the mass spectrum of a compound is found by putting a small amount of the compound into a vacuum and bombarding it with electrons. The molecules of the compound are split into various fragments and the molecular weight of the fragments are recorded together with their abundance. The largest fragment consists of molecules of the compound itself which were not fragmented. A typical mass spectrum is given in Table 7.1, and graphed in Figure 7.1. The relative abundances are normalized so that the maximum abundance is 1000.

TABLE 7.1 Mass Spectrum of Diphenyl-Acetylene

m/e*	Relative Abundance	m/e*	Relative Abundance	m/e*	Relative Abundance
38	1.6	87	13.0	151	79.0
39	10.0	88	36.0	152	84.0
40	0.5	89	74.0	153	10.0
50	10.0	90	2.1	154	0.6
51	23.0	91	1.3	163	7.0
52	6.7	102	10.0	164	2.0
53	0.6	103	11.0	165	0.7
63	36.0	115	5.4	174	5.3
64	3.8	126	27.0	175	23.0
65	4.1	127	8.9	176	140.0
74	18.0	128	6.4	177	100.0
75	24.0	129	0.8	178	1000.0
76	100.0	139	23.0	179	150.0
77	11.0	140	4.1	180	11.0
78	2.2	150	39.0	181	0.5

*m/e = molecular weight

A trained chemist can interpret a mass spectrum in terms of the chemical bonds involved and often reconstruct the original molecule. According to McLafferty (1973, preface): "The fragment ions indicate the pieces of which the molecule is composed, and the interpreter attempts to deduce how these pieces fit together in the original molecular structure. In recent years such corre-

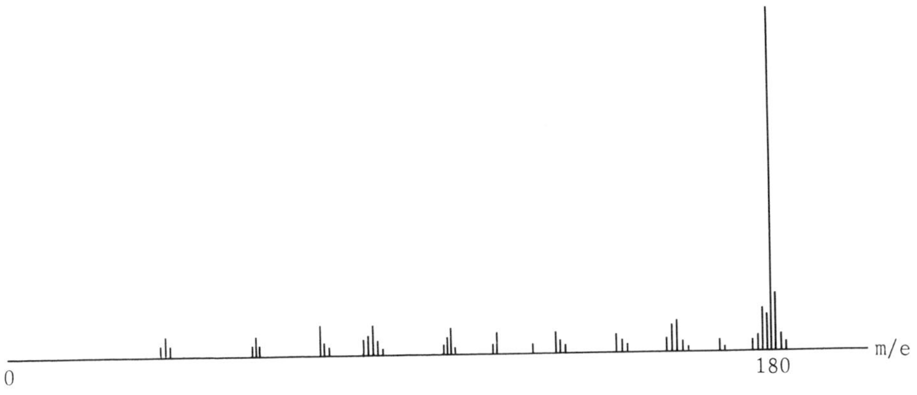

FIGURE 7.1 Graph of mass spectrum of diphenyl-acetylene.

lations have been achieved for the spectra of a variety of complex molecules."

The EPA, as part of its regulatory function, collects numerous samples of air and water containing unknown compounds and tries to determine the presence of toxic substances. Although mass spectra are not difficult to measure, a competent spectral interpretation is costly and time consuming.

Instead of attempting to reconstruct the complete molecular structure, a more modest goal was considered as possible for an automatized screening device: Construct computerized algorithms to recognize the presence of certain key elements in the unknown compound.

The available data base is large. The chemical information system maintained by EPA-NIH currently contains about 33,000 mass spectra corresponding to identified molecular structures. The compounds range in molecular weight from under 100 to over 1000. This suggests that one possible identification method is to match the mass spectrum of an unknown compound against the library and select the compound whose spectrum is most similar to the unknown. This has difficulties. First, the unknown compound may not be in the library. Second, mass spectra are not precisely replicable. Depending on the apparatus and the user, mass spectra measurements of the same compound can vary in the relative abundance measurements. We have heard estimates of variability ranging from 10 percent to 25 percent. Third, such searches are computationally expensive.

At any rate, the EPA decided to fund a tree structured classification approach. The first tree, detecting chlorine, had a test sample misclassification rate of 8 percent and is implemented in the chemical information system. The bromine tree, described shortly, has a test sample misclassification rate of 6 percent and has also been implemented. Because of the nature of the tree structure, both algorithms produce classifications in a fraction of a second (CPU time).

7.3.2 *Data Reduction and Question Formulation*

The critical element of the bromine tree was the construction of a set of questions designed to recognize bromine hallmarks. In particular, bromine has two important footprints:

1. For every 100 atoms of bromine occurring naturally at atomic weight 79, there are 98 isotopic atoms of weight 81.
2. Small fragments containing any halogens in the compound (chlorine and bromine predominantly) tend to split off easily.

The first step in this process was a reduction in dimensionality of the data. For any spectrum, denote by H_m the relative fragment abundance at molecular weight m. Then H_m was defined to be a *peak* or local maximum if

$$H_m = \max(H_{m-5}, H_{m-4}, \ldots, H_{m+4}, H_{m+5}).$$

In a spectrum the peaks are the important landmarks, and what is usually seen in a graph of the spectrum are peaks surrounded on either side by small relative abundances corresponding to the main fragment at the peak plus or minus one or more hydrogen atoms.

However, if there is an isotopic structure, then the cluster of abundances around the peak may reflect that structure. For example, if a fragment of weight m contains one bromine atom, then $H_{m+2}/H_m \simeq .98$. Therefore, corresponding to every peak location seven ratios were computed

$$R_1 = \frac{H_{m-6}}{H_m}, \quad R_2 = \frac{H_{m-4}}{H_m}, \quad R_3 = \frac{H_{m-2}}{H_m}, \quad R_4 = \frac{H_{m+2}}{H_m},$$

$$R_5 = \frac{H_{m+4}}{H_m}, \quad R_6 = \frac{H_{m+6}}{H_m}, \quad R_7 = \frac{H_{m+8}}{H_m}.$$

If a fragment contains one bromine atom, the theoretical value for the ratios $\mathbf{R} = (R_1, \ldots, R_7)$ is

$$\mathbf{R}_1^* = (0, 0, 0, .98, 0, 0, 0).$$

However, if it contains two bromine atoms, the most frequently occurring combination is one 79 bromine atom and one isotopic bromine atom. The theoretical ratio vector then is

$$\mathbf{R}_3^* = (0, 0, .51, .49, 0, 0, 0).$$

If bromine occurs in combination with chlorine, then since chlorine (weight 35) has an isotope of weight 37 that occurs 24.5 percent of the time, there is a different theoretical ratio vector. The data base showed significant numbers of compounds with one, two, three, four bromine atoms and one or two bromine atoms in combination with one or two chlorine atoms. This gives eight theoretical ratio vectors. However, we reasoned that with any error, two nearly equal abundances might be shifted; that is, with a small experimental error, a fragment containing a single bromine atom might have its peak at the isotopic weight. For this reason, the theoretical shifted ratio $\mathbf{R}_2^*$,

$$\mathbf{R}_2^* = (0, 0, 1.0, 0, 0, 0, 0),$$

was added together with another shifted ratio $\mathbf{R}_5^*$, where adjacent abundances were nearly equal.

The set of theoretical ratios are listed in Table 7.2. Corresponding to every theoretical ratio a set of weights was defined. These are given in Table 7.3.

TABLE 7.2

	Bromine-Chlorine	Theoretical Ratios						
R_1^*	1 Br	0	0	0	.98	0	0	0
R_2^*	(shifted R_1^*)	0	0	1.0	0	0	0	0
R_3^*	2 Br	0	0	.51	.49	0	0	0
R_4^*	3 Br	0	0	.34	.98	.32	0	0
R_5^*	(shifted R_4^*)	0	.34	1.0	.32	0	0	0
R_6^*	4 Br	0	.17	.68	.65	.16	0	0
R_7^*	1 Br, 1 Cl	0	0	.77	.24	0	0	0
R_8^*	2 Br, 1 Cl	0	0	.44	.70	.14	0	0
R_9^*	1 Br, 2 Cl	0	0	.61	.46	.07	0	0
R_{10}^*	2 Br, 2 Cl	0	0	.38	.90	.32	.04	0

TABLE 7.3 Weights

W_1	0	0	1/3	1/3	1/3	0	0
W_2	0	1/3	1/3	1/3	0	0	0
W_3	0	1/4	1/4	1/4	1/4	1/4	0
W_4	0	1/5	1/5	1/5	1/5	1/5	0
W_5	1/5	1/5	1/5	1/5	1/5	0	0
W_6	1/6	1/6	1/6	1/6	1/6	1/6	0
W_7	0	1/4	1/4	1/4	1/4	0	0
W_8	0	1/5	1/5	1/5	1/5	1/5	0
W_9	0	1/5	1/5	1/5	1/5	1/5	0
W_{10}	0	1/6	1/6	1/6	1/6	1/6	1/6

For each peak location and corresponding adjacent abundance ratios $\mathbf{R} = (R_1, \ldots, R_7)$, a distance D_i from $\mathbf{R}$ to the theoretical ratio $\mathbf{R}_i^*$ was defined by

$$D_i = \sum_{j=1}^{7} W_{ij} |R_j - R^*_{i,j}|.$$

Then D was set equal to $\min_i D_i$ and B was defined as the number of bromine atoms in the theoretical ratio closest to **R**. Thus, D gave an overall measure of how closely the abundances adjacent to a given peak mimicked clusters of abundances reflecting isotopic structure of bromine, possibly combined with chlorine.

If a peak occurred at weight m, the loss L was defined as

$$L = \text{molecular weight of compound} - m$$

and the height of the peak denoted by H. The final data reduction consisted of associating an (L, H, D, B) vector to each peak in the spectrum. Over the data base, the number of peaks ranged from a low of 2 to over 30.

For each reduced spectrum, associate with each L its rank RL among the losses; that is, if there are N peaks with losses arranged so that

$$L_1 \leq L_2 \leq \cdots \leq L_H,$$

then the value of RL corresponding to L_1 is 1, to L_2 is 2, and so on. Similarly, define the rank height RH to be 1 for the largest H, 2 for the second highest H, and so on.

The set of questions used were in two groups. The first group was aimed at the isotopic structure:

Q_1: Are there $\geq k$ peaks, $k = 1, 2, 3, 4$, such that for every peak
(a) $D \leq d$, $d = .01, .02, .04, .08, .16, .32$,
(b) $RL \leq r$, $r = k + 1, k + 2, k + 3, k + 4, k + 8$,
(c) $RH \leq h$, $h = k + 1, k + 2, k + 3, k + 4, k + 8$?

Therefore, Q_1 contained 600 questions.

The second set of questions (modified slightly to simplify the presentation) is aimed at the losses:

Q_2: Is there a peak with loss L, $L = 0, 1, \ldots, 250$ and $RL \leq r$, $r = 1, 2, 3, 4, 5, \infty$?

Q_2 contains 1500 questions. The Boolean combinations of splits in Q_2 were constructed (see Section 5.2.3). Oring was used instead of anding, so these splits were of the form s_1 or s_2....

A third set of 15 questions was aimed at the total number of peaks in the spectrum but did not produce any splits in the tree construction.

7.3.3 *Tree Construction and Results*

Out of about 33,000 compounds, only 873 contained bromine. An independent test sample approach was used, and the cases given in Table 7.4 were selected at random.

TABLE 7.4

	Bromine	Nonbromine	Ratios
Learning sample	582	5820	2
Test sample	291	2910	1
Ratios	1	10	

The decision to make the nonbromine set 10 times as large as the bromine was somewhat arbitrary. The reasoning was that the nonbromines represented a much larger diversity of compounds than the bromines; therefore, the sample size selected should be large enough to reflect this diversity. On the other hand, using all of about 32,000 nonbromine spectra seemed redundant and wasteful.

In view of the purpose of the project, we wanted to roughly equalize the bromine and nonbromine misclassification rates. Therefore, equal priors were used, together with the Gini splitting criterion. Tree pruning was done as described in Chapter 3.

The initial tree T_1 contained 50 nodes. The graph of the misclassification rates on pruning up is illustrated in Figure 7.2.

The minimum value of R^{ts} = 5.8 percent occurred with 13 terminal nodes. This tree is pictured in Figure 7.3. The test sample

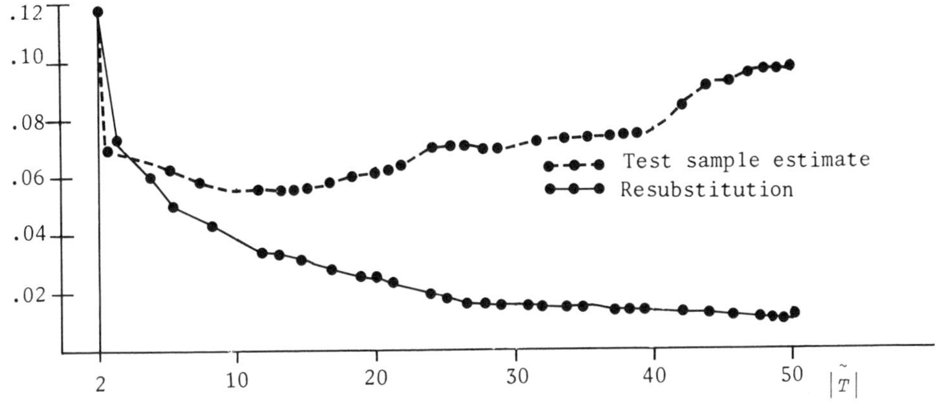

FIGURE 7.2

bromine misclassification rate was 7.2 percent and the nonbromine 4.3 percent. Standard errors of R^{ts} ranged from 1.0 percent for 50 nodes to a low of 0.8 percent around the 13-terminal-node range. As might be expected, the standard error for the bromine misclassification rate alone was higher, about 1.5 percent in the 13-node range, compared with .4 percent for the nonbromines.

Over 75 percent of the bromines ended in one terminal node defined by isotopic structure questions only. About 80 percent of the nonbromines were in a terminal node defined by an initial split on isotopic structure and then whittled down by a sequence of questions about loss locations.

As a final verification, the remaining 23,000 nonbromines were run through the optimal tree. The resulting misclassification rate was 4.4 percent.

We believe that the drop from 8 percent on the chlorine tree to 6 percent on the bromine was due to experience gained in formulating an informative set of questions. For example, Boolean combinations were not used in the chlorine tree. The procedure is still unrefined, and further development of the question set could result in a significant drop in misclassification.

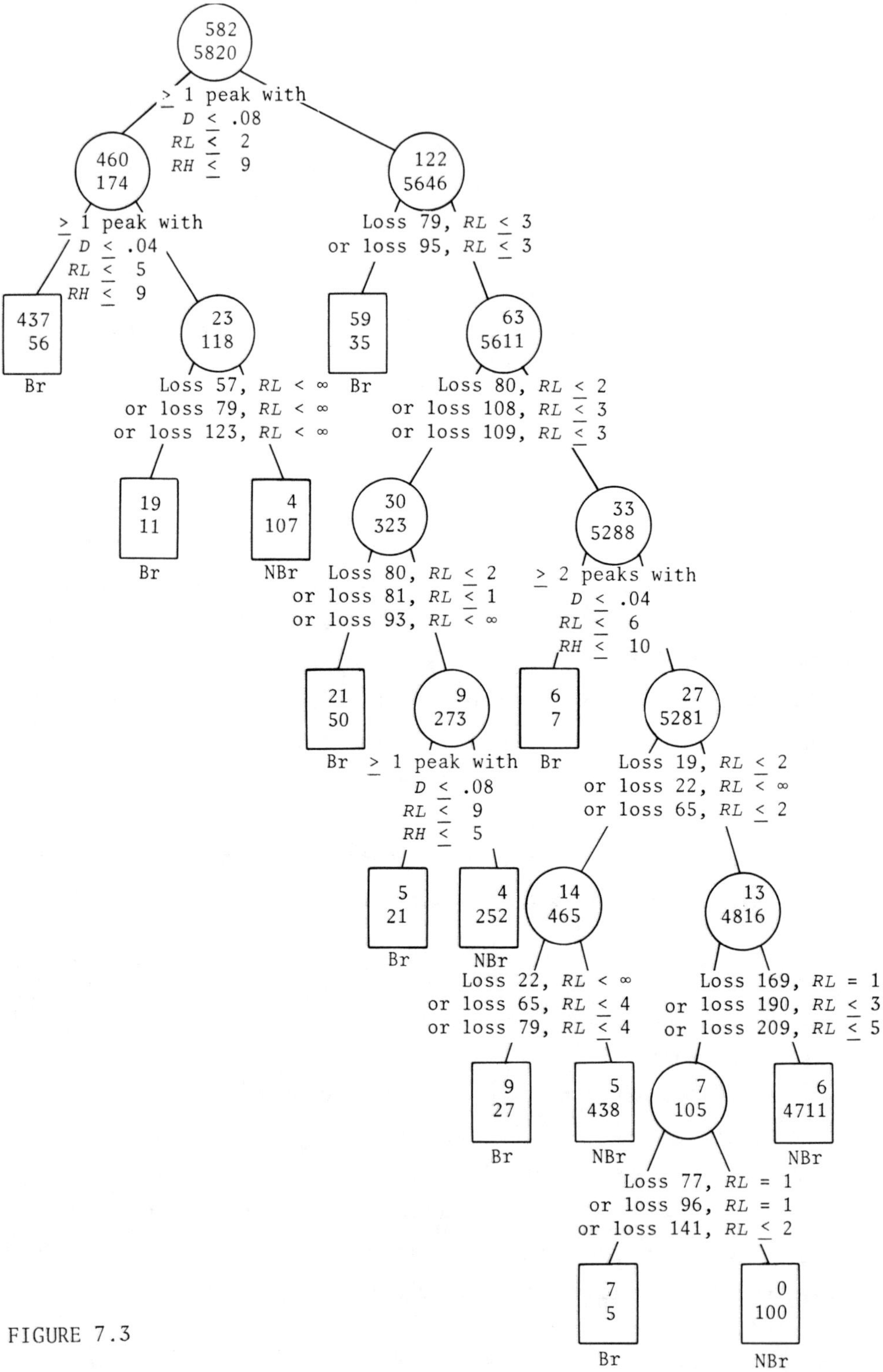
582
5820
≥ 1 peak with
D ≤ .08
RL ≤ 2
RH ≤ 9
460
174
122
5646
≥ 1 peak with
D ≤ .04
RL ≤ 5
RH ≤ 9
Loss 79, RL ≤ 3
or loss 95, RL ≤ 3
437
56
Br
23
118
59
35
Br
63
5611
Loss 57, RL < ∞
or loss 79, RL < ∞
or loss 123, RL < ∞
Loss 80, RL ≤ 2
or loss 108, RL ≤ 3
or loss 109, RL ≤ 3
19
11
Br
4
107
NBr
30
323
33
5288
Loss 80, RL ≤ 2
or loss 81, RL ≤ 1
or loss 93, RL < ∞
≥ 2 peaks with
D ≤ .04
RL ≤ 6
RH ≤ 10
21
50
Br
9
273
6
7
Br
27
5281
≥ 1 peak with
D ≤ .08
RL ≤ 9
RH ≤ 5
Loss 19, RL ≤ 2
or loss 22, RL < ∞
or loss 65, RL ≤ 2
5
21
Br
4
252
NBr
14
465
13
4816
Loss 22, RL < ∞
or loss 65, RL ≤ 4
or loss 79, RL ≤ 4
Loss 169, RL = 1
or loss 190, RL ≤ 3
or loss 209, RL ≤ 5
9
27
Br
5
438
NBr
7
105
6
4711
NBr
Loss 77, RL = 1
or loss 96, RL = 1
or loss 141, RL ≤ 2
7
5
Br
0
100
NBr

FIGURE 7.3

Hopefully, the detail in which this example was presented does not obscure the point: The hierarchical tree structure gives the analyst considerable flexibility in formulating and testing a large number of questions designed to extract information from nonstandard data structures.

There is flexibility not only in handling nonstandard data structures but also in using information. For example, we have designed tree structured programs that make use of a "layers of information" approach. The set of questions is divided into subsets of S_1, S_2, ... The tree first uses the questions in S_1 until it runs out of splits that significantly decrease impurity. Then it starts using the splits in S_2, and so on.

In general, our experience has been that once one begins thinking in terms of free structures, many avenues open up for new variations and uses.

8

REGRESSION TREES

8.1 INTRODUCTION

The tree structured approach in regression is simpler than in classification. The same impurity criterion used to grow the tree is also used to prune the tree. Besides this, there are no priors to deal with. Each case has equal weight.

Use of a stepwise optimal tree structure in least squares regression dates back to the Automatic Interaction Detection (AID) program proposed by Morgan and Sonquist (1963). Its development is traced in Sonquist and Morgan (1964), Sonquist (1970), Sonquist, Baker, and Morgan (1973), Fielding (1977), and Van Eck (1980). The major difference between AID and CART lies in the pruning and estimation process, that is, in the process of "growing an honest tree." Less important differences are that CART does not put any restrictions on the number of values a variable may take and contains the same additional features as in classification—variable combinations, predictive association measure to handle missing data and assess variable importance, and subsampling. On the other hand, CART does not contain some of the options available in AID, such as a limited look ahead.

In this chapter, we first start with an example of a regression tree grown on some well-known data (Section 8.2). In Section 8.3 some standard background for the regression problem is discussed. Then the basic splitting criterion and tree growing methodology are described in Section 8.4. Pruning and use of a test sample or cross-validation to get right-sized trees are covered in Section 8.5. In Section 8.6 simulated data are generated from a known model and used to illustrate the pruning and estimation procedure. There are two issues in cross-validation that are discussed in Section 8.7. Many features of classification trees carry over to the regression context. These are described in Sections 8.8, 8.9, and 8.10, using both a real data set and simulated data for illustrations.

In place of least squares, other error criteria can be incorporated into the tree structured framework. In particular, least absolute deviation regression is implemented in CART. This is described and illustrated in Section 8.11.

8.2 AN EXAMPLE

For their 1978 paper, Harrison and Rubinfeld gathered data about Boston housing values to see if there was any effect of air pollution concentration (NOX) on housing values. The data consisted of 14 variables measured for each of 506 census tracts in the Boston area.

These variables, by tract, are as follows:

y: median value of homes in thousands of dollars (MV)
x_1: crime rate (CRIM)
x_2: percent land zoned for lots (ZN)
x_3: percent nonretail business (INDUS)
x_4: 1 if on Charles River, 0 otherwise (CHAS)
x_5: nitrogen oxide concentration, pphm (NOX)
x_6: average number of rooms (RM)
x_7: percent built before 1940 (AGE)
x_8: weighted distance to employment centers (DIS)

x_9: accessibility to radial highways (RAD)
x_{10}: tax rate (TAX)
x_{11}: pupil/teacher ratio (P/T)
x_{12}: percent black (B)
x_{13}: percent lower-status population (LSTAT)

Using various transformations of both the dependent variable MV and independent variables, Harrison and Rubinfeld fitted the data with a least squares regression equation of the form

$$\begin{aligned}\log(\mathrm{MV}) = a_1 &+ a_2(\mathrm{RM})^2 + a_3(\mathrm{AGE}) + a_4 \log(\mathrm{DIS}) \\ &+ a_5 \log(\mathrm{RAD}) + a_6(\mathrm{TAX}) + a_7(\mathrm{P/T}) \\ &+ a_8(\mathrm{B} - 63)^2 + a_9 \log(\mathrm{LSTAT}) + a_{10}(\mathrm{TAX}) \qquad (8.1) \\ &+ a_{11}(\mathrm{ZN}) + a_{12}(\mathrm{INDUS}) + a_{13}(\mathrm{CHAS}) + a_{14}(\mathrm{NOX})^b + \varepsilon,\end{aligned}$$

where b is a parameter to be estimated. The resubstitution estimate for the proportion of variance explained is .81.

These data became well known when they were extensively used in the book *Regression Diagnostics*, by Belsley, Kuh, and Welsch (1980). A regression tree was grown on these data using the methods outlined in this chapter. The tree is pictured in Figure 8.1.

The number within each node is the average of the dependent variable MV over all tracts (cases) in the node. For instance, over all 506 tracts, the average MV is 22.5 (thousands of dollars). Right underneath each node is the question that split the node. The numbers in the lines are the number of cases going right and left. So, for example, 430 tracts had RM $\leq$ 6.9 and 76 did not.

To use this tree as a predictor, data (x_1, ..., x_{13}) for a tract is dropped down the tree until it comes to rest in a terminal node. Then the predicted value of MV is the average of the MV's for that terminal node.

It is interesting to track the actions of the tree. The split RM $\leq$ 6.9 separates off 430 tracts with the low average MV of 19.9 from 76 tracts with a high average of 37.2. Then the left branch is split on LSTAT $\leq$ 14, with 255 tracts having less than 14 percent lower-status population and 175 more than 14 percent. The 255 tracts going left are then split on DIS $\leq$ 1.4. The 5 tracts

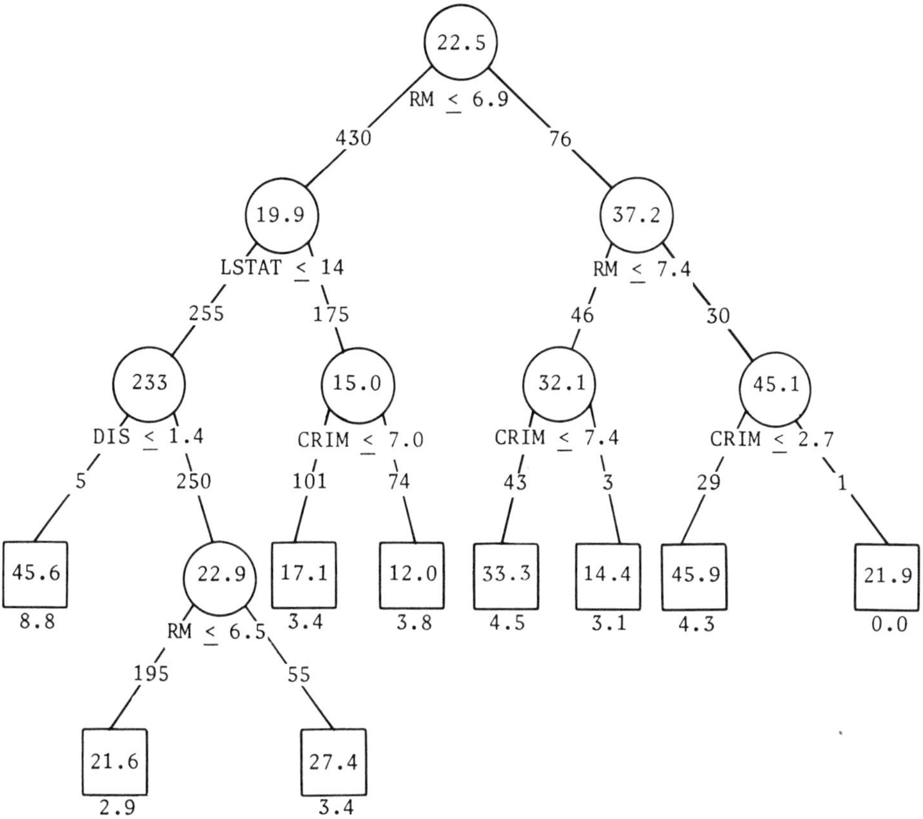

FIGURE 8.1

with DIS $\leq$ 1.4 have a high average MV of 45.6. These 5 are high housing cost inner city tracts.

The other branches can be similarly followed down and interpreted. We think that the tree structure, in this example, gives results more easy to interpret than equation (8.1). The cross-validated estimate of the proportion of variance explained is .73 ± .04. This is difficult to compare with the Harrison-Rubinfeld result, since they used log(MV) as the dependent variable and resubstitution estimation.

Only 4 of the 13 variables appear in splits; RM, LSTAT, DIS, and CRIM. In particular, the air pollution variable (NOX) does not appear. It is the fourth best surrogate to an LSTAT split at a node and either the first or second best surrogate to CRIM splits at three other nodes.

The variable importances (in order) are in Table 8.1.

TABLE 8.1

Variable	Importance
LSTAT	100
RM	91
P/T	38
DIS	27
NOX	25
CRIM	25
INDUS	25
TAX	21
AGE	17
RAD	15
B	13
ZN	11
CHAS	1

The numbers beneath the terminal nodes in Figure 8.1 are the standard deviations of the MV values in the terminal node. The leftmost terminal node, for example, contains five MV values with an average of 45.6 and a standard deviation of 8.8. Clearly, these five MV values have a wide range. The standard deviations in the other nodes give similar indications of how tightly (or loosely) the MV values are clustered in that particular node.

This tree was grown using the techniques outlined in the rest of this chapter. It will be used along the way to illustrate various points. Since we do not know what "truth" is for these data, simulated data from a known model will be introduced and worked with in Section 8.6.

8.3 LEAST SQUARES REGRESSION

In regression, a case consists of data $(\mathbf{x}, y)$ where $\mathbf{x}$ falls in a measurement space X and y is a real-valued number. The variable y is usually called the *response* or *dependent* variable. The variables in $\mathbf{x}$ are variously referred to as the *independent* variables, the *predictor variables*, the *carriers*, etc. We will stay with the terminology of $\mathbf{x}$ as the *measurement vector* consisting of measured variables and y as the response variable.

A prediction rule or predictor is a function $d(\mathbf{x})$ defined on X taking real values; that is, $d(\mathbf{x})$ is a real-valued function on X.

Regression analysis is the generic term revolving around the construction of a predictor $d(\mathbf{x})$ starting from a learning sample $\mathcal{L}$. Construction of a predictor can have two purposes: (1) to predict the response variable corresponding to future measurement vectors as accurately as possible; (2) to understand the structural relationships between the response and the measured variables.

For instance, an EPA sponsored project had as its goal the prediction of tomorrow's air pollution levels in the Los Angeles basin based on today's meteorology and pollution levels. Accuracy was the critical element.

In the example given concerning the Boston housing data, accuracy is a secondary and somewhat meaningless goal. It does not make sense to consider the problem as one of predicting housing values in future Boston census tracts. In the original study, the point of constructing a predictor was to get at the relationship between NOX concentration and housing values.

Suppose a learning sample $\mathcal{L}$ consisting of N cases $(\mathbf{x}_1, y_1)$, $\ldots$, $(\mathbf{x}_N, y_N)$ was used to construct a predictor $d(\mathbf{x})$. Then the question arises of how to measure the accuracy of this predictor. If we had a very large test sample $(\mathbf{x}_1', y_1')$, $\ldots$, $(\mathbf{x}_{N_2}', y')$ of size N_2, the accuracy of $d(\mathbf{x})$ could be measured as the average error

$$\frac{1}{N_2} \sum_{n=1}^{N_2} |y'_n - d(\mathbf{x}'_n)|$$

in $d(\mathbf{x}'_n)$ as a predictor of y'_n, $n = 1, \ldots, N_2$. This measure leads to *least absolute deviation* regression (see Section 8.11).

But for reasons having to do with ease of computations, the measure of accuracy classically used in regression is the averaged squared error,

$$\frac{1}{N_2} \sum_{n=1}^{N_2} (y'_n - d(\mathbf{x}'_n))^2.$$

The methodology revolving about this measure is *least squares regression*.

To define accuracy in the mean squared error sense, a theoretical framework is needed. Assume that the random vector $(\mathbf{X}, Y)$ and the learning sample $\mathcal{L}$ are independently drawn from the same underlying distribution.

DEFINITION 8.2. *Define the mean squared error $R^*(d)$ of the predictor d as*

$$R^*(d) = E(Y - d(\mathbf{X}))^2.$$

That is, $R^*(d)$ is the expected squared error using $d(\mathbf{X})$ as a predictor of Y, where the expectation is taken holding $\mathcal{L}$ fixed.

Previously, $R^*(d)$ was used to denote the misclassification rate of a classifier d. We use the same notation here in the regression context to have a uniform notation for the measure of accuracy of a predictor, regardless of whether we are predicting a class label or an ordered response.

Using the preceding definition, the optimal (or Bayes) predictor has a simple form.

PROPOSITION 8.3. *The predictor d_B which minimizes $R^*(d)$ is*

$$d_B(\mathbf{x}) = E(Y|\mathbf{X} = \mathbf{x}).$$

In other words, $d_B(\mathbf{x})$ is the conditional expectation of the response, given that the measurement vector is $\mathbf{x}$. The surface $y = d_B(\mathbf{x})$ is often referred to as the regression surface of Y on $\mathbf{X}$.

The proof of Proposition 8.3 is simple conditional expectation manipulation. The key element is the following elementary lemma.

LEMMA 8.4. *The constant a which minimizes*

$$E(Y - a)^2$$

is $E(Y)$.

See Section 9.1 for a complete proof.

8.3.1 Error Measures and Their Estimates

Given a learning sample $\mathcal{L}$ consisting of $(\mathbf{x}_1, y_1), \ldots, (\mathbf{x}_N, y_N)$, again one wants to use $\mathcal{L}$ both to construct a predictor $d(\mathbf{x})$ and to estimate its error $R^*(d)$. There are several ways to estimate R^*. The usual (and worst) is the resubstitution estimate

$$R(d) = \frac{1}{N} \sum_n (y_n - d(\mathbf{x}_n))^2. \tag{8.5}$$

Test sample estimates $R^{ts}(d)$ are gotten by randomly dividing $\mathcal{L}$ into $\mathcal{L}_1$ and $\mathcal{L}_2$ and using $\mathcal{L}_1$ to construct d and $\mathcal{L}_2$ to form

$$R^{ts}(d) = \frac{1}{N_2} \sum_{(\mathbf{x}_n, y_n) \in \mathcal{L}_2} (y_n - d(\mathbf{x}_n))^2. \tag{8.6}$$

The V-fold cross-validation estimate $R^{CV}(d)$ comes from dividing $\mathcal{L}$ in V subsets $\mathcal{L}_1, \ldots, \mathcal{L}_V$, each containing (as nearly as possible) the same number of cases. For each v, $v = 1, \ldots, V$, apply the same construction procedure to the learning sample $\mathcal{L} - \mathcal{L}_v$, getting the predictor $d^{(v)}(\mathbf{x})$. Then set

$$R^{CV}(d) = \frac{1}{N} \sum_v \sum_{(\mathbf{x}_n, y_n) \in \mathcal{L}_v} (y_n - d^{(v)}(\mathbf{x}_n))^2. \tag{8.7}$$

The rationale for R^{ts} and R^{CV} is the same as that given in Section 1.4.

In classification, the misclassification rate has a natural and intuitive interpretation. But the mean squared error of a predictor does not. Furthermore, the value of $R^*(d)$ depends on the scale in which the response is measured. For these reasons, a normalized measure of accuracy which removes the scale dependence is often used. Let

$$\mu = E(Y).$$

Then

$$R^*(\mu) = E(Y - \mu)^2$$

is the mean squared error using the constant μ as a predictor of Y, which is also the variance of Y.

DEFINITION 8.8. *The relative mean squared error $RE^*(d)$ in $d(\mathbf{x})$ as a predictor of Y is*

$$RE^*(d) = R^*(d)/R^*(\mu).$$

The idea here is that μ is the baseline predictor for Y if nothing is known about $\mathbf{X}$. Then judge the performance of any predictor d based on $\mathbf{X}$ by comparing its mean squared error to that of μ.

The relative error is always nonnegative. It is usually, but not always, less than 1. Most sensible predictors $d(\mathbf{x})$ are more accurate than μ, and $RE^*(d) < 1$. But on occasion, some construction procedure may produce a poor predictor d with $RE^*(d) \geq 1$.

Let

$$\bar{y} = \frac{1}{N} \sum_n y_n$$

and

$$R(\bar{y}) = \frac{1}{N} \sum_n (y_n - \bar{y})^2.$$

Then the resubstitution estimate $RE(d)$ for $RE^*(d)$ is $R(d)/R(\bar{y})$. Using a test sample, the estimate is

$$RE^{ts}(d) = R^{ts}(d)/R^{ts}(\bar{y}).$$

The cross-validated estimate $RE^{CV}(d)$ is $R^{CV}(d)/R(\bar{y})$.

In least squares linear regression, if $d(\mathbf{x})$ is the best linear predictor, then the quantity

$$1 - RE(d)$$

is called the *proportion of the variance explained by d*. Furthermore, it can be shown that if ρ is the sample correlation between the values y_n and $d(\mathbf{x}_n)$, $n = 1, \ldots, N$, then

$$\rho^2 = 1 - RE(d).$$

The ρ^2 value of .81 was reported in the original linear regression study of the Boston data.

In general, $R(d)$ is not a variance and it does not make sense to refer to $1 - RE(d)$ as the proportion of variance explained. Neither is $1 - RE(d)$ equal to the square of the sample correlation between the y_n and $d(\mathbf{x}_n)$ values.

Therefore, we prefer to use the terminology *relative error* and the estimates of $RE^*(d)$ as a measure of accuracy rather than $1 - RE^*(d)$. The value .73 for "the proportion of variance explained" given by the tree predictor grown on the Boston data is the estimate $1 - RE^{CV}(d)$.

8.3.2 Standard Error Estimates

Standard error estimates for R^{ts} and R^{CV} are covered in Sections 11.4 and 11.5. Understanding the concept behind their derivation is important to their interpretation. For this reason, we give the derivation in the test sample setting.

Let the learning sample consist of N_1 cases independently selected from an underlying probability distribution, and suppose that the learning sample is used to construct a predictor $d(\mathbf{x})$. The test sample consists of N_2 cases drawn independently from the same distribution. Denote these by $(\mathbf{X}_1, Y_1)$, $(\mathbf{X}_2, Y_2)$, ..., $(\mathbf{X}_{N_2}, Y_{N_2})$. Then an unbiased estimate for $R^*(d)$ is

$$R^{ts} = \frac{1}{N_2} \sum_{n=1}^{N_2} (Y_n - d(\mathbf{X}_n))^2. \tag{8.9}$$

Since the individual terms in (8.9) are independent (holding the learning sample fixed), the variance of R^{ts} is the sum of the variances of the individual terms. All cases have the same distribution, so the variance of each term equals the variance of the first term. Thus, the standard deviation of R^{ts} is

$$\frac{1}{\sqrt{N_2}} \left\{E(Y_1 - d(\mathbf{X}_1))^4 - \left[E(Y_1 - d(\mathbf{X}_1))^2\right]^2\right\}^{1/2}.$$

Now use the sample moment estimates

$$E(Y_1 - d(\mathbf{X}_1))^4 \simeq \frac{1}{N_2} \sum_{n=1}^{N_2} (Y_n - d(\mathbf{X}_n))^4$$

and

$$E(Y_1 - d(\mathbf{X}_1))^2 \cong \frac{1}{N_2} \sum_{n=1}^{N_2} (Y_n - d(\mathbf{X}_n))^2 = R^{ts}$$

to estimate the standard error of R^{ts} by

$$SE(R^{ts}) = \frac{1}{\sqrt{N_2}} \left[\frac{1}{N_2} \sum_{n=1}^{N_2} (Y_n - d(\mathbf{X}_n))^4 - (R^{ts})^2\right]^{1/2}.$$

Since sample fourth moments can be highly variable, less credence should be given to the SE's in regression than in classification.

The measure RE^* is a ratio, and its estimates RE^{ts} and RE^{CV} are ratio estimators with more complicated standard error formulas (see Sections 11.4 and 11.5).

Proportionally, the standard error of RE^{ts}, say, may be larger than that of R^{ts}. This is because the variability of $R^{ts}(d)/R^{ts}(\bar{y})$ is affected by the variability of both the numerator and the denominator and by the interaction between them. At times, the variability of the denominator is the dominant factor.

8.3.3 *Current Regression Methods*

In regression, predictors have usually been constructed using a parametric approach. The assumption is made that

$$E(Y|\mathbf{X} = \mathbf{x}) = d(\mathbf{x}, \theta),$$

where d has known functional form depending on $\mathbf{x}$ and a finite set of parameters $\theta = (\theta_1, \theta_2, \ldots)$. Then θ is estimated as that parameter value $\hat{\theta}$ which minimizes $R(d(\mathbf{x}, \hat{\theta}))$; that is,

$$R(d(\mathbf{x}, \hat{\theta})) = \min_{\theta} R(d(\mathbf{x}, \theta)).$$

For instance, in linear regression the assumptions are that $\mathbf{x} = (x_1, \ldots, x_M)$ with $x_1, \ldots, x_M$ ordered variables and that

$$E(Y|\mathbf{X} = \mathbf{x}) = b_0 + b_1 x_1 + \cdots + b_M x_M,$$

where the coefficients $b_0, \ldots, b_M$ are to be estimated. Assuming further that the error term is normally distributed $N(0, \sigma^2)$ and independent from case to case leads to an elegant inferential theory.

But our focus is on data sets whose dimensionality requires some sort of variable selection. In linear regression the common practice is to use either a stepwise selection or a best subsets algorithm. Since variable selection invalidates the inferential model, stepwise or best subsets regression methods have to be viewed as heuristic data analysis tools.

However, linear regression with variable selection can be a more powerful and flexible tool than discriminant analysis in classification. The assumptions necessary for good performance are much less stringent, its behavior has been widely explored, diagnostic tools for checking goodness of fit are becoming available, and robustifying programs are flourishing.

Therefore, tree structured regression as a competitor to linear regression should be looked at in somewhat a different light than tree structured classification and used in those problems where its distinctive characteristics are desirable.

Two nonparametric regression methods, nearest neighbor and kernel, have been studied in the literature (see the bibliography in Collomb, 1980). Their drawbacks are similar to the analogous methods in classification. They give results difficult to interpret and use.

8.4 TREE STRUCTURED REGRESSION

A tree structured predictor is similar to a tree structured classifier. The space X is partitioned by a sequence of binary splits into terminal nodes (see Figure 8.2). In each terminal node t, the predicted response value $y(t)$ is constant.

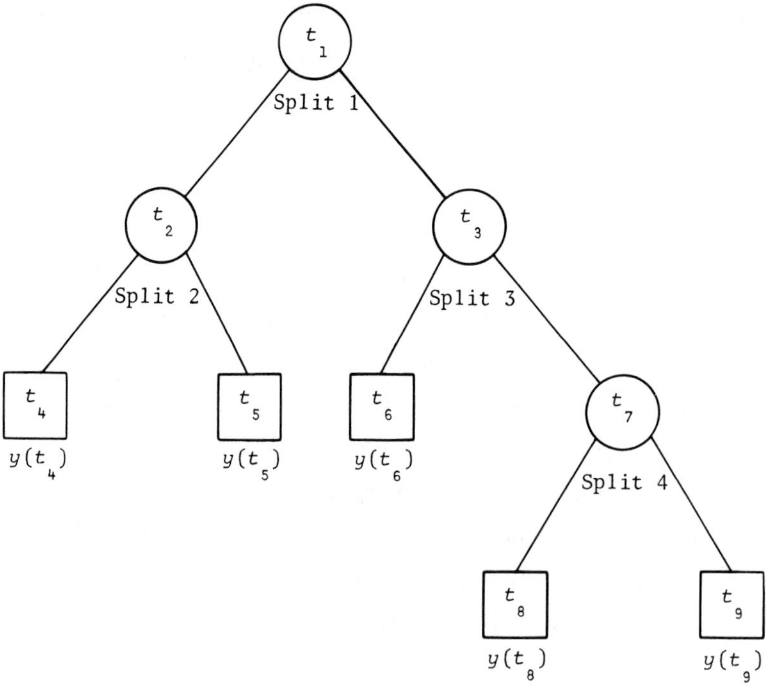

FIGURE 8.2

Since the predictor $d(\mathbf{x})$ is constant over each terminal node, the tree can be thought of as a histogram estimate of the regression surface (see Figure 8.3).

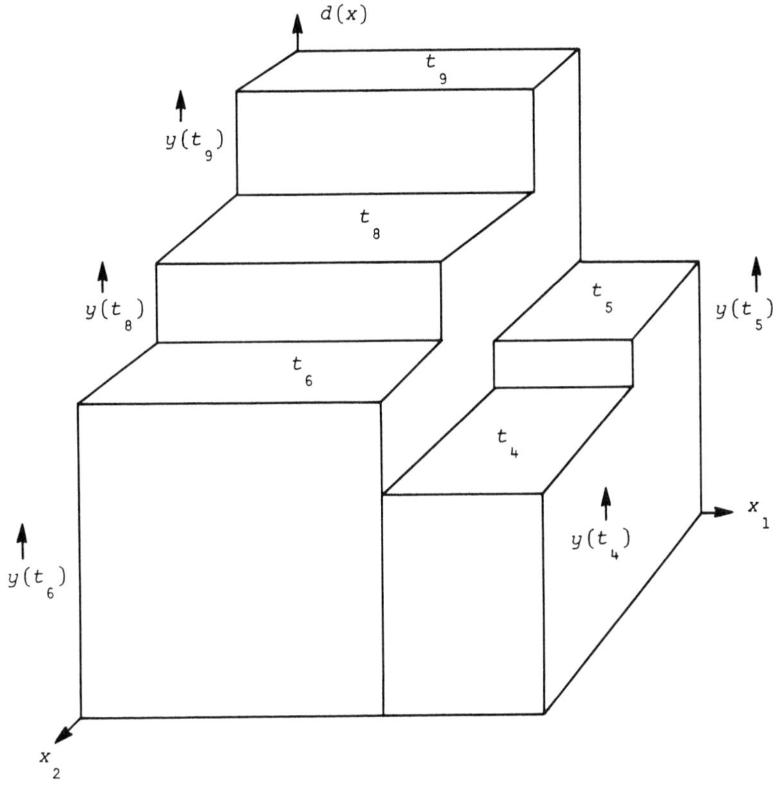

FIGURE 8.3

Starting with a learning sample $\mathcal{L}$, three elements are necessary to determine a tree predictor:

1. A way to select a split at every intermediate node
2. A rule for determining when a node is terminal
3. A rule for assigning a value $y(t)$ to every terminal node t

It turns out, as in classification, that the issue of the node assignment rule is easiest to resolve.

We start with the resubstitution estimate for $R^*(d)$, that is,

$$R(d) = \frac{1}{N} \sum_n (y_n - d(\mathbf{x}_n))^2.$$

Then take $y(t)$ to minimize $R(d)$.

PROPOSITION 8.10. *The value of $y(t)$ that minimizes $R(d)$ is the average of y_n for all cases $(\mathbf{x}_n, y_n)$ falling into t; that is, the minimizing $y(t)$ is*

$$\bar{y}(t) = \frac{1}{N(t)} \sum_{\mathbf{x}_n \in t} y_n,$$

where the sum is over all y_n such that $\mathbf{x}_n \in t$ and $N(t)$ is the total number of cases in t.

The proof of Proposition 8.10 is based on seeing that the number a which minimizes $\sum_n (y_n - a)^2$ is

$$a = \frac{1}{N} \sum_n y_n .$$

Similarly, for any subset $y_{n'}$ of the y_n, the number which minimizes $\sum_{n'} (y_{n'} - a)^2$ is the average of the $y_{n'}$.

From now on, we take the predicted value in any node t to be $\bar{y}(t)$. Then, using the notation $R(T)$ instead of $R(d)$,

$$R(T) = \frac{1}{N} \sum_{t \in \tilde{T}} \sum_{\mathbf{x}_n \in t} (y_n - \bar{y}(t))^2. \tag{8.11}$$

Set

$$R(t) = \frac{1}{N} \sum_{\mathbf{x}_n \in t} (y_n - \bar{y}(t))^2$$

so (8.11) can be written as

$$R(T) = \sum_{t \in \tilde{T}} R(t). \tag{8.12}$$

These expressions have simple interpretations. For every node t, $\sum_{x_n \in t} (y_n - \bar{y}(t))^2$ is the within node sum of squares. That is, it is the total squared deviations of the y_n in t from their average. Summing over $t \in \tilde{T}$ gives the total within node sum of squares, and dividing by N gives the average.

Given any set of splits S of a current terminal node t in $\tilde{T}$,

DEFINITION 8.13. *The best split* $\mathcal{s}^*$ *of* t *is that split in* S *which most decreases* $R(T)$.

More precisely, for any split $\mathcal{s}$ of t into t_L and t_R, let

$$\Delta R(\mathcal{s}, t) = R(t) - R(t_L) - R(t_R).$$

Take the best split $\mathcal{s}^*$ to be a split such that

$$\Delta R(\mathcal{s}^*, t) = \max_{\mathcal{s} \in S} \Delta R(\mathcal{s}, t).$$

Thus, a regression tree is formed by iteratively splitting nodes so as to maximize the decrease in $R(T)$. In classification trees, choosing the best splits to be the ones that minimized the resubstitution misclassification rate had undesirable properties. Alternative criteria had to be found.

The natural regression criterion is stouter. There are no similar problems in defining the best splits to be those that minimize the resubstitution error measure. Using this criterion, the best split at a node is that split on the **x** variables which most successfully separates the high response values from the low ones. This can be seen operating in the Boston data tree (Figure 8.1). At each intermediate node t, one of $\bar{y}(t_L)$, $\bar{y}(t_R)$ is considerably lower than $\bar{y}(t)$ and the other higher.

An alternative form of the criterion is interesting. Let $p(t) = N(t)/N$ be the resubstitution estimate for the probability that a case chosen at random from the underlying theoretical distribution falls into node t. Define

$$s^2(t) = \frac{1}{N(t)} \sum_{x_n \in t} (y_n - \bar{y}(t))^2 \tag{8.14}$$

so that $R(t) = s^2(t)p(t)$, and

$$R(T) = \sum_{t \in \tilde{T}} s^2(t)p(t). \tag{8.15}$$

Note that $s^2(t)$ is the sample variance of the y_n values in the node t. Then the best split of t minimizes the weighted variance

$$p_L s^2(t_L) + p_R s^2(t_R),$$

where p_L and p_R are the proportion of cases in t that go left and right, respectively.

We noted in Chapter 3 that in two-class problems, the impurity criterion given by $p(1|t)p(2|t)$ is equal to the node variance computed by assigning the value 0 to every class 1 object, and 1 to every class 2 object. Therefore, there is a strong family connection between two-class trees and regression trees.

8.5 PRUNING AND ESTIMATING

8.5.1 *Pruning*

In the original AID program, a node was declared terminal if

$$\max_{\mathcal{s}} \Delta R(\mathcal{s}, t) \leq .006R(t_1).$$

Then the resubstitution estimates $R(T)$ or $1 - RE(T)$ were used as measures of accuracy. The difficulties are the same as in classification. The trees grown are not the right size and the estimates are overly optimistic. The error measure $R(t)$ has the property that for *any* split of t into t_L, t_R,

$$R(t) \geq R(t_L) + R(t_R).$$

Again, the more splitting done, the better $RE(T)$ looks. For example, Figure 8.4 is a graph of $RE(T)$ versus $|\widetilde{T}|$ for the Boston data.

The method we use to select a tree is exactly the same as that used to select a classification tree. First, a large tree

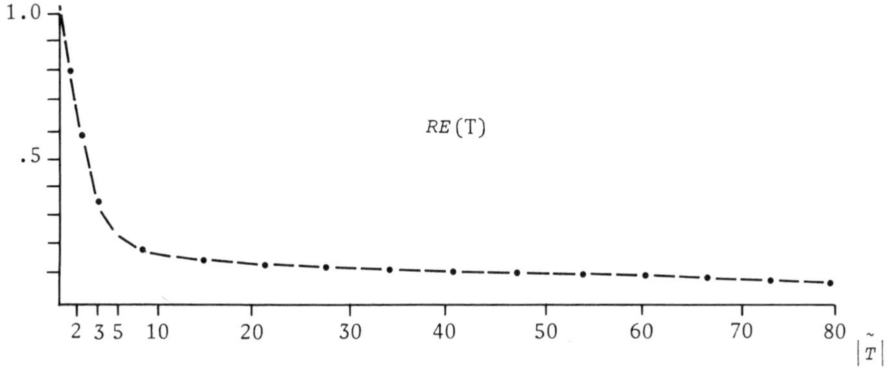

FIGURE 8.4

T_{max} is grown by successively splitting so as to minimize $R(T)$, until for every $t \in \widetilde{T}_{max}$, $N(t) \leq N_{min}$. Usually, N_{min} is taken as 5. These starting trees are usually much larger than classification starting trees. In classification, splitting stops if the node is pure or $N(t) \leq N_{min}$. The corresponding, but less frequently satisfied, purity condition for regression trees is that all y values in a node are the same. In the Boston tree sequence, T_{max} has 78 terminal nodes.

Define the error-complexity measure $R_\alpha(T)$ as

$$R_\alpha(T) = R(T) + \alpha|\widetilde{T}|.$$

Now minimal error-complexity pruning is done exactly as minimal cost-complexity pruning in classification. The result is a decreasing sequence of trees

$$T_1 \succ T_2 \succ \cdots \succ \{t_1\}$$

with $T_1 \preceq T_{max}$ and a corresponding increasing sequence of α values

$$0 = \alpha_1 < \alpha_2 < \cdots$$

such that for $\alpha_k \leq \alpha < \alpha_{k+1}$, T_k is the smallest subtree of $T_{\max}$ minimizing $R_\alpha(T)$.

8.5.2 Estimating $R^*(T_k)$ and $RE^*(T_k)$

To select the right sized tree from the sequence $T_1 \succ T_2 \succ \cdots$, honest estimates of $R(T_k)$ are needed. To get test sample estimates, the cases in $\mathcal{L}$ are randomly divided into a learning sample $\mathcal{L}_1$ and a test sample $\mathcal{L}_2$. The learning sample $\mathcal{L}_1$ is used to grow the sequence $\{T_k\}$ of pruned trees. Let $d_k(\mathbf{x})$ denote the predictor corresponding to the tree T_k. If $\mathcal{L}_2$ has N_2 cases, define

$$R^{ts}(T_k) = \frac{1}{N_2} \sum_{(x_n, y_n) \in \mathcal{L}_2} (y_n - d_k(\mathbf{x}_n))^2 .$$

In practice, we have generally used cross-validation except with large data sets. In V-fold cross-validation $\mathcal{L}$ is randomly divided into $\mathcal{L}_1, \ldots, \mathcal{L}_V$ such that each subsample $\mathcal{L}_v$, $v = 1, \ldots, V$, has the same number of cases (as nearly as possible).

Let the vth learning sample be $\mathcal{L}^{(v)} = \mathcal{L} - \mathcal{L}_v$, and repeat the tree growing and pruning procedure using $\mathcal{L}^{(v)}$. For each v, this produces the trees $T^{(v)}(\alpha)$ which are the minimal error-complexity trees for the parameter value α.

Grow and prune using all of $\mathcal{L}$, getting the sequences $\{T_k\}$ and $\{\alpha_k\}$. Define $\alpha'_k = \sqrt{\alpha_k \alpha_{k+1}}$. Denote by $d_k^{(v)}(\mathbf{x})$ the predictor corresponding to the tree $T^{(v)}(\alpha'_k)$. The cross-validation estimates $R^{CV}(T_k)$ and $RE^{CV}(T_k)$ are given by

$$R^{CV}(T_k) = \frac{1}{N} \sum_{v=1}^{V} \sum_{(\mathbf{x}_n, y_n) \in \mathcal{L}_v} (y_n - d_k^{(v)}(\mathbf{x}_n))^2$$

and

$$RE^{CV}(T_k) = R^{CV}(T_k)/R(\bar{y}).$$

For the Boston data, $RE^{CV}(T_k)$ is plotted against $|\tilde{T}_k|$ in Figure 8.5 and compared to the resubstitution estimates.

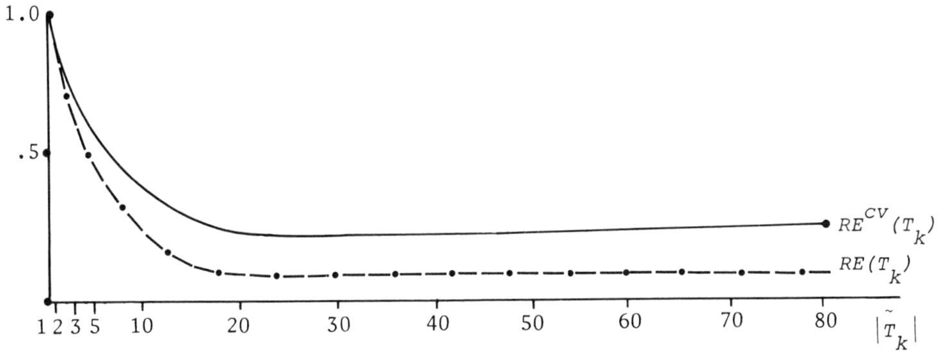

FIGURE 8.5

8.5.3 Tree Selection

In regression, the sequence $T_1 > \cdots > \{t_1\}$ tends to be larger than in classification. In the Boston example, there were 75 trees in the sequence.

The pruning process in regression trees usually takes off only two terminal nodes at a time. This contrasts with classification, where larger branches are pruned, resulting in a smaller sequence of pruned subtrees.

The mechanism is illustrated by the following hypothetical classification example. Figure 8.6 illustrates a current branch of the tree starting from an intermediate node. There are two classes with equal priors and the numbers in the nodes are the class populations.

If the two leftmost terminal nodes are pruned, the result is as illustrated in Figure 8.7. There are a total of 100 misclassified in this branch. But in the top node, there are also 100 misclassified. Therefore, the top node, by itself, is a smaller branch having the same misclassification rate as the three-node configuration in Figure 8.7. In consequence, if the two leftmost nodes in Figure 8.6 were pruned, the entire branch would be pruned.

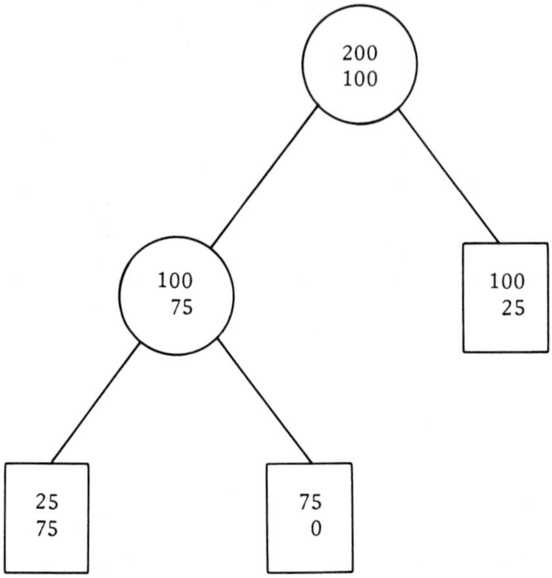

FIGURE 8.6

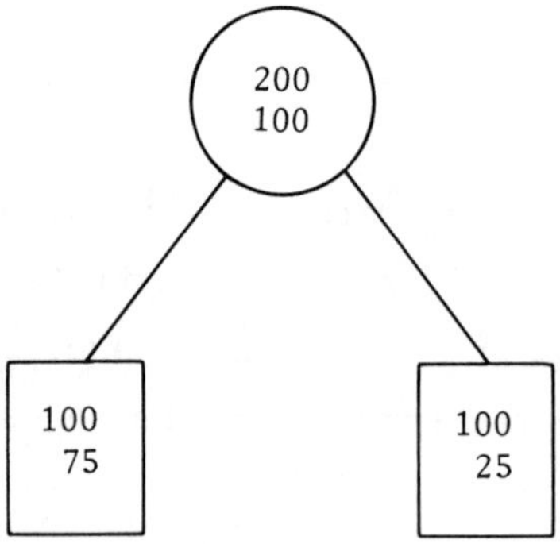

FIGURE 8.7

In classification, a split almost always decreases the impurity $I(T)$, but, as just illustrated, may not decrease $R(T)$. In regression trees, a split almost always decreases $R(T)$. Since we are pruning up on $R(T)$, in general only two terminal nodes at a time will be pruned.

Not only is the sequence of trees generally larger in regression, but the valley containing the minimum value of $R^{CV}(T_k)$ also tends to be flatter and wider.

In the Boston example, $R^{CV}(T_1) = 18.8$. The minimum occurred at tree 32 with $R^{CV}(T_{32}) = 18.6$ and $|\tilde{T}_{32}| = 49$. For $1 \leq k \leq 67$, $18.6 \leq R^{CV}(T_k) \leq 21.6$. The SE estimates for R^{CV} in this range were about 2.8. Clearly, the selection of any tree from T_1 to T_{67} on the basis of R^{CV} is somewhat arbitrary.

In keeping with our philosophy of selecting the smallest tree commensurate with accuracy, the 1 SE rule was used. That is, the T_k selected was the smallest tree such that

$$R^{CV}(T_k) \leq R^{CV}(T_{k_0}) + \text{SE},$$

where

$$R^{CV}(T_{k_0}) = \min_k R^{CV}(T_k)$$

and SE is the standard error estimate for $R^{CV}(T_{k_0})$. Tree 66 was selected by this rule, with $R^{CV}(T_{66}) = 21.1$ and $|\tilde{T}_{66}| = 9$.

If the analyst wants to select the tree, Section 11.6 gives a method for restricting attention to a small subsequence of the main sequence of pruned trees.

8.6 A SIMULATED EXAMPLE

To illustrate the pruning and estimation process outlined in Section 8.5, as well as other features, we use some simulated data.

The data were generated from this model: Take $X_1, \ldots, X_{10}$ independent and

$$P(X_1 = -1) = P(X_1 = 1) = 1/2$$

$$P(X_m = -1) = P(X_m = 0) = P(X_m = 1) = 1/3, \; m = 2, \ldots, 10.$$

Let Z be independent of $X_1, \ldots, X_{10}$ and normally distributed with mean zero and variance 2. Then if $X_1 = 1$, set

$$Y = 3 + 3X_2 + 2X_3 + X_4 + Z,$$

if $X_1 = -1$, set

$$Y = -3 + 3X_5 + 2X_6 + X_7 + Z.$$

Variables X_8, X_9, X_{10} are noise.

This example consists of two distinct regression equations with the choice of equation dictated by the binary variable X_1. The other variables are ordered three-valued variables. The best predictor is

$$d_B(\mathbf{x}) = \begin{cases} 3 + 3x_2 + 2x_3 + x_4 \text{ if } x_1 = 1 \\ -3 + 3x_5 + 2x_6 + x_7 \text{ if } x_1 = -1 \end{cases}$$

with $RE^*(d_B) = .10$ and $R^*(d_B) = 2.0$.

The learning sample $\mathcal{L}$ consisted of 200 cases generated from the model. The test sample consists of an additional 5000 cases. In the initial run of this example, $N_{\min}$ was set equal to 1. This produced, with pruning, a sequence of 180 subtrees. Table 8.2 gives a partial summary of RE, $RE^{CV} \pm$ SE, and RE^{ts}.

For the trees with 11 or more terminal nodes, the cross-validated estimates are quite accurate. In the smaller trees, the discrepancies are generally large, and the resubstitution estimate is more accurate. This problem will be discussed later.

The loss of accuracy in the smaller trees is not critical, since accuracy is maintained in the range containing the best trees. Tree 168 is selected by the 1 SE rule, with $RE^{CV} = .17 \pm .02$ and $RE^{ts} = .17$. It has 13 terminal nodes and is displayed in Figure 8.8. The numbers in the lines connecting the nodes are the popula-

TABLE 8.2

Tree No.	$\lvert\tilde{T}\rvert$	RE	$RE^{CV} \pm$ SE	RE^{ts}‡
1	200	.00	.25 ± .03	.25
49	150	.00	.25 ± .03	.25
96	100	.01	.25 ± .03	.25
135	50	.04	.21 ± .02	.22
157	25	.08	.19 ± .02	.18
162	20	.10	.18 ± .02	.19
164	18	.11	.18 ± .02	.19
165	16	.12	.18 ± .02	.17
166	15	.12	.18 ± .02	.18
167*	14*	.13	.17 ± .02	.18
168†	13†	.13	.17 ± .02	.17
169	12	.14	.19 ± .02	.17
170	11	.16	.24 ± .03	.19
171	10	.18	.27 ± .03	.21
172	9	.20	.27 ± .03	.22
173	8	.21	.27 ± .03	.24
174	7	.24	.29 ± .03	.25
175	6	.27	.32 ± .03	.26
176	5	.31	.43 ± .04	.30
177	4	.36	.42 ± .04	.33
178	3	.47	.60 ± .04	.43
179	2	.60	.60 ± .04	.55
180	1	1.00	1.00	1.00

*Lowest RE^{CV} tree.

†1 SE tree.

‡The SE for the RE^{ts} estimate is about .005

tions going to the left and right nodes. The numbers in the nodes are the learning sample node averages. In the terminal nodes, the number above the line is the learning sample node average. The number below the line is the test set node average.

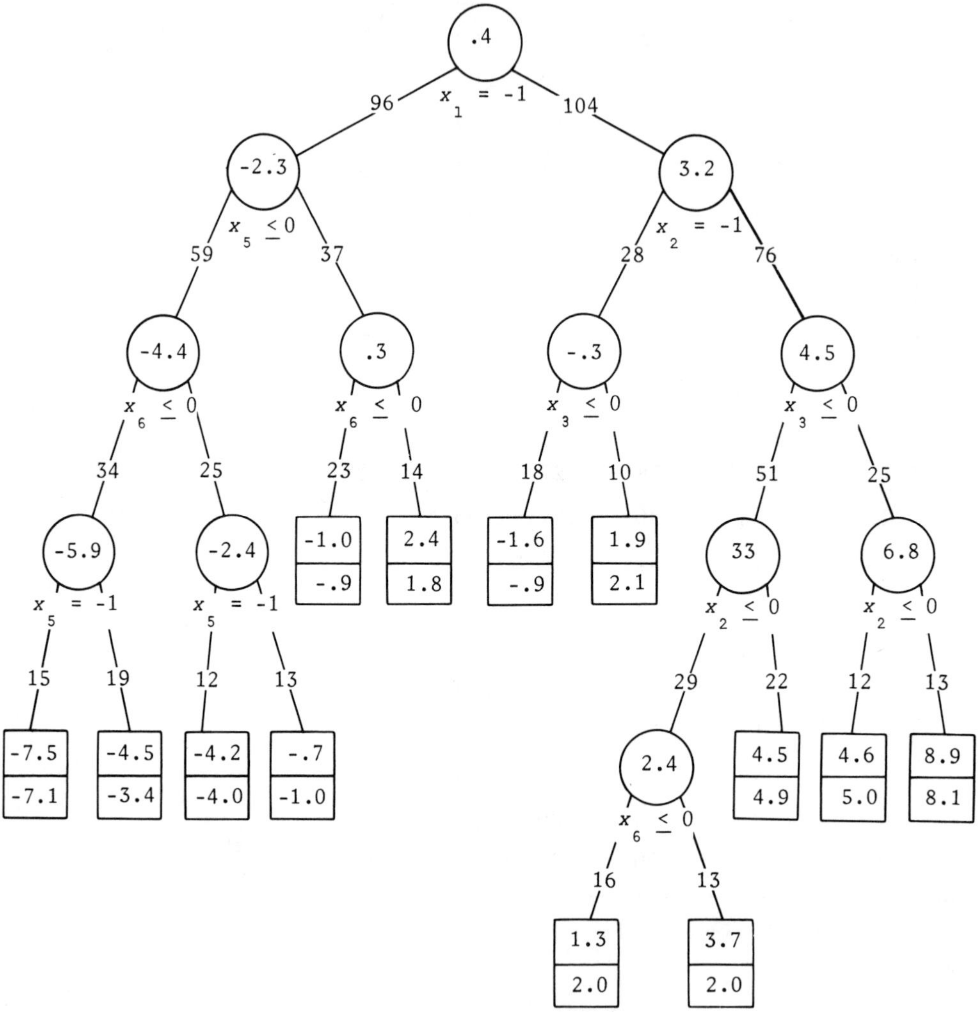

FIGURE 8.8

With one exception, the tree follows the structure of the model generating the data. The first split is on the binary variable x_1 and separates the 96 cases generated from the equation

$$Y = -3 + 3X_5 + 2X_6 + X_7 + Z$$

from the 104 cases generated from

$$Y = 3 + 3X_2 + 2X_3 + X_4 + Z.$$

Then the left side of the tree splits repeatedly on x_5 and x_6, the right side on x_2 and x_3.

On the right side, however, there is one noisy split on x_6, which produces the two lowest terminal nodes. These two also have the largest discrepancies between the learning and test sample node averages. This split illustrates once more the caution that must be used in interpreting data structure from the tree output.

To further check the cross-validation procedure, four replicate data sets were generated from the same model but with different random number seeds. A summary of the 1 SE trees is given in Table 8.3.

TABLE 8.3

Data Set	$\lvert\tilde{T}\rvert$	RE	RE^{cv}	RE^{ts}
1	12	.13	.20 ± .02	.20
2	16	.11	.18 ± .02	.17
3	12	.17	.22 ± .03	.18
4	16	.10	.16 ± .02	.16

8.7 TWO CROSS-VALIDATION ISSUES

8.7.1 The Small Tree Problem

The data generated in the example was also run with 2, 5, 25, and 50 cross-validations. Partial results are given in Table 8.4.

TABLE 8.4

$\lvert\tilde{T}\rvert$	RE^{ts}	$RE^{cv}(2)$	$RE^{cv}(5)$	$RE^{cv}(10)$	$RE^{cv}(25)$	$RE^{cv}(50)$
75	.22	.30	.26	.23	.24	.25
50	.21	.28	.25	.21	.23	.24
25	.18	.23	.22	.19	.21	.21
20	.19	.22	.21	.18	.20	.20
15	.18	.23	.20	.18	.19	.18
10	.21	.36	.32	.27	.28	.28
8	.24	.39	.32	.27	.27	.26
6	.26	.41	.36	.32	.32	.31
5	.30	.41	.41	.43	.43	.43
4	.34	.62	.41	.42	.44	.46
3	.43	.62	.60	.60	.62	.63
2	.55	.86	.60	.60	.60	.61
1	1.00	1.00	1.00	1.00	1.00	1.00

Using two cross-validations produces a serious loss of accuracy. With five cross-validations, the reduction in accuracy is still apparent. But 10, 25, and 50 cross-validations give comparable results, with 10 being slightly better.

The lack of accuracy in the smaller trees was noted in classification, but to a lesser extent. In regression trees the effect can be more severe. Analysis of the problem leads to better understanding of how cross-validation works in tree structures. There are some aspects of the problem that are specific to this particular simulated data set. Instead of discussing these, we focus on two aspects that are more universal.

Table 8.5 lists the estimates of RE^* and the corresponding complexity parameter α_k for the 13 smallest trees.

The cross-validated estimate of RE^* for the three terminal node tree uses the cross-validation trees corresponding to the value of $\alpha = \sqrt{499 \cdot 467} = 483$. Of these 10 cross-validation trees,

TABLE 8.5

$\lvert\tilde{T}\rvert$	α_k	RE^{cv}	RE^{ts}
1	1652	1.00	1.00
2	499	.60	.55
3	467	.60	.43
4	203	.42	.34
5	172	.43	.30
6	112	.32	.26
7	103	.29	.25
8	78	.27	.24
9	78	.27	.22
10	57	.27	.21
11	39	.24	.19
12	29	.19	.17
13	25	.17	.17

7 have two terminal nodes, and 3 have three terminal nodes. Thus, the majority of the cross-validation trees have accuracies comparable to the two terminal node tree. Largely because of this, the tenfold cross-validated estimate of RE^* for the three terminal node tree is the same as that for the two terminal node tree.

There are two reasons for the disparity in the number of terminal nodes between the main tree and the cross-validation trees. First, the three terminal node main tree is optimal only over the comparatively narrow range (467,499). For $\alpha = 483$, the two terminal node main tree has cost-complexity almost as small as the three terminal node tree. Second, since the cross-validation trees are grown on a subset of the data, for the same number of terminal nodes they will tend to have lower resubstitution error rates than the main tree. The combination of these two factors is enough to swing the balance from three terminal nodes in the main tree to two terminal nodes in some of the cross-validation trees.

In general, whenever there is a tree T_k in the main tree sequence that is optimal over a comparatively narrow α-range (α_k, α_{k+1}), we can expect that some of the cross-validation trees have fewer terminal nodes than the main tree. The cross-validated RE^* estimate will then be biased upward toward the RE^* values corresponding to T_{k+1}. If the RE^* value for T_{k+1} is considerably larger than that for T_k, the bias may be large. Thus, the effect is more pronounced in the smaller trees where RE^* is rapidly increasing.

Another potential source of bias is unbalanced test samples. Suppose that in tenfold cross-validation on 200 cases, a test sample contains, say, 10 of the 20 highest y values, together with 10 "typical" values. Suppose that the tree with two terminal nodes grown on the remaining 180 cases generally sends high y values right and lower ones left. Because of the absence of 10 high y values, the mean of the right node will be smaller than if all cases were available. Then when the test sample is run through the tree, the sum of squares will be inflated.

This bias is also reduced as the tree grows larger. Suppose the tree is grown large enough so that the remaining 10 of the 20 highest y values are mostly split off into a separate node t. At this point the absence of the 10 high response values, assuming they would also fall into t, does not affect the mean of any node except t.

Furthermore, assuming that the 10 cases in the test sample are randomly selected from the 20 original high y value cases, the average within-node sum of squares resulting when these cases drop into t is an unbiased estimate of the average within-node sum of squares for t.

To summarize, sources of bias in small trees are cross-validation trees that have fewer terminal nodes than the corresponding main tree and unbalanced test samples. The former might be remedied by selecting the cross-validation trees to have, as nearly as possible, the same number of terminal nodes as the main tree; the latter by stratifying the cases by their y values and selecting the test samples by combining separate samples from each stratum. Both of these have been tried. The results are summarized in Table 8.6. The 3rd column of this table gives the original cross-validation results. The 4th and 5th columns give cross-validation estimates using trees with the same number of terminal nodes as the main tree. The 5th column estimate uses stratified test samples, the 4th column estimates do not.

TABLE 8.6

$\lvert\tilde{T}\rvert$	RE^{ts}	RE^{cv} (orig.)	RE^{cv} (unstrat.)	RE^{cv} (strat.)
2	.55	.60	.61	.60
3	.43	.60	.54	.55
4	.34	.42	.42	.40
5	.30	.43	.38	.35
6	.26	.32	.32	.32
7	.25	.29	.28	.27

Using cross-validation trees with the same number of nodes as the main tree and stratifying the test sets reduce the bias in the estimates. Even so, a marked upward bias remains. To some extent, this seems to be data-set dependent. When other seeds were used to generate data from the same model, the small tree effect was usually present but not so pronounced.

At any rate, in the examples we have examined, the 1 SE tree has always been in the range where the cross-validation estimates are reasonably accurate. (Glick, 1978, has some interesting comments on the "leave-one-out" estimate as used in linear discrimination, which may be relevant to the preceding discussion.)

8.7.2 Stratification and Bias

Since there was some indication that stratification gave more accurate estimates in the preceding example, it was implemented in

CART and tested on 10 more replicates generated from our simulation model.

The idea is this: In tenfold cross-validation, the cases are ordered by their y values and then put into bins corresponding to this ordering.

Thus, for example, the first bin consists of the cases having the 10 lowest y values. The second bin contains the 10 next lowest y values and so on. Then each of the 10 test samples is constructed by drawing one case at random (without replacement) from each bin.

The comparison of the unstratified and stratified cross-validation estimates is given in Table 8.7.

TABLE 8.7

	Unstratified			Stratified		
Data Set	$\|\tilde{T}\|$	$RE^{CV} \pm$ SE	R^{ts}	$\|\tilde{T}\|$	$RE^{CV} \pm$ SE	R^{ts}
5	16	.20 ± .02	.17	19	.19 ± .02	.18
6	10	.26 ± .03	.22	9	.22 ± .03	.24
7	14	.16 ± .02	.18	14	.16 ± .02	.18
8	10	.28 ± .03	.21	10	.24 ± .03	.21
9	14	.14 ± .02	.20	14	.14 ± .02	.20
10	11	.24 ± .03	.20	11	.24 ± .03	.20
11	11	.19 ± .02	.18	11	.18 ± .02	.18
12	12	.17 ± .02	.17	12	.17 ± .02	.17
13	15	.14 ± .02	.16	14	.15 ± .02	.17
14	12	.21 ± .02	.20	13	.19 ± .02	.18

As Table 8.7 shows, the stratified estimates never do worse than the unstratified. On the 8th data set stratification is considerably more accurate. It is difficult to generalize from this single example, but our current thinking is that stratification is the preferred method.

Note that on the 9th data set, the cross-validation estimate is about 3 SE's lower than the test sample estimate. Both in regression and classification, we have noted that in SE units, the cross-validation estimates tend to differ more from the 5000 test sample estimates than predicted on the basis of classical normal theory. This problem is currently being studied and appears complex.

8.8 STANDARD STRUCTURE TREES

As in classification, we say that the data have *standard structure* if the measurement space $\mathcal{X}$ is of fixed dimensionality M, $\mathbf{x} = (x_1, \ldots, x_M)$, where the variables may be either ordered or categorical.

The standard set of splits then consists of all splits of the form {is $x_m < c$?} on ordered variables and {is $x_m \in S$?} for categorical variables where S is any subset of the categories.

Categorical variables in standard structure regression can be handled using a result similar to Theorem 4.5. If $x_m \in \{b_1, \ldots, b_L\}$ is categorical, then for any node t, define $\bar{y}(b_\ell)$ as the average over all y_n in the node such that the mth coordinate of $\mathbf{x}_n$ is b_ℓ. Order these so that

$$\bar{y}(b_{\ell_1}) \leq \bar{y}(b_{\ell_2}) \leq \cdots \leq \bar{y}(b_{\ell_L}).$$

PROPOSITION 8.16. *The best split on x_m in node t is one of the $L - 1$ splits*

$$x_m \in \{b_{\ell_1}, \ldots, b_{\ell_h}\}, \; h = 1, \ldots, L - 1.$$

This reduces the search for the best subset of categories from $2^{L-1} - 1$ to $L - 1$ subsets. The proof is in Section 9.4.

Procedures that carry over without any modification are

Variable combinations
Surrogate splits
Missing value algorithm
Variable importance

In addition, exploratory trees and subsampling carry over for any data structures. Actually, subsampling is simpler in regression because there are no classes. If the node population $N(t)$ is greater than the threshold population N_0, select a random sample of size N_0 from the $N(t)$ cases and use the sample to split the node.

In regression, use of linear combination splits does not have the appeal that it does in classification. In fact, the linear combination algorithm run on both the Boston and the simulated data did not produce any significant decrease in either the cross-validated or the test sample estimates of the relative error. The reason appears to be that whether using linear combinations or univariate splits, what is produced is a flat-topped histogram-type approximation to the regression surface.

A promising alternative for improving accuracy is to grow a small tree using only a few of the most significant splits. Then do multiple linear regression in each of the terminal nodes. Obviously, this would be well tailored to the simulated data set. But the trade-off, in general, is a more complicated tree structure. See Friedman (1979) for a related method.

8.9 USING SURROGATE SPLITS

The definition of surrogate splits and their use in missing values and variable importance carries over, in entirety, to regression trees. They are just as useful here as in classification.

8.9.1 Missing Value Examples

In the Boston data, 5 percent, 10 percent, and 25 percent of the measurement variables were deleted at random. The expected number of deletions per case are .7, 1.3, and 3.3, respectively, and the expected percentages of complete cases are 51 percent, 25 percent, and 2 percent. The corresponding trees were grown and cross-validated and the 1 SE trees selected. The results are given in Table 8.8.

TABLE 8.8

% Data Missing	$\|\tilde{T}\|$	RE^{CV}
0	9	.22
5	16	.22
10	12	.31
25	6	.35

A similar experiment was carried out with the simulated data. The test samples used had the corresponding proportion of data randomly deleted. Table 8.9 has the results for the 1 SE trees.

TABLE 8.9

% Data Missing	$\|\tilde{T}\|$	RE^{CV}	RE^{ts}
0	13	.17 ± .02	.17
5	7	.36 ± .04	.32
10	6	.40 ± .05	.43
25	4	.60 ± .06	.64

As in classification, the question arises as to whether the loss in accuracy is due to the tree constructed or to the increased difficulty in classifying cases with variables missing. Two experiments were carried out. In the first, complete test samples were dropped down the trees constructed with incomplete data. In the second, incomplete test samples were dropped down the 1 SE tree constructed on complete data (see Table 8.10).

TABLE 8.10

Experiment 1		Experiment 2	
% Learning Data Missing	R^{ts}	% Test Data Missing	R^{ts}
0	.17	0	.17
5	.24	5	.27
10	.28	10	.36
25	.33	25	.65

The situation is similar to the digit recognition problem. The measurement variables are independent. There are no meaningful surrogate splits. Tree construction holds up fairly well. But the deletion of a few variables in the test set, in particular, one or more of x_1, x_2, x_3, x_5, x_6, can throw the predicted response value far off.

8.9.2 Variable Importance

The variable importances are defined as in classification, using ΔR instead of ΔI. The results for the Boston data are given in Section 8.2. Only four variables appear in splits in the 1 SE tree, but a number of other variables are ranked as high as, or higher than, two of the variables appearing in splits (DIS, CRIM).

In particular, P/T is the third-ranking variable, after LSTAT and RM. The variables NOX, INDUS, and TAX have variable importances about equal to those at DIS, CRIM. The importances for the other five variables AGE, RAD, B, ZN, and CHAS taper off slowly.

From the structure of the model generating the simulated data, one would clearly rank the variable in importance as

x_1	most important
x_2, x_5	tied, second most important
x_3, x_6	tied, third most important
x_4, x_7	tied, fourth most important
x_8, x_9, x_{10}	tied, least important

The graph of the outputted variable importances is shown in Figure 8.9. The values track our expectations.

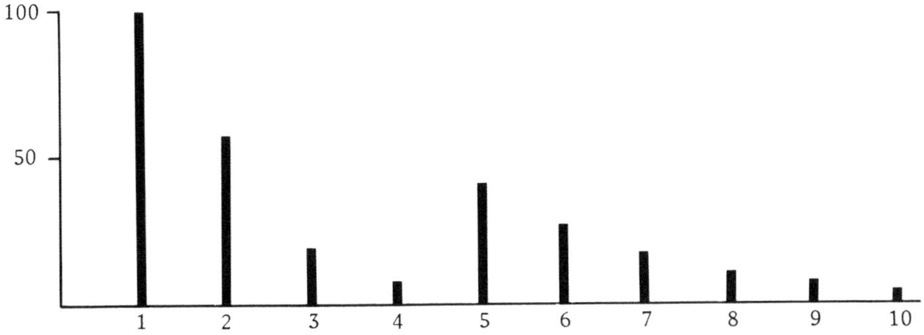

FIGURE 8.9

8.10 INTERPRETATION

The same facilities for tree interpretation exist for regression trees as for classification trees; exploratory trees can be rapidly grown and the cross-validation output can be examined.

8.10.1 Tree Stability

The results of repeated runs with new seeds on the regression model are in contrast to the two classification examples. In the latter, the tree structures were unstable due to redundancy or high correlations in the data. The regression model has much more clear cut structure. The variables are independent, with an obvious order of importance. The first few splits in the original data set and in the four replicated data sets were similar. Figure 8.10 shows the variables split on in the five data sets.

The variable importances are also fairly stable. Table 8.11 gives the rankings of the nonnoise variables in the five runs.

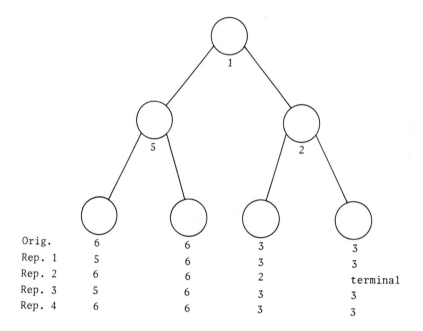

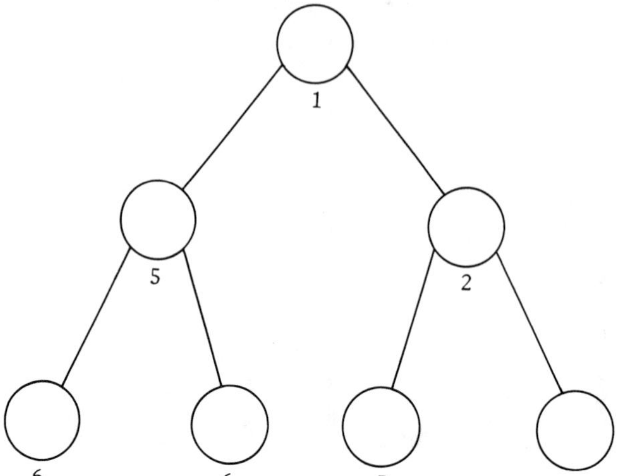

Orig.	6	6	3	3
Rep. 1	5	6	3	3
Rep. 2	6	6	2	terminal
Rep. 3	5	6	3	3
Rep. 4	6	6	3	3

FIGURE 8.10

TABLE 8.11 Variable Importance Ranking

Variable	Original Data	Rep. 1	Rep. 2	Rep. 3	Rep. 4
x_1	1	1	1	1	1
{ x_2	2	2	2	3	3
{ x_5	3	3	3*	2	2
{ x_3	5	5	3*	5	4
{ x_6	4	4	3*	4	5
{ x_4	7	6*	8	7	6
{ x_7	6	6*	7	6	7

*Importances rounding off to the same two digits have tied ranks.

If the structure of the data is clear cut, then the tree tends to give a stable picture of the structure. Instabilities in the tree reflect correlated variables, alternative prediction rules, and noise.

8.10.2 Robustness and Outliers

Tree structured regression is quite robust with respect to the measurement variables, but less so with respect to the response variable. As usual, with least squares regression, a few unusually high or low y values may have a large influence on the residual sum of squares.

However, the tree structure may treat these outliers in a way that both minimizes their effect and signals their presence. It does this by isolating the outliers in small nodes. If there are one or a few cases in a node whose y values differ significantly from the node mean, then a significant reduction in residual sum of squares can be derived from splitting these cases off.

As an illustration, look at the rightmost terminal node in Figure 8.1. It contains only one case, with a y value of 21.9. The parent node had a node mean of 45.1 with a standard deviation of 6.1. The residual sum of squares was decreased about 50 percent by splitting the one case to the right. A similar occurrence can be seen in the third terminal node from the right.

To the extent that the tree structure has available splits that can isolate outliers, it is less subject to distortion from them than linear regression.

8.10.3 Within-Node Standard Deviations

To illustrate a point, assume that the data are generated from a model of the form

$$Y = g(\mathbf{X}) + \varepsilon,$$

where ε is a random error term independent of $\mathbf{X}$ with mean zero. Denote the variance of ε at $\mathbf{X} = \mathbf{x}$ by $\sigma^2(\mathbf{x})$. If $\sigma^2(\mathbf{x})$ is constant, then the model is called homoscedastic. In general, the error variance is different over different parts of the measurement space.

Lack of homoscedasticity may have an unfortunate effect on tree structures. For instance, in a given node t, the within-node variance $s^2(t)$ may be large compared to other nodes, even though $\bar{y}(t)$ is a good approximation to the regression surface over t. Then the search will be on for a noisy split on some variable, resulting in a $p_L s^2(t_L) + p_R s^2(t_R)$ value considerably lower than $s^2(t)$. If such a split can be found, then it may be retained even when the tree is pruned upward.

Because the successive splits try to minimize the within-node variances, the resubstitution estimates $s^2(t)$ will tend to be biased low. Table 8.12 gives the node means and standard deviations computed using resubstitution and the 5000-case test sample for the 13 terminal nodes of the tree (Figure 8.8) grown on the simu-

TABLE 8.12

		Resubstitution		Test Sample	
t	$N(t)$	Mean	Standard Deviation	Mean	Standard Deviation
1	15	-7.5	1.7	-7.1	1.9
2	19	-4.5	1.6	-3.9	1.9
3	12	-4.2	1.2	-4.0	1.6
4	13	- .7	1.8	-1.0	1.6
5	23	-1.0	1.7	- .9	1.9
6	14	2.4	1.1	1.8	1.6
7	18	-1.6	1.8	- .9	1.9
8	10	1.9	1.6	2.1	1.7
9	16	1.3	1.3	2.0	2.0
10	13	3.7	1.8	2.0	2.0
11	22	4.5	1.6	4.9	1.9
12	12	4.6	2.2	5.0	1.6
13	13	8.9	1.8	8.1	1.6

lated data. The nodes are ordered from left to right. The systematic downward bias is apparent.

It would be tempting to think of $\bar{y}(t) \pm 2s(t)$ as a 95 percent confidence interval in the following sense. Drop a large independent test sample down the tree. Of all the cases falling into terminal node t, about 95 percent of the corresponding y values are in the range $\bar{y}(t) \pm 2s(t)$.

This is not valid and does not even hold up as a heuristic. First of all, as already noted, the $s^2(t)$ are generally biased low. An adjustment on the $s^2(t)$ similar to the adjustment on $r(t)$ in classification has been explored, but the improvement was marginal. Second, the dominant error often is not in the downward bias of $s^2(t)$, but instead is that $\bar{y}(t)$ is a poor estimate of the true node mean. To see this, look at nodes 9 and 10 in Table 8.12. Finally, use of $\bar{y}(t) \pm 2s(t)$ implicitly invokes the unwarranted assumption that the node distribution of y values is normal.

The smaller the terminal nodes, the more potentially biased are the estimates $\bar{y}(t)$, $s^2(t)$. In situations where it is important that those individual node estimates be accurate, it is advisable to pick smaller trees having larger terminal nodes. For instance, selecting the tree in the sequence with 12 terminal nodes instead of 13 would have eliminated the noisy split leading to the terminal nodes 9 and 10 in Table 8.12. The single parent node of 9 and 10 with a population of 29 cases becomes terminal in the smaller tree. It has node mean 2.4 and standard deviation 1.9 compared with test sample estimates of 2.0 and 2.0.

8.11 LEAST ABSOLUTE DEVIATION REGRESSION

8.11.1 Background

As mentioned in Section 8.3, if a learning sample $\mathcal{L}$ is used to construct a predictor $d(\mathbf{x})$, then based on a large test sample of size N_2 consisting of $(\mathbf{x}_1, y_1')$, ..., $(\mathbf{x}'_{N_2}, y'_{N_2})$, a natural measure of the accuracy of $d(\mathbf{x})$ is

$$\frac{1}{N_2} \sum_{n=1}^{N_2} |y'_n - d(\mathbf{x}'_n)|.$$

Use of this type of measure leads to least absolute deviation (LAD) regression. (For a recent review of linear least absolute deviation regression methods see Narula and Wellington, 1982.)

To give a theoretical definition of accuracy in the least deviation sense, assume that $(\mathbf{X}, Y)$ and the learning sample cases are independently drawn from the same distribution.

DEFINITION 8.17. *Define the absolute error $R^*(d)$ of the predictor d as*

$$R^*(d) = E|Y - d(\mathbf{X})|.$$

Recall that the median $\nu(Y)$ of a random variable Y, or of the distribution of Y, is defined as a fixed number having the property that

$$P(Y \geq \nu(Y)) \geq .5$$
$$P(Y \leq \nu(Y)) \geq .5.$$

Unlike means or expected values, the median of a distribution may not be unique.

PROPOSITION 8.18. *Any predictor d_B of the form*

$$d_B(\mathbf{x}) = \nu(Y|\mathbf{X} = \mathbf{x})$$

minimizes the absolute error $R^*(d)$.

That is, $d_B(\mathbf{x})$ is a median of the conditional distribution of Y given $\mathbf{X} = \mathbf{x}$. If these medians are not unique, there are a number of minimizing predictors. The key to the proof of Proposition 8.18 is

LEMMA 8.19. *To minimize*

$$E|Y - a|$$

take a equal to any median value of Y.

Let $\nu = \nu(Y)$. Then

$$R^*(\nu) = E|Y - \nu|$$

is the absolute error using the constant ν as a predictor. In analogy with Definition 8.8, use:

DEFINITION 8.20. *The relative absolute error $R^*(d)$ in $d(\mathbf{x})$ as a predictor of Y is*

$$RE^*(d) = R^*(d)/R^*(\nu).$$

The resubstitution estimate for $R^*(d)$ is

$$R(d) = \frac{1}{N} \sum_n |Y_n - d(\mathbf{x}_n)|.$$

Based on the test sample $(\mathbf{x}_1', y_1), \ldots, (\mathbf{x}_{N_2}', y_{N_2})$, the test sample estimate is

$$R^{ts}(d) = \frac{1}{N_2} \sum_{n=1}^{N_2} |y_n' - d(\mathbf{x}_n')|;$$

and the cross-validated estimate $R^{CV}(d)$ is defined analogously to (8.7).

Let $\hat{\nu}$ be a sample median of $y_1, \ldots, y_N$, and set

$$R(\hat{\nu}) = \frac{1}{N} \sum_n |y_n - \hat{\nu}|.$$

Then the resubstitution estimate for the absolute relative error is $R(d)/R(\hat{\nu})$. The test sample estimate is $R^{ts}(d)/R^{ts}(\hat{\nu})$ and the cross-validated estimate is $R^{CV}(d)/R(\hat{\nu})$.

Standard error estimates are derived in much the same way as in least squares regression. However, they depend only on sample second moments and first absolute moments. Therefore, they should be much less variable and more believable than the corresponding least squares estimates.

Since LAD regression was not implemented and tested until late in the course of this monograph, the results in the later chapters, particularly the consistency results in Chapter 12, do not address the least absolute deviation criterion. We are virtually certain that with minor modifications the consistency results can be extended to the LAD context.

8.11.2 Tree Structured LAD Regression

To specify how tree structured least absolute deviation regression works, given a learning set $\mathcal{L}$, three elements are again necessary:

1. *A way to select a split at every intermediate node*
2. *A rule for determining when a node is terminal*
3. *A rule for assigning a value* $y(t)$ *to every terminal node* t

To resolve (3), start with the resubstitution estimate for $R^*(d)$, i.e.,

$$R(d) = \frac{1}{N} \sum_n |y_n - d(\mathbf{x}_n)|.$$

Then select $y(t)$ to minimize $R(d)$.

PROPOSITION 8.21. *Any sample median* $\nu(t)$ *of the* y_n *values corresponding to all cases* $(\mathbf{x}_n, y_n)$ *falling into* t *minimizes* $R(d)$.

The proof of (8.21) is based on the simple observation that $\sum_n |y_n - a|$ is minimized when a is any sample median of the $\{y_n\}$.

In this section, the predicted value in any node t is henceforth taken as any sample median $\nu(t)$ of the y values in the node. Then

$$R(T) = \frac{1}{N} \sum_{t \in \tilde{T}} \sum_{\mathbf{x}_n \in t} |y_n - \nu(t)|. \tag{8.22}$$

Set

$$R(t) = \frac{1}{N} \sum_{\mathbf{x}_n \in t} |y_n - \nu(t)|;$$

then

$$R(T) = \sum_{t \in \tilde{T}} R(t). \tag{8.23}$$

As in Definition 8.13, given a set of splits S of a node t, the best split s^* is that one which most decreases $R(T)$. Alternatively, the best split s^* in S is any minimizer of

$$R(s, t) = R(t) - R(t_L) - R(t_R).$$

More simply, consider all splits of t into t_L, t_R given by S. Then the best split maximizes

$$\sum_{\mathbf{x}_n \in t} |y_n - \nu(t)| - \sum_{\mathbf{x}_n \in t_L} |y_n - \nu(t_L)| - \sum_{\mathbf{x}_n \in t_R} |y_n - \nu(t_R)|;$$

equivalently, it minimizes

$$\sum_{\mathbf{x}_n \in t_L} |y_n - \nu(t_L)| + \sum_{\mathbf{x}_n \in t_R} |y_n - \nu(t_R)|.$$

Thus LAD regression iteratively attempts to minimize the sum of the absolute deviations from the node medians.

To get a form analagous to (8.15), define the *average deviation* $\bar{d}(t)$ for the node t as

$$\bar{d}(t) = \frac{1}{N(t)} \sum_{\mathbf{x}_n \in t} |y_n - \nu(t)|. \tag{8.24}$$

Then

$$R(T) = \sum_{t \in \tilde{T}} \bar{d}(t) p(t)$$

and the best split of t minimizes the weighted sum of average deviations

$p_L\bar{d}(t_L) + p_R\bar{d}(t_R)$.

Tree selection is done either by using a test sample or by cross-validation to prune upward. The trees selected in this section were picked out of the pruned sequence by using the 1 SE rule on the cross-validated estimates of $R^*(d)$.

8.11.3 *The CART Implementation on Standard Data Structures*

For x_1, say, an ordered variable taking many values, there are a large number of potential splits on x_1 at the larger nodes of the tree. The obvious way to evaluate any split on x_1 of a node t into t_L and t_R is to order the y values in t_L and t_R, compute $\nu(t_L)$ and $\nu(t_R)$, and then compute the absolute deviations from $\nu(t_L)$ and $\nu(t_R)$ in each node. The result is a very slow-running program.

Instead, CART uses a fast update algorithm devised by Hal Forsey. This algorithm was incorporated into CART by Padraic Neville. The resulting LAD regression program runs about the same magnitude of speed as ls (least squares) regression. For instance, on the Boston housing data with tenfold cross-validation, ls regression takes 707 CPU seconds on a VAX 11/750 to construct the final tree (Figure 8.1). On the same data and machine, LAD regression requires 1100 CPU seconds, or 56 percent more running time. The update algorithm does not work on categorical variables. On these, CART directly sorts and evaluates the medians and deviations for all possible splits. Therefore, the presence of categorical variables taking many possible values will result in longer running times.

8.11.4 *Examples*

The LAD regression program was run on the Boston housing data and on the simulated data described in Section 8.6. The tree selected for the Boston data is shown in Figure 8.11. The number inside each node circle or rectangle is a median of the y values in the

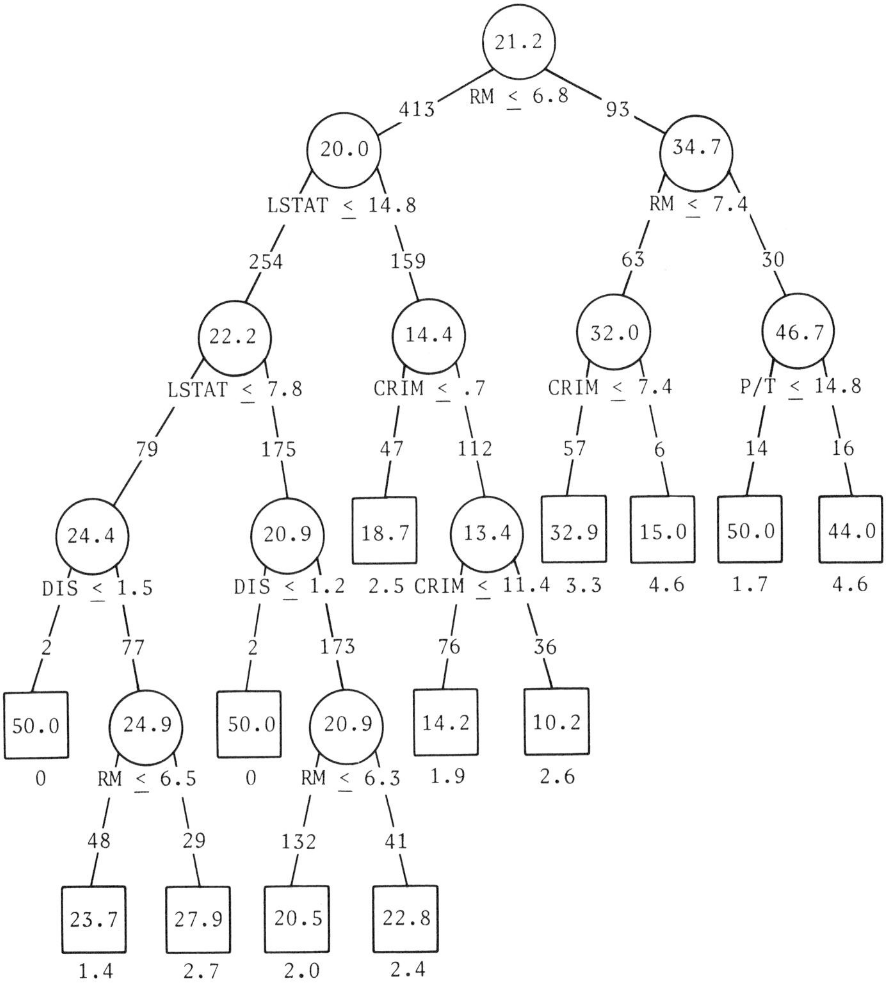
21.2
RM ≤ 6.8
413
93
20.0
34.7
LSTAT ≤ 14.8
RM ≤ 7.4
254
159
63
30
22.2
14.4
32.0
46.7
LSTAT ≤ 7.8
CRIM ≤ .7
CRIM ≤ 7.4
P/T ≤ 14.8
79
175
47
112
57
6
14
16
24.4
20.9
18.7
13.4
32.9
15.0
50.0
44.0
DIS ≤ 1.5
DIS ≤ 1.2
2.5
CRIM ≤ 11.4
3.3
4.6
1.7
4.6
2
77
2
173
76
36
50.0
24.9
50.0
20.9
14.2
10.2
0
RM ≤ 6.5
0
RM ≤ 6.3
1.9
2.6
48
29
132
41
23.7
27.9
20.5
22.8
1.4
2.7
2.0
2.4

FIGURE 8.11

node. The number underneath each terminal node is the average absolute deviation in the node. Otherwise, the notation is the same as in Figure 8.1. The cross-validation estimates RE^* and R^* are .44 ± .03 and 2.9 ± .2, respectively, based on stratified test samples. The results using unstratified samples are very similar.

Variable importance in LAD regression is computed similarly to ls regression, but based on changes in sums of absolute deviations instead of sums of squares. The resulting importances are given in Table 8.13 and compared with the ls importances.

TABLE 8.13

	Importance	
Variable	(LAD)	(ls)
LSTAT	100	100
RM	86	91
INDUS	59	25
CRIM	53	25
P/T	45	38
TAX	43	21
NOX	41	25
DIS	41	27
AGE	34	17
RAD	25	15
B	21	13
ZN	10	11
CHAS	3	1

It is difficult to decide which tree is "best." Relative error in LAD regression is not directly comparable to relative error in ls regression. The latter is the ratio of the mean squared error of the regression to the original mean squared error. If any comparisons are made, they should be between the relative error in LAD regression and ratios of *root* mean squared errors in ls regression;

equivalently, the square root of the relative error in ls regression. Even this is not quite cricket, since by the Schwarz inequality, for any set of numbers $a_1, \ldots, a_N$

$$\frac{1}{N}\sum_{n=1}^{N} |a_i| \leq \sqrt{\frac{1}{N}\sum_{n=1}^{N} a_i^2}$$

It follows that if the medians are about equal to the means, the mean absolute deviation will usually be less than the root mean squared error.

The RE^{CV} in LAD regression for the Boston data is .44. The $\sqrt{RE^{CV}}$ for ls regression is .52. But, as we pointed out in the preceding paragraph, the implications of this difference are not clear.

There are two interesting contrasts in the tree structures. There are four inner city tracts with high MV values (all equal to 50). The ls regression puts these four together with another tract having MV value 28 into one terminal node. The LAD regression isolates these four into two terminal nodes, each having two of the tracts.

In the last split on the right, there is an intermediate node in both trees containing the same 30 tracts. These tracts are characterized by RM > 7.4. The ls tree splits by finding one outlier with low MV having a higher value of CRIM. The LAD tree, not weighing outliers as heavily as least squares, splits on P/T with fourteen low P/T and high MV going left and sixteen lower MV tracts going right.

In terms of rankings, there is one significant change in the variable importances. In the LAD tree INDUS is the third most important variable by a considerable margin over P/T, which is the third ranked by a goodly margin in the ls tree. Interestingly enough, INDUS is never split on in the LAD tree. It achieves its ranking by showing up often as a surrogate for the splits on RM.

LAD regression was also run on data generated from the model specified in Section 8.1. On the original data set, the LAD tree selected is exactly the same as the tree selected by 1s regression and pictured in Figure 8.8. The RE^{CV} value is .41 ± .03 and RE^{ts} is .41 ± .01. For the 1s tree $\sqrt{RE^{CV}}$ is also .41.

To check the accuracy of the cross-validation estimates, LAD regression was run on the first five data sets used in Table 8.7. Stratified test samples were used. The results are given in Table 8.14.

TABLE 8.14

Data Set	$\lvert\tilde{T}\rvert$	$RE^{CV} \pm SE$	RE^{ts}
5	14	.44 ± .03	.44
6	9	.46 ± .03	.48
7	15	.42 ± .03	.41
8	9	.49 ± .03	.45
9	12	.39 ± .03	.43

These estimates, in SE units, are generally closer to the test sample estimates than the 1s estimates are to their corresponding test sample estimates. In particular, the estimate for data set 9 is not unusually low in terms of SE units. But generalizing from the results on only six data sets is folly, and we plan more extensive testing and comparison.

8.12 OVERALL CONCLUSIONS

Regression trees have been used by the authors in fields as diverse as air pollution, criminal justice, and the molecular structure of toxic substances. Its accuracy has been generally competitive with linear regression. It can be much more accurate on nonlinear problems but tends to be somewhat less accurate on problems with good linear structure.

Our philosophy in data analysis is to look at the data from a number of different viewpoints. Tree structured regression offers an interesting alternative for looking at regression type problems. It has sometimes given clues to data structure not apparent from a linear regression analysis. Like any other tool, its greatest benefit lies in its intelligent and sensible application.

9

BAYES RULES AND PARTITIONS

In the remainder of the book, tree structured procedures will be developed and studied in a general framework, which includes regression, classification, and class probability estimation as special cases. The reader is presumed to be familiar with the motivation and intuition concerning tree structured procedures. But otherwise the material in the following chapters can be read independently of that given earlier. In Chapters 9 and 10 the joint distribution of $\mathbf{X}$ and Y is assumed to be known. Additional topics that arise when tree structured procedures are applied to a learning sample will be dealt with in Chapter 11. Chapter 12 is devoted to a mathematical study of the consistency of tree structured procedures as the size of the learning sample tends to infinity.

9.1 BAYES RULE

Let $\mathcal{X}$ denote the set of possible measurement vectors, let $\mathbf{X}$ denote an $\mathcal{X}$-valued random variable whose distribution is denoted by $P(d\mathbf{x})$, and let Y denote a real-valued response (or classification) variable. Let A denote the set of possible "actions," and let $L(y, a)$ denote the "loss" if y is the actual value of Y and a is the action taken. A decision rule d is an A-valued function on X: action $a = d(\mathbf{x})$ is prescribed whenever $\mathbf{x}$ is the observed value of $\mathbf{X}$. The risk $R(d)$ of such a rule is defined to be the expected loss when the rule is used; namely, $R(d) = EL(Y, d(\mathbf{X}))$.

In the regression problem, A is the real line and $L(y, a) = (y - a)^2$. Thus, $R(d) = E[(Y - d(\mathbf{X}))^2]$ is the mean square error of $d(\mathbf{X})$ viewed as a predictor of Y.

In the classification problem as treated in previous chapters, the possible values of Y are restricted to the finite set $\{1, \ldots, J\}$; $A = \{1, \ldots, J\}$; $L(y, a) = C(a|y)$ is the cost of classifying a class y object as a class a object; and $R(d)$ is the expected cost of using the classification rule d. Even in the classification context it is worthwhile to consider the added generality obtained by allowing the set of actions to differ from the set of classes. For example, $\{1, \ldots, J\}$ might correspond to possible diseases and A to possible treatments.

In the class probability estimation problem the values of Y are again restricted to $\{1, \ldots, J\}$. Now A is the set of J-tuples $(a_1, \ldots, a_J)$ of nonnegative numbers that sum to 1; and $L(y, a)$ is defined by

$$L(y, a) = \sum_j (\psi_j(y) - a_j)^2 \quad \text{for} \quad a = (a_1, \ldots, a_J),$$

where $\psi_j(y) = 1$ or $\psi_j(y) = 0$ according as $y = j$ or $y \neq j$. The risk of a decision rule $d = (d_1, \ldots, d_J)$ is given by

$$R(d) = EL(y, d(\mathbf{X})) = \sum_j E[(\psi_j(Y) - d_j(\mathbf{X}))^2].$$

The quantity $d_j(\mathbf{X})$ can be viewed as a predictor of $\psi_j(Y)$, with $R(d)$ denoting the sum of the corresponding mean square errors of prediction.

A Bayes rule d_B is any rule d that minimizes $R(d)$. To find such a Bayes rule, observe that $R(d) = EE[L(Y, d(\mathbf{X}))|\mathbf{X}]$ and hence that

$$R(d) = \int E[L(Y, d(\mathbf{x}))|\mathbf{X} = \mathbf{x}]P(d\mathbf{x});$$

that is, $R(d)$ is the integral of $E[L(Y, d(\mathbf{x}))|\mathbf{X} = \mathbf{x}]$ with respect to the distribution of $\mathbf{X}$. Thus, d is a Bayes rule if for each $\mathbf{x} \in X$, $a = d(\mathbf{x})$ minimizes $E[L(Y, a)|\mathbf{X} = \mathbf{x}]$. Also, the risk $R(d_B)$ of a Bayes rule d_B can be written as

$$R(d_B) = \int \min_a E[L(Y, a)|\mathbf{X} = \mathbf{x}]P(d\mathbf{x}).$$

(In general, min must be replaced by inf.)

In the regression problem, let $\mu(\mathbf{x}) = E[Y|\mathbf{X} = \mathbf{x}]$, $\mathbf{x} \in X$, denote the regression function of Y on $\mathbf{X}$ and observe that $E[Y - \mu(\mathbf{x})| \mathbf{X} = \mathbf{x}] = 0$. Thus,

$$\begin{aligned} E[L(Y, a)|\mathbf{X} = \mathbf{x}] &= E[(Y - a)^2|\mathbf{X} = \mathbf{x}] \\ &= E[(Y - \mu(\mathbf{x}) + \mu(\mathbf{x}) - a)^2|\mathbf{X} = \mathbf{x}] \\ &= E[(Y - \mu(\mathbf{x}))^2|\mathbf{X} = \mathbf{x}] + (\mu(\mathbf{x}) - a)^2; \end{aligned}$$

consequently,

$$\min_a E[L(Y, a)|\mathbf{X} = \mathbf{x}] = E[(Y - \mu(\mathbf{x}))^2|\mathbf{X} = \mathbf{x}]$$

and $\mu(\mathbf{x})$ is the unique value of a that minimizes $E[L(Y, a)|\mathbf{X} = \mathbf{x}]$. Therefore, the Bayes rule is given by $d_B(\mathbf{x}) = \mu(\mathbf{x}) = E[Y|\mathbf{X} = \mathbf{x}]$. Observe also that for any rule d,

$$R(d) = R(d_B) + E[(d_B(\mathbf{X}) - d(\mathbf{X}))^2].$$

In the classification problem, set $P(j|\mathbf{x}) = P(Y = j|\mathbf{X} = \mathbf{x})$. Then

$$E[L(Y, i)|\mathbf{X} = \mathbf{x}] = \sum_j L(j, i)P(j|\mathbf{x}) = \sum_j C(i|j)P(j|\mathbf{x}).$$

Thus, to obtain a Bayes rule, choose $d_B(\mathbf{x})$ to be, say, the smallest value of i that minimizes $\sum_j C(i|j)P(j|\mathbf{x})$. The risk of the Bayes rule is given by

$$R(d_B) = \int \min_i \left[\sum_j C(i|j)P(j|\mathbf{x}) \right] P(d\mathbf{x}).$$

To find the Bayes rule for the class probability estimation problem, observe that $P(j|\mathbf{x}) = E[\psi_j(Y)|\mathbf{X} = \mathbf{x}]$ and hence that $E[\psi_j(Y) - P(j|\mathbf{x})|\mathbf{X} = \mathbf{x}] = 0$. It is also easily seen that

$$E[(\psi_j(Y) - P(j|\mathbf{x}))^2|\mathbf{X} = \mathbf{x}] = \text{Var}[\psi_j(Y)|\mathbf{X} = \mathbf{x}]$$
$$= P(j|\mathbf{x})(1 - P(j|\mathbf{x})).$$

Thus, it follows as in the proof of the corresponding result for the regression problem that

$$E[L(Y, a)|\mathbf{X} = \mathbf{x}] = \sum_j P(j|\mathbf{x})(1 - p(j|\mathbf{x})) + \sum_j (P(j|\mathbf{x}) - a_j)^2$$

for $a = (a_1, \ldots, a_J)$. Consequently, the Bayes rule is given by $d_B(\mathbf{x}) = (P(1|\mathbf{x}), \ldots, P(j|\mathbf{x}))$, and it has risk

$$R(d_B) = \int \sum_j P(j|\mathbf{x})(1 - P(j|\mathbf{x}))P(d\mathbf{x}).$$

For any rule d,

$$R(d) = R(d_B) + \sum_j E[(P(j|\mathbf{X}) - d_j(\mathbf{X}))^2].$$

9.2 BAYES RULE FOR A PARTITION

A subset t of X is called a node if $P(t) > 0$, P being the distribution of $\mathbf{X}$. Let $\tilde{T}$ be a partition of X into a finite number of disjoint nodes (which, in applications to tree structured rules, are the terminal nodes of a tree T). The *partition function* τ associated with $\tilde{T}$ is the function from X to $\tilde{T}$ that assigns to each $\mathbf{x}$ the node $t \in \tilde{T}$ containing $\mathbf{x}$, so that $\tau(\mathbf{x}) = t$ if, and only if, $\mathbf{x} \in t$. A decision rule d is said to correspond to the partition $\tilde{T}$ if it is constant on the terminal nodes of T, that is, if for each node $t \in \tilde{T}$ there is a $\nu(t) \in \alpha$ such that $d(\mathbf{x}) = \nu(t)$ for all $\mathbf{x} \in t$. The decision rule d is then given explicitly by $d(\mathbf{x}) = \nu(\tau(\mathbf{x}))$. The risk of such a rule is given by

$$R(d) = \sum_{\widetilde{T}} E[L(Y, \nu(t))|\mathbf{X} \in t]P(t).$$

A Bayes rule corresponding to the partition $\widetilde{T}$ is a rule of the preceding form having the smallest possible risk. Thus, d is a Bayes rule corresponding to $\widetilde{T}$ if, and only if, $d(\mathbf{x}) = \nu(\tau(\mathbf{x}))$, where for each $t \in \widetilde{T}$, $a = \nu(t)$ minimizes $E[L(Y, a)|\mathbf{X} \in t]$. From now on, given a node t, let $\nu(t)$ be a value of a that minimizes $E[L(Y, a)| \mathbf{X} \in t]$. Also, set

$$r(t) = E[L(Y, \nu(t))|\mathbf{X} \in t] = \min_a E[L(Y, a)|\mathbf{X} \in t]$$

and $R(t) = P(t)r(t)$.

Let $d_{\widetilde{T}}$ be a Bayes rule corresponding to $\widetilde{T}$. It is convenient to abbreviate $R(d_{\widetilde{T}})$ to $R(\widetilde{T})$. Then

$$R(\widetilde{T}) = \sum_{\widetilde{T}} P(t)r(t) = \sum_{\widetilde{T}} R(t).$$

In the regression problem given a node t, set $\mu(t) = E(Y|\mathbf{X} \in t)$ and

$$\sigma^2(t) = \text{Var}(Y|\mathbf{X} \in t) = E[(Y - \mu(t))^2|\mathbf{X} \in t].$$

The unique Bayes rule corresponding to a partition $\widetilde{T}$ is given by $d_{\widetilde{T}}(\mathbf{x}) = \mu(\tau(\mathbf{x}))$. Also, $r(t) = \sigma^2(t)$.

In the classification problem, given a node t, set $P(j|t) = P(Y = j|\mathbf{X} \in t)$. Here a Bayes rule corresponding to $\widetilde{T}$ is obtained by choosing $\nu(t)$ to be a value (for example, the smallest value) of i that minimizes $\sum_j C(i|j)P(j|t)$ and setting $d_{\widetilde{T}}(\mathbf{x}) = \nu(\tau(\mathbf{x}))$. Moreover,

$$r(t) = \min_i \sum_j C(i|j)P(j|t).$$

In the class probability estimation problem, the unique Bayes rule $d_{\widetilde{T}}$ corresponding to $\widetilde{T}$ is given by $d_{\widetilde{T}}(\mathbf{x}) = \nu(\tau(\mathbf{x}))$, where $\nu(t) = (P(1|t), \ldots, P(J|t))$. Also,

$$r(t) = \sum_j P(j|t)(1 - P(j|t)).$$

THEOREM 9.1. *Let t be a node and let $\tilde{T}^{(t)}$ be a collection of nodes that forms a partition t. Then $R(t) \geq \sum_{\tilde{T}^{(t)}} R(s)$, with equality holding if, and only if,*

$$r(s) = E[L(Y, \nu(t))|\mathbf{X} \in s] \text{ for all } s \in \tilde{T}^{(t)}.$$

PROOF. It is easily seen that

$$P(t) = \sum_{\tilde{T}^{(t)}} P(s) \tag{9.2}$$

and

$$\begin{aligned} R(t) = P(t)r(t) &= P(t)E[L(Y, \nu(t))|\mathbf{X} \in t] \\ &= \sum_{\tilde{T}^{(t)}} P(s)E[L(Y, \nu(t))|\mathbf{X} \in s]. \end{aligned} \tag{9.3}$$

Consequently,

$$R(t) - \sum_{\tilde{T}^{(t)}} R(s) = \sum_{\tilde{T}^{(t)}} P(s)\{E[L(Y, \nu(t))|\mathbf{X} \in s] - r(s)\},$$

which yields the desired result.

Let $\tilde{T}$, $\tilde{T}'$ be two partitions of X into disjoint nodes. Then $\tilde{T}'$ is a *refinement* of $\tilde{T}(\tilde{T}' \succcurlyeq \tilde{T})$ if for any pair of nodes $t \in \tilde{T}$, $s \in \tilde{T}'$, either s is a subset of t or s and t are disjoint. (Recall that if T and T' are trees having root X, then the corresponding collections $\tilde{T}$ and $\tilde{T}'$ of terminal nodes are each partitions of X. Observe that if $T' \succcurlyeq T$, as defined in Section 3.2, then $\tilde{T}' \succcurlyeq \tilde{T}$; that is, $\tilde{T}'$ is a refinement of $\tilde{T}$.)

THEOREM 9.4. *Let T and T' be partitions of X into disjoint nodes such that $\tilde{T}' \succcurlyeq \tilde{T}$. Then $R(\tilde{T}') \leq R(\tilde{T})$, with equality holding if and only if $r(s) = E[L(Y, \nu(t))|\mathbf{X} \in s]$ for all pairs $t \in \tilde{T}$, $s \in \tilde{T}'$ such that $s \subset t$.*

PROOF. Given $t \in \tilde{T}$, let $\tilde{T}^{(t)}$ be the partition of t defined by $\tilde{T}^{(t)} = \{s \in \tilde{T}' : s \subset t\}$. It follows from (9.2) and (9.3) that

$$R(\tilde{T}) - R(\tilde{T}') = \sum_{\tilde{T}} \sum_{\tilde{T}^{(t)}} P(s)\{E[L(Y, \nu(t))|\mathbf{X} = s] - r(s)\},$$

which yields the desired result.

9.3 RISK REDUCTION SPLITTING RULE

Let $\widetilde{T}$ be a fixed partition and let $t \in \widetilde{T}$ also be fixed. Consider a split $\mathcal{s}$ of t into two disjoint nodes t_L and t_R. Set

$$p_L = P(\mathbf{X} \in t_L | \mathbf{X} \in t) = \frac{P(t_L)}{P(t)}$$

and

$$p_R = P(\mathbf{X} \in t_R | \mathbf{X} \in t) = \frac{P(t_R)}{P(t)},$$

and observe that $p_L + p_R = 1$. Let $\widetilde{T}'$ be the modification to $\widetilde{T}$ obtained by replacing t by the pair t_L, t_R. The risk reduction $\Delta R(\mathcal{s}, t) = R(\widetilde{T}) - R(\widetilde{T}')$ due to the split is given by

$$\begin{aligned}\Delta R(\mathcal{s}, t) &= R(t) - R(t_L) - R(t_R) \\ &= P(t)[r(t) - p_L r(t_L) - p_R r(t_R)].\end{aligned}$$

The relative risk reduction

$$\Delta R(\mathcal{s}|t) = \frac{R(\widetilde{T}) - R(\widetilde{T}')}{P(t)}$$

due to the split is given by

$$\Delta R(\mathcal{s}|t) = r(t) - p_L r(t_L) - p_R r(t_R).$$

The *risk reduction splitting rule* is to choose $\mathcal{s}$ to maximize $\Delta R(\mathcal{s}|t)$. The following result follows immediately from either Theorem 9.1 or Theorem 9.4.

THEOREM 9.5. *Let $\mathcal{s}$ be a split of t into t_L, t_R. Then $\Delta R(\mathcal{s}|t) \geq 0$, with equality holding if, and only if,*

$$r(t_L) = E[L(Y, \nu(t))|\mathbf{X} \in t_L] \text{ and } r(t_R) = E[L(Y, \nu(t))|\mathbf{X} \in t_R].$$

Consider a split $\mathcal{s}$ of t into t_L, t_R. According to Theorem 9.5, the risk of the Bayes rule for the modified partition cannot be more than the risk for the original partition; it is strictly smaller unless every choice of the action a that is optimal for t is also optimal for t_L and t_R.

In the regression problem,

$$\Delta R(s|t) = \sigma^2(t) - P_L\sigma^2(t_L) - P_R\sigma^2(t_R),$$

which can be rewritten as

$$\Delta R(s|t) = E[(Y - \mu(t))^2|\mathbf{X} \in t] - P_L E[(Y - \mu(t_L))^2|\mathbf{X} \in t_L] - P_R E[(Y - \mu(t_R))^2|\mathbf{X} \in t_R].$$

Observe that $\mu(t) = P_L\mu(t_L) + P_R\mu(t_R)$ and that

$$E[(Y - \mu(t))^2|\mathbf{X} \in t] = P_L E[(Y - \mu(t))^2|\mathbf{X} \in t_L] + P_R E[(Y - \mu(t))^2|\mathbf{X} \in t_R].$$

Consequently,

$$\begin{aligned}\Delta R(s|t) &= P_L(\mu(t_L) - \mu(t))^2 + P_R(\mu(t_R) - \mu(t))^2 \\ &= P_L\mu^2(t_L) + P_R\mu^2(t_R) - \mu^2(t) \\ &= P_L P_R(\mu(t_L) - \mu(t_R))^2.\end{aligned}$$

Analogous results hold in the class probability estimation problem. That is,

$$P(j|t) = P_L P(j|t_L) + P_R P(j|t_R) \quad \text{for} \quad 1 \leq j \leq J$$

and

$$\begin{aligned}\Delta R(s|t) &= \sum_j P(j|t)(1 - P(j|t)) - P_L \sum_j P(j|t_L)(1 - P(j|t_L)) \\ &\quad - P_R \sum_j P(j|t_R)(1 - P(j|t_R)) \\ &= P_L \sum_j P^2(j|t_L) + P_R \sum_j P^2(j|t_R) - \sum_j P^2(j|t) \\ &= P_L \sum_j (P(j|t_L) - P(j|t))^2 + P_R \sum_j (P(j|t_R) - P(j|t))^2 \\ &= P_L P_R \sum_j (P(j|t_L) - P(j|t_R))^2.\end{aligned}$$

In the special case $J = 2$,

$$\Delta R(\delta|t) = 2[P(1|t)(1 - P(1|t)) - p_L P(1|t_L)(1 - P(1|t_L)) - p_R P(1|t_R)(1 - P(1|t_R))]$$

$$= 2[p_L P^2(1|t_L) + p_R P^2(1|t_R) - P^2(1|t)]$$

$$= 2[p_L(P(1|t_L) - P(1|t))^2 + p_R(P(1|t_L) - P(1|t))^2]$$

$$= 2p_L p_R(P(1|t_L) - P(1|t_R))^2.$$

9.4 CATEGORICAL SPLITS

Consider splits of t into t_L, t_R based on the mth coordinate of $\mathbf{x}$, which is assumed here to be a categorical variable whose possible values range over a finite set B. Then

$$t_L = \{\mathbf{x} \in t : x_m \in B_1\} \quad \text{and} \quad t_R = \{\mathbf{x} \in t : x_m \in B_2\},$$

where B_1, B_2 is a partition of B. Suppose that the splitting rule is equivalent to choosing the partition B_1, B_2 to minimize $p_L\phi(\mu(t_L)) + p_R\phi(\mu(t_R))$, where ϕ is a concave function on an interval containing the range of $\mu(\cdot)$. (This is true for the risk reduction splitting rule in the regression problem with $\phi(y) = -y^2$; see Section 4.2 for other examples.) A minimizing partition will be called optimal. In finding an optimal partition of B, it is assumed that $P(\mathbf{X} \in t, X_m = b) > 0$ for all $b \in B$ (exceptional b's can be ignored). It is also assumed that $E(Y|\mathbf{X} \in t, X_m = b)$ is not constant in $b \in B$, for otherwise every partition is optimal.

The motivation for the next result in terms of computational efficiency is given in Section 4.2.2. In the special case that $\phi(y) = -y^2$, the result is due to Fisher (1958).

THEOREM 9.6. *There is an optimal partition B_1, B_2 of B such that*

$$E(Y|\mathbf{X} \in t, X_m = b_1) < E(Y|\mathbf{X} \in t, X_m = b_2)$$

for $b_1 \in B_1$ and $b_2 \in B_2$.

PROOF. The following proof contains some clever simplifications due to P. Feigin. It can be assumed that $t = X$ (otherwise condition on $\mathbf{X} \in t$). Set $q_b = P(X_m = b)$ for $b \in B$. By assumption, B contains at least two points and $q_b > 0$ for $b \in B$. Set $y_b = E(Y|X_m = b)$ for $b \in B$. Given a subset B_1 of B, set $q(B_1) = \sum_{B_1} q_b$ and

$$\bar{y}(B_1) = \begin{cases} \frac{1}{q(B_1)} \sum_{B_1} y_b q_b & \text{if } B_1 \text{ is nonempty} \\ \bar{y}(B) & \text{otherwise.} \end{cases}$$

Since ϕ is a concave function on an interval,

$$\phi(\lambda y_1 + (1 - \lambda)y_2) \geq \lambda\phi(y_1) + (1 - \lambda)\phi(y_2)$$

for y_1, y_2 in the interval and $0 \leq \lambda \leq 1$. Let Ψ be defined on subsets of B by

$$\Psi(B_1) = q(B_1)\phi(\bar{y}(B_1)) + q(B_2)\phi(\bar{y}(B_2)),$$

where $B_2 = B - B_1$. Note here that $\Psi(B_1) = \Psi(B_2)$. By the concavity of ϕ,

$$\Psi(B_1) \leq \phi(q(B_1)\bar{y}(B_1) + q(B_2)\bar{y}(B_2)) = \phi(\bar{y}(B)) = \Psi(B)$$

for all subsets B_1 of B.

If $t_L = \{\mathbf{x} \in X : x_m \in B_1\}$ and $t_R = \{\mathbf{x} \in X : x_m \in B_2\}$, then

$$\Psi(B_1) = p_L\phi(\mu(t_L)) + p_R\phi(\mu(t_R)).$$

Thus, B_1, B_2 is an optimal partition of B if and only if B_1 minimizes $\Psi(B_1)$, in which case B_1 will be called an optimal subset of B. It can be assumed that $\Psi(B_1) < \Psi(B)$ if B_1 is optimal (otherwise every partition of B is optimal). Thus, in finding an optimal subset B_1 of B, B_1 can be allowed to range over *all* subsets of B, including B itself and the empty set.

Since there are only finitely many subsets of B, there is an optimal one. If B_1 is an optimal subset and $B_2 = B - B_1$, then $q(B_1) > 0$, $q(B_2) > 0$, and $\bar{y}(B_1) \neq \bar{y}(B_2)$ (for otherwise $\Psi(B_1) = \Psi(B)$). The optimal subset B_1 can be chosen so that $\bar{y}(B_1) < \bar{y}(B_2)$

(otherwise replace B_1 by B_2). Let B_1 be a subset of B such that B_1 is optimal, $\bar{y}(B_1) < \bar{y}(B_2)$, and, subject to these restrictions, $Y_1 - Y_2$ is made as small as possible, where $Y_1 = \max_{B_1} y_b$ and $Y_2 = \min_{B_2} y_b$. To verify the statement of the theorem, it suffices to show that $Y_1 < Y_2$ or, equivalently, to show that the assumption $Y_1 \geq Y_2$ leads to a contradiction.

To this end, assume that $Y_1 \geq Y_2$. Set $A_1 = \{b \in B_1 : y_b = Y_1\}$, $A_2 = \{b \in B_2 : y_b = Y_2\}$, $A_3 = \{b \in B_1 : y_b < Y_1\}$, and $A_4 = \{b \in B_2 : y_b > Y_2\}$. Also set $Q_i = q(A_i)$ for $i = 1, 2, 3, 4$, $v = Q_2/(Q_1 + Q_2)$, $Y_3 = \bar{y}(A_3)$, and $Y_4 = \bar{y}(A_4)$. (Note that $Q_1 > 0$ and $Q_2 > 0$.) It will be shown that

$$\Psi(B_1) \geq v\Psi(A_3) + (1 - v)\Psi(A_4). \tag{9.7}$$

Since B_1 is optimal, it follows from (9.7) that A_3 is optimal; also, $Q_3 > 0$, so

$$Y_3 < \frac{Q_1Y_1 + Q_3Y_3}{Q_1 + Q_3} < Y_1$$

and hence

$$Y_3 < \frac{Q_1Y_1 + \dfrac{Q_2 + Q_4}{Q_1 + Q_3}(Q_1Y_1 + Q_3Y_3)}{Q_1 + Q_2 + Q_4}. \tag{9.8}$$

Now.

$$\frac{Q_1Y_1 + Q_3Y_3}{Q_1 + Q_3} = \bar{y}(B_1) < \bar{y}(B_2) = \frac{Q_2Y_2 + Q_4Y_4}{Q_2 + Q_4},$$

so it follows from (9.8) that

$$\bar{y}(A_3) = Y_3 < \frac{Q_1Y_1 + Q_2Y_2 + Q_4Y_4}{Q_1 + Q_2 + Q_4} = \bar{y}(A_1 \cup A_2 \cup A_4) = \bar{y}(B - A_3).$$

Finally,

$$\max_{A_3} y_b - \min_{B-A_3} y_b < Y_1 - Y_2.$$

Therefore, the assumption that $Y_1 - Y_2$ is as small as possible subject to the indicated restrictions has been contradicted.

Set $Q_{ij} = Q_i + Q_j$, $Q_{ijk} = Q_i + Q_j + Q_k$, $Y_{ij} = (Q_iY_i + Q_jY_j)/Q_{ij}$, and $Y_{ijk} = (Q_iY_i + Q_jY_j + Q_kY_k)/Q_{ijk}$. It remains to verify that

$$Q_{13}\phi(Y_{13}) + Q_{24}\phi(Y_{24}) \geq v\bigl(Q_3\phi(Y_3) + Q_{124}\phi(Y_{124})\bigr) + (1 - v)\bigl(Q_4\phi(Y_4) + Q_{123}\phi(Y_{123})\bigr), \tag{9.9}$$

which is equivalent to (9.7).

It is easily seen that $Q_{13} = vQ_3 + (1 - v)Q_{123}$ and $Q_{24} = (1 - v)Q_4 + vQ_{124}$. Define Y'_{13} and Y'_{24} by

$$Q_{13}Y'_{13} = vQ_3Y_3 + (1 - v)Q_{123}Y_{123} = Q_{13}Y_{13} - Q_{12}v(1 - v)(Y_1 - Y_2)$$

and

$$Q_{24}Y'_{24} = (1 - v)Q_4Y_4 + vQ_{124}Y_{124} = Q_{24}Y_{24} + Q_{12}v(1 - v)(Y_1 - Y_2).$$

It follows from the concavity of ϕ that

$$Q_{13}\phi(Y'_{13}) \geq vQ_3\phi(Y_3) + (1 - v)Q_{123}\phi(Y_{123}) \tag{9.10}$$

and

$$Q_{24}\phi(Y'_{24}) \geq (1 - v)Q_4\phi(Y_4) + vQ_{124}\phi(Y_{124}). \tag{9.11}$$

Next it will be shown that

$$Q_{13}\phi(Y_{13}) + Q_{24}\phi(Y_{24}) \geq Q_{13}\phi(Y'_{13}) + Q_{24}\phi(Y'_{24}). \tag{9.12}$$

By assumption, $Y_1 \geq Y_2$ and $Y_{13} < Y_{24}$. If $Y_1 = Y_2$, then $Y'_{13} = Y_{13}$ and $Y'_{24} = Y_{24}$, so (9.12) is trivially satisfied. Otherwise, $Y_1 > Y_2$ and hence

$$Y'_{13} < Y_{13} < Y_{24} < Y'_{24}.$$

Thus, by the concavity of ϕ,

$$\frac{\phi(Y_{13}) - \phi(Y'_{13})}{Y_{13} - Y'_{13}} \geq \frac{\phi(Y'_{24}) - \phi(Y_{24})}{Y'_{24} - Y_{24}}.$$

Consequently,

$$Q_{13}\bigl(\phi(Y_{13}) - \phi(Y'_{13})\bigr) = Q_{12}v(1 - v)(Y_1 - Y_2)\frac{\phi(Y_{13}) - \phi(Y'_{13})}{Y_{13} - Y'_{13}}$$

$$\geq Q_{12}v(1 - v)(Y_1 - Y_2)\frac{\phi(Y_{24}') - \phi(Y_{24})}{Y_{24}' - Y_{24}}$$

$$= Q_{24}(\phi(Y_{24}') - \phi(Y_{24})),$$

so (9.12) again holds. Since (9.9) follows from (9.10) - (9.12), the proof of the theorem is complete.

10

OPTIMAL PRUNING

In Section 10.1 a convenient terminology for discussing trees is introduced. This terminology is used in Section 10.2 to state and verify the mathematical properties of optimal pruning. In Section 10.3 an explicit optimal pruning algorithm is described and illustrated. Optimal pruning is treated in this chapter as a self-contained concept. Its connection with tree structured statistical procedures will be reestablished in Chapter 11.

10.1 TREE TERMINOLOGY

For purposes of this chapter, a *tree* can be defined as consisting of a finite nonempty set T of positive integers and two functions left(·) and right(·) from t to $T \cup \{0\}$, which together satisfy the following two properties:

(i) For each $t \in T$, either left(t) = right(t) = 0 or left(t) > t and right(t) > t;

(ii) For each $t \in T$, other than the smallest integer in T, there is exactly one $s \in T$ such that either t = left(s) or t = right(s).

For simplicity, T itself will be referred to as a tree and, as before, each element of T will be referred to as a node. The correspondence between nodes of T and subsets of X is irrelevant in this chapter.

Figure 10.1 shows a tree that arose in the discussion of the digit recognition example in Section 2.6.1. Table 10.1 determines the same tree by specifying the values of $\ell(t)$ = left(t) and $r(t)$ = right(t). The reader is encouraged to apply to this tree the various definitions that follow.

The triple T, left(·), right(·) can always be chosen (as they were in the example in Figure 10.1) so that for each node t either left(t) = right(t) = 0 or left(t) > 0 and right(t) = left(t) + 1. Under this restriction, the tree is determined by specifying left(·).

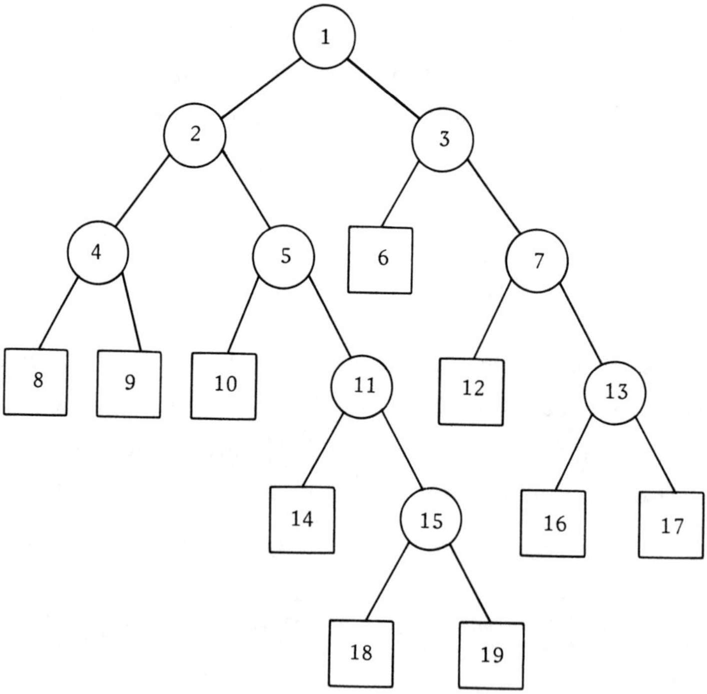

FIGURE 10.1

TABLE 10.1

t	$\ell(t)$	$r(t)$
1	2	3
2	4	5
3	6	7
4	8	9
5	10	11
6	0	0
7	12	13
8	0	0
9	0	0
10	0	0
11	14	15
12	0	0
13	16	17
14	0	0
15	18	19
16	0	0
17	0	0
18	0	0
19	0	0

The minimum element of a tree T is called the *root* of T, denoted by root(T). If s, $t \in T$ and t = left(s) or t = right(s), then s is called the *parent* of t. The root of T has no parent, but every other node in t has a unique parent. Let parent($\cdot$) be the function from T to $T \cup \{0\}$ defined so that parent(root(T)) = 0 and parent(t) is the parent of t for $t \neq$ root(T). A node t is called a *terminal node* if it is not a parent, that is, if left(t) = right(t) = 0. Let $\widetilde{T}$ denote the collection of terminal nodes of T. The elements in $T - \widetilde{T}$ are called *nonterminal nodes*.

A node s is called an *ancestor* of a node t if s = parent(t) of s = parent(parent(t)) or .... A node t is called a *descendant* of s if s is an ancestor of t. If s is an ancestor of t, there is a unique "path" $s_0, \ldots, s_m$ from s to t (that is, such that $s_0 = s$, s_{k-1} = parent(s_k) for $1 \leq k \leq m$, and $s_m = t$). Let $\ell(s, t) = m$ denote the *length* of this path. Also set $\ell(t, t) = 0$.

Given a nonempty subset T_1 of T, define left$_1(\cdot)$ and right$_1(\cdot)$ from T_1 to $T_1 \cup \{0\}$ by

$$\text{left}_1(t) = \begin{cases} \text{left}(t) \text{ if left}(t) \in T_1 \\ 0 \text{ otherwise} \end{cases}$$

and

$$\text{right}_1(t) = \begin{cases} \text{right}(t) \text{ if right}(t) \in T_1 \\ 0 \text{ otherwise.} \end{cases}$$

Then T_1 is called a *subtree* of T if the triple T_1, left$_1(\cdot)$, right$_1(\cdot)$ forms a tree. If T_1 is a subtree of T, and T_2 is a subtree of T_1, then T_2 is a subtree of T. Given $t \in T$, the collection T_t consisting of t and all its descendants is called the *branch of t stemming from t*. It is a subtree of t. The branch $(T_1)_t$ of a subtree T_1 of T is denoted by T_{1t}.

The tree T is said to be *trivial* if any of the following equivalent conditions is satisfied: $|T| = 1$; $|\widetilde{T}| = 1$; $T = \{\text{root}(T)\}$; $T - \widetilde{T}$ is empty. (Recall that $|S|$ denotes the number of elements in the set S.) Otherwise, T is said to be nontrivial.

Given a nontrivial tree T, set t_1 = root(T), t_L = left(t_1), t_R = right(t_1), $T_L = T_{t_L}$, and $T_R = T_{t_R}$. Then T_L is called the *left primary branch of T* and T_R is called the *right primary branch*. Observe that $\{t_1\}$, T_L, T_R are nonempty disjoint sets whose union is T and that $\widetilde{T}_L$, $\widetilde{T}_R$ are nonempty disjoint sets whose union is $\widetilde{T}$; in particular,

$$|T| = 1 + |T_L| + |T_R| \tag{10.1}$$

and

$$|\widetilde{T}| = |\widetilde{T}_L| + |\widetilde{T}_R|. \tag{10.2}$$

Properties of trees are typically proved by induction based on the observation that the primary branches of a nontrivial tree are trees having fewer terminal nodes than the original one. For example, it follows easily from (10.1), (10.2), and induction that

$$|T| = 2|\widetilde{T}| - 1. \tag{10.3}$$

A subtree T_1 of T is called a *pruned subtree* of T if $\text{root}(T_1) = \text{root}(T)$; this is denoted by $T_1 \preccurlyeq T$ or $T \succcurlyeq T_1$. The notation $T_1 \prec T$ (or $T \succ T_1$) is used to indicate that $T_1 \preccurlyeq T$ and $T_1 \neq T$. Observe that $\preccurlyeq$ is transitive; that is, if $T_1 \preccurlyeq T$ and $T_2 \preccurlyeq T_1$, then $T_2 \preccurlyeq T$. Similarly, $\prec$ is transitive. An arbitrary subset T_1 of T is a subtree of T having root t if, and only if, it is a pruned subtree of T_t.

The definitions given here differ slightly from those in Chapter 3. There T_1 was called a pruned subtree of T if $T_1 \prec T$. Also, pruning was given a "bottom-up" definition. This definition is very suggestive, but the "top-down" definition $(\text{root}(T_1) = \text{root}(T))$ given here will be more convenient in using induction to establish the validity of the optimal pruning algorithm. The two definitions are obviously equivalent to each other.

Let T be a nontrivial tree having root t_1 and primary branches T_L and T_R. Let T' be a nontrivial pruned subtree of T and let T'_L and T'_R, respectively, denote its left and right primary branches. Then T'_L is a pruned subtree of T_L and T'_R is a pruned subtree of T_R. Moreover, every pair of pruned subtrees of T_L, T_R arises uniquely in this manner. Observe that $\widetilde{T}'_L$ and $\widetilde{T}'_R$ are disjoint sets whose union is $\widetilde{T}'$. In particular,

$$|\widetilde{T}'| = |\widetilde{T}'_L| + |\widetilde{T}'_R|. \tag{10.4}$$

Let T'' be a second nontrivial pruned subtree of T. Then $T'' \preccurlyeq T'$ if, and only if, $T''_L \preccurlyeq T'_L$ and $T''_R \preccurlyeq T'_R$. The following result is now easily proved by induction.

THEOREM 10.5. *Let T_1 and T_2 be pruned subtrees of T. Then T_2 is a pruned subtree of T_1 if, and only if, every nonterminal node of T_2 is also a nonterminal node of T_1.*

Let $\#(T)$ temporarily denote the number of pruned subtrees of a tree T. If T is trivial, then $\#(T) = 1$. Otherwise,

$$\#(T) = \#(T_L)\#(T_R) + 1. \tag{10.6}$$

It is obvious from (10.6) that the maximum number of pruned subtrees of a tree having m terminal nodes increases rapidly with m. In particular, let T_n be the tree all of whose terminal nodes have exactly n ancestors, so that $|\widetilde{T}| = 2^n$. It follows from (10.6) that $\#(T_{n+1}) = (\#(T_n))^2 + 1$. Consequently, $\#(T_1) = 2$, $\#(T_2) = 5$, $\#(T_3) = 26$, $\#(T_4) = 677$, and so forth. Now $(\#(T_n))^{1/|\widetilde{T}_n|}$ is easily seen to be increasing in n and to converge rapidly to a number $b \doteq 1.5028368$. It is left as an interesting curiosity for the reader to show that $\#(T_n) = [b^{|\widetilde{T}_n|}]$ for *every* $n \geq 1$, where $[c]$ denotes the largest integer no larger than c. P. Feigin came up with a short elegant proof of this result.

10.2 OPTIMALLY PRUNED SUBTREES

(This material should be read in conjunction with the more intuitive discussion in Section 3.2.)

Let T_0 be a fixed nontrivial tree (for example, the tree $T_{\max}$ in earlier chapters) and let $R(t)$, $t \in T_0$, be fixed real numbers. Given a real number α, set $R_\alpha(t) = R(t) + \alpha$ for $t \in T_0$. Given a subtree T of T_0, set

$$R(T) = \sum_{\widetilde{T}} R(t)$$

and

$$R_\alpha(T) = \sum_{\widetilde{T}} R_\alpha(t) = R(T) + \alpha|\widetilde{T}|.$$

If T is a trivial tree having root t_1, then $R(T) = R(t_1)$ and $R_\alpha(T) = R_\alpha(t_1)$.

A pruned subtree T_1 of T is said to be an *optimally pruned subtree* of T (with respect to α) if

$$R_\alpha(T_1) = \min_{T' \preccurlyeq T} R_\alpha(T').$$

Since there are only finitely many pruned subtrees of T, there is clearly an optimal one, but not necessarily a unique one. An optimally pruned subtree T_1 of T is said to be the *smallest optimally pruned subtree* of T if $T' \succcurlyeq T_1$ for every optimally pruned subtree T' of T. There is clearly at most one smallest optimally pruned subtree of T (with respect to α); when it exists, it is denoted by $T(\alpha)$.

Let T' be a nontrivial pruned subtree of T and let T'_L and T'_R be its two primary branches. Then

$$R_\alpha(T') = R_\alpha(T'_L) + R_\alpha(T'_R).$$

Theorem 10.7 now follows easily by induction. (In this result, $T_L(\alpha)$ and $T_R(\alpha)$ denote the smallest optimally pruned subtrees of T_L and T_R, respectively.)

THEOREM 10.7. *Every tree T has a unique smallest optimally pruned subtree $T(\alpha)$. Let T be a nontrivial tree having root t_1 and primary branches T_L and T_R. Then*

$$R_\alpha(T(\alpha)) = \min[R_\alpha(t_1)), R_\alpha(T_L(\alpha) + R_\alpha(T_R(\alpha))].$$

If $R_\alpha(t_1) \le R_\alpha(T_L(\alpha)) + R_\alpha(T_R(\alpha))$, then $T(\alpha) = \{t_1\}$; otherwise, $T(\alpha) = \{t_1\} \cup T_L(\alpha) \cup T_R(\alpha)$.

The next result follows immediately from the transitivity of $\preccurlyeq$.

THEOREM 10.8. *If $T(\alpha) \preccurlyeq T' \preccurlyeq T$, then $T(\alpha) = T'(\alpha)$.*

Increasing α leads to a greater penalty for complex trees and presumably to the same or smaller $T(\alpha)$. Statement (i) of the next result is to this effect. The validity of the result is highly dependent on the structure of the collection of pruned subtrees; for example, the analogous result for all subsets regression is false.

THEOREM 10.9. (i) *If* $\alpha_2 \geq \alpha_1$, *then* $T(\alpha_2) \preccurlyeq T(\alpha_1)$.
(ii) *If* $\alpha_2 > \alpha_1$ and $T(\alpha_2) \prec T(\alpha_1)$, *then*

$$\alpha_1 < \frac{R(T(\alpha_2)) - R(T(\alpha_1))}{|\widetilde{T}(\alpha_1)| - |T(\alpha_2)|} \leq \alpha_2.$$

PROOF. It is easily seen that if $T(\alpha_1)$ is trivial, then $T(\alpha_2)$ is trivial for $\alpha_2 \geq \alpha_1$. Statement (i) now follows from Theorem 10.7 by induction. If $\alpha_2 > \alpha_1$ and $T(\alpha_2) \prec T(\alpha_1)$, then

$$R(T(\alpha_2)) + \alpha_2|\widetilde{T}(\alpha_2)| \leq R(T(\alpha_1)) + \alpha_2|\widetilde{T}(\alpha_1)|$$

and

$$R(T(\alpha_1)) + \alpha_1|\widetilde{T}(\alpha_1)| < R(T(\alpha_2)) + \alpha_1|\widetilde{T}(\alpha_2)|,$$

which together yield the conclusion to (ii).

THEOREM 10.10. *If* $R_\alpha(t) \geq R_\alpha(T_t)$ *for all* $t \in T - \widetilde{T}$, *then* $R_\alpha(T(\alpha)) = R_\alpha(T)$ *and*

$$T(\alpha) = \{t \in T : R_\alpha(s) > R_\alpha(T_s) \textit{ for all ancestors } s \textit{ of } t\}.$$

PROOF. The theorem will be proved by induction. It is clearly true when $|\widetilde{T}| = 1$. Suppose it is true for all trees having fewer than n terminal nodes, where $n \geq 2$. Let T be a tree having n terminal nodes, root t_1, and primary branches T_L and T_R. By hypothesis,

$$T_L(\alpha) = \{t \in T_L : R_\alpha(s) > R_\alpha(T_s) \text{ for all ancestors } s \in T_L \text{ of } t\}$$

and

$$T_R(\alpha) = \{t \in T_R : R_\alpha(s) > R_\alpha(T_s) \text{ for all ancestors } s \in T_R \text{ of } t\}; \text{ also}$$

$$R_\alpha(T_L(\alpha)) = R_\alpha(T_L) \quad \text{and} \quad R_\alpha(T_R(\alpha)) = R_\alpha(T_R).$$

Consequently,

$$R_\alpha(T) = R_\alpha(T_L) + R_\alpha(T_R) = R_\alpha(T_L(\alpha)) + R_\alpha(T_R(\alpha)).$$

Thus, it follows from Theorem 10.7 that if $R_\alpha(t_1) = R_\alpha(T)$, then $T(\alpha) = \{t_1\}$; and if $R_\alpha(t_1) > R_\alpha(T)$, then $T(\alpha) = \{t_1\} \cup T_L(\alpha) \cup T_R(\alpha)$ and $R_\alpha(T(\alpha)) = R_\alpha(T_L(\alpha)) + R_\alpha(T_R(\alpha))$. In either case the conclusion of the theorem is satisfied. Therefore, the theorem is valid by induction.

Given a nontrivial tree T, set

$$g(t, T) = \frac{R(t) - R(T_t)}{|\widetilde{T}_t| - 1} \quad \text{for } t \in T - \widetilde{T}.$$

It is easily seen that for each $t \in T - \widetilde{T}$ and real number α, the following two statements are valid:

(i) $g(t, T) \geq \alpha$ if, and only if, $R_\alpha(t) \geq R_\alpha(T_t)$,
(ii) $g(t, T) > \alpha$ if, and only if, $R_\alpha(t) > R_\alpha(T_t)$.

THEOREM 10.11. *Given a nontrivial tree T, set $\alpha_1 = \min_{t \in T - \widetilde{T}} g(t, T)$. Then T is the unique optimally pruned subtree of itself wrt α for $\alpha < \alpha_1$; T is an optimally pruned subtree of itself wrt α_1, but not the smallest one; and T is not an optimally pruned subtree of itself wrt α for $\alpha > \alpha_1$. Set $T_1 = T(\alpha_1)$. Then*

$$T_1 = \{t \in T : g(s, T) > \alpha_1 \text{ for all ancestors } s \text{ of } t\}. \tag{10.12}$$

Let $t \in T_1 - \widetilde{T}_1$. Then

$$\begin{aligned} g(t, T_1) &> g(t, T) \quad \text{if } T_{1t} \prec T_t, \\ &= g(t, T) \quad \text{otherwise.} \end{aligned} \tag{10.13}$$

PROOF. It follows immediately from Theorem 10.10 that T is the unique optimally pruned subtree of itself wrt α for $\alpha < \alpha_1$; that T is an optimally pruned subtree of itself wrt α_1, but not the smallest one; and that (10.12) holds. In particular,

$$R(T_1) + \alpha_1|\widetilde{T}_1| = R(T) + \alpha_1|\widetilde{T}|,$$

but $|\widetilde{T}_1| < |\widetilde{T}|$. Let $\alpha > \alpha_1$. Then

$$R(T_1) - R(T) = \alpha_1(|\widetilde{T}| - |\widetilde{T}_1|) < \alpha(|\widetilde{T}| - |\widetilde{T}_1|) \quad \text{and hence}$$

$$R_\alpha(T(\alpha)) \leq R_\alpha(T_1) < R_\alpha(T).$$

Consequently, T is not an optimally pruned subtree of itself wrt α. Let $t \in T_1 - \widetilde{T}_1$. If $T_{1t} = T_t$, then $g(t, T_1) = g(t, T)$ by definition. Suppose $T_{1t} < T_t$. By Theorem 10.10, T_{1t} is the smallest optimally pruned subtree of T_t wrt α_1. Choose $\alpha_2 > \alpha_1$ so that $\{t\}$ is the smallest optimally pruned subtree of T_t wrt α_2. Since T_t is the unique optimally pruned subtree of itself wrt α for $\alpha < \alpha_1$, it follows from Theorem 10.9 (ii) that

$$\frac{R(t) - R(T_{1t})}{|\widetilde{T}_{1t}| - 1} > \alpha_1 \geq \frac{R(T_{1t}) - R(T_t)}{|\widetilde{T}_t| - |\widetilde{T}_{1t}|}.$$

Consequently,

$$\begin{aligned} R(t) - R(T_t) &= R(t) - R(T_{1t}) + R(T_{1t}) - R(T_t) \\ &< (R(t) - R(T_{1t}))\left(1 + \frac{|\widetilde{T}_t| - |T_{1t}|}{|\widetilde{T}_{1t}| - 1}\right) \\ &= (R(t) - R(T_{1t}))\left(\frac{|\widetilde{T}_t| - 1}{|\widetilde{T}_{1t}| - 1}\right) \end{aligned}$$

and hence

$$g(t, T_1) = \frac{R(t) - R(T_{1t})}{|\widetilde{T}_{1t}| - 1} > \frac{R(t) - R(T_t)}{|\widetilde{T}_t| - 1} = g(t, T).$$

Therefore, (10.13) holds and the proof of the theorem is complete.

Recall that T_0 is the nontrivial tree mentioned at the beginning of this section. Set $\alpha_1 = \min_{t \in T_0 - \widetilde{T}_0} g(t, T_0)$ and

$$T_1 = \{t \in T_0 : g(s, T_0) > \alpha_1 \text{ for all ancestors } s \text{ of } t\}. \quad (10.14)$$

Then $T_1 \prec T_0$. By Theorem 10.11, $T_0(\alpha) = T_0$ for $\alpha < \alpha_1$ and $T_0(\alpha_1) = T_1$. If T_1 is the trivial tree, then $T_0(\alpha) = T_1$ for $\alpha \geq \alpha_1$ by Theorem 10.9(i).

Suppose instead that T_1 is nontrivial. Set $\alpha_2 = \min_{t \in T_1 - \widetilde{T}_1} g(t, T_1)$ and

$$T_2 = \{t \in T_1 : g(s, T_1) > \alpha_2 \text{ for all ancestors } s \text{ of } t\}.$$

Then $T_2 \prec T_1$, and $\alpha_2 > \alpha_1$ by (10.13) and (10.14). It follows from Theorem 10.11 that $T_1(\alpha) = T_1$ for $\alpha < \alpha_2$ and $T_1(\alpha_2) = T_2$. If $\alpha_1 \leq \alpha < \alpha_2$, then $T_0(\alpha) \preccurlyeq T_0(\alpha_1) = T_1 \preccurlyeq T_0$ by Theorem 10.9(i) and hence $T_0(\alpha) = T_1(\alpha) = T_1$ by Theorem 10.8. Similarly, $T_0(\alpha_2) \preccurlyeq T_0(\alpha_1) = T_1 \preccurlyeq T_0$ and hence $T_0(\alpha_2) = T_1(\alpha_2) = T_2$. If T_2 is trivial, then $T_0(\alpha) = T_2$ for $\alpha \geq \alpha_2$.

Otherwise, the preceding process can be repeated. Indeed, it can be repeated until a trivial tree is obtained. Thus, there is a positive integer K, real numbers α_k, $1 \leq k \leq K$, and trees T_k, $1 \leq k \leq K$, such that

$$-\infty < \alpha_1 < \cdots < \alpha_K < \infty;$$

$$T_0 \succ T_1 \cdots \succ T_K = \{\text{root } (T_0)\};$$

$$\alpha_{k+1} = \min_{t \in T_k - \widetilde{T}_k} g(t, T_k), \quad 0 \leq k < K; \tag{10.15}$$

$$T_{k+1} = \{t \in T_k : g(s, T_k) > \alpha_{k+1} \text{ for all ancestors } s \text{ of } t\}, \quad 0 \leq k < K; \tag{10.16}$$

and

$$T_0(\alpha) = \begin{cases} T_0, & \alpha < \alpha_1, \\ T_k, & 1 \leq k < K \text{ and } \alpha_k \leq \alpha < \alpha_{k+1}, \\ T_K, & \alpha \geq \alpha_K. \end{cases} \tag{10.17}$$

By definition,

$$g(t, T_k) = \frac{R(t) - R(T_{kt})}{|\widetilde{T}_{kt}| - 1}, \quad t \in T_k - \widetilde{T}_k. \tag{10.18}$$

Formulas (10.15), (10.16), and (10.18) together yield an algorithm for determining K, the α_k's, and the T_k's and hence for determining $T_0(\alpha)$, $-\infty < \alpha < \infty$, from (10.17).

Let $0 \leq k < K$. Then T_k is an optimally pruned subtree of itself wrt α_{k+1}, but $T_k(\alpha_{k+1}) = T_{k+1}$; in particular, $R_{\alpha_{k+1}}(T_k) =$

$R_{\alpha_{k+1}}(T_{k+1})$. It follows by elementary algebra that

$$\alpha_{k+1} = \frac{R(T_{k+1}) - R(T_k)}{|\widetilde{T}_k| - |\widetilde{T}_{k+1}|}, \quad 0 \leq k < K. \tag{10.19}$$

Formula (10.23) in the next result is more convenient for determining $T_0(\alpha)$, $-\infty < \alpha < \infty$, than is (10.17).

THEOREM 10.20. *Let* $g_k(t)$, $t \in T_0 - \widetilde{T}_0$ *and* $0 \leq k \leq K - 1$, *be defined recursively by*

$$g_0(t) = g(t, T_0), \quad t \in T_0 - \widetilde{T}_0, \tag{10.21}$$

and, for $1 \leq k \leq K - 1$,

$$g_k(t) = \begin{cases} g(t, T_k), & t \in T_k - \widetilde{T}_k \\ g_{k-1}(t) & \text{otherwise.} \end{cases} \tag{10.22}$$

Then for $-\infty < \alpha < \infty$,

$$T_0(\alpha) = \{t \in T_0 : g_{K-1}(s) > \alpha \text{ for all ancestors } s \text{ of } t\}. \tag{10.23}$$

PROOF. It suffices to show that if $0 \leq k \leq K - 1$ and $\alpha \leq \alpha_{k+1}$, then

$$T_0(\alpha) = \{t \in T_0 : g_k(s) > \alpha \text{ for all ancestors } s \text{ of } t\} \tag{10.24}$$

(for then (10.23) holds for $\alpha \leq \alpha_K$; since $T_0(\alpha_K)$ is trivial and $T_0(\alpha) \preccurlyeq T_0(\alpha_K)$ for $\alpha \geq \alpha_K$ by Theorem 10.9(i), (10.23) holds for $-\infty < \alpha < \infty$).

It follows from Theorem 10.11 that (10.24) is valid for $k = 0$ and $\alpha \leq \alpha_1$. Suppose now that $1 \leq k \leq K - 1$ and

$$T_0(\alpha) = \{t \in T_0 : g_{k-1}(s) > \alpha \text{ for all ancestors } s \text{ of } t\}, \quad \alpha \leq \alpha_k. \tag{10.25}$$

Then, in particular

$$T_k = T_0(\alpha_k) = \{t \in T_0 : g_{k-1}(s) > \alpha_k \text{ for all ancestors } s \text{ of } t\} \tag{10.26}$$

and hence

$$g_{k-1}(s) > \alpha_k, \quad s \in T_k - \widetilde{T}_k. \tag{10.27}$$

Now $T_k - \widetilde{T}_k \subset T_{k-1} - \widetilde{T}_{k-1}$, so by (10.13) and (10.22),

$$g_k(s) = \begin{cases} g(s, T_k) \geq g_{k-1}(s) & \text{for} \quad s \in T_k - \widetilde{T}_k \\ g_{k-1}(s) \text{ otherwise;} \end{cases} \tag{10.28}$$

thus it follows from (10.27) that for $s \in T_0 - \widetilde{T}_0$ and $\alpha \leq \alpha_k$, $g_k(s) > \alpha$ if, and only if, $g_{k-1}(s) > \alpha$. Hence (10.25) implies (10.24) for $\alpha \leq \alpha_k$. Suppose now that $\alpha_k < \alpha \leq \alpha_{k+1}$. Then

$$T_0(\alpha) = T_k(\alpha) = \{t \in T_k : g(s, T_k) > \alpha \text{ for all ancestors } s \text{ of } t\},$$

so it follows from (10.22) that

$$T_0(\alpha) = \{t \in T_k : g_k(s) > \alpha \text{ for all ancestors } s \text{ of } t\}. \tag{10.29}$$

Suppose $t \notin T_k$. By (10.26) there is an ancestor s of t such that $g_{k-1}(s) \leq \alpha_k$; $s \notin T_k - \widetilde{T}_k$ by (10.27), so $g_k(s) = g_{k-1}(s) \leq \alpha_k < \alpha$ by (10.28). Thus, it follows from (10.29) that (10.24) holds for $\alpha_k < \alpha \leq \alpha_{k+1}$ and hence for $\alpha \leq \alpha_{k+1}$. By induction, (10.24) holds for the indicated range of k and α. This completes the proof of the theorem.

The next theorem clearly follows from the previous one; for an application, see Algorithm 11.1.

THEOREM 10.30. *Let $t \in \widetilde{T}_0$ and let $-\infty < \alpha < \infty$. If $g_{K-1}(s) > \alpha$ for all ancestors s of t, then $t \in \widetilde{T}_0(\alpha)$. Otherwise, let s be the first node of T_0 along the path from the root of T_0 to t for which $g_{K-1}(s) \leq \alpha$. Then s is the unique ancestor of t in $\widetilde{T}_0(\alpha)$.*

Recall that if s is an ancestor of t, then $\ell(s, t)$ is the length of the path from s to t. Given $-\infty < \alpha < \infty$ and $t \in T_0$, set

$$S_\alpha(t) = \min[R(s) - \alpha(\ell(s, t) + 1) : s = t \text{ or } s \text{ is an ancestor of } t].$$

THEOREM 10.31. *Suppose that* $R(t) \geq 0$ *for all* $t \in T_0$. *Then* $S_\alpha(t) > 0$ *for* $-\infty < \alpha < \infty$ *and every nonterminal node* t *of* $T_0(\alpha)$.

PROOF. Let t be a nonterminal node of $T_0(\alpha)$ and let s be t or an ancestor of t. Now $T_0(\alpha)$ is the unique optimally pruned subtree of itself wrt α; so

$$\frac{R(s) - R(T_{0s}(\alpha))}{|\widetilde{T}_{0s}(\alpha)| - 1} > \alpha$$

by Theorem 10.10 and hence $R(s) > \alpha(|\widetilde{T}_{0s}(\alpha)| - 1)$. It is easily shown by induction that $|\widetilde{T}_{0s}(\alpha)| \geq \ell(s, t) + 2$ and therefore that $R(s) > \alpha(\ell(s, t) + 1)$. This yields the desired conclusion.

THEOREM 10.32. *Suppose that* $R(t) \geq 0$ *for all* $t \in T_0$. *Given a real number* α, *set*

$$T_{\text{suff}}(\alpha) = \{\mu \in T_0 : S_\alpha(t) > 0 \text{ for all ancestors } t \text{ of } u\}.$$

Then $T_0(\alpha) \preccurlyeq T_{\text{suff}}(\alpha) \preccurlyeq T_0$.

PROOF. It is easily seen that $T_{\text{suff}}(\alpha)$ is a subtree of T_0, and it clearly contains the root of T_0. Thus, $T_{\text{suff}}(\alpha) \preccurlyeq T_0$. Let v be a nonterminal node of $T_0(\alpha)$. Then $S_\alpha(v) > 0$ by Theorem 10.31. Moreover, if t is an ancestor of v, it is a nonterminal node of $T_0(\alpha)$; so $S_\alpha(t) > 0$ by the same theorem. Consequently, v is a nonterminal node of $T_{\text{suff}}(\alpha)$. In other words, every nontrivial terminal node of $T_0(\alpha)$ is also a nonterminal node of $T_{\text{suff}}(\alpha)$. Therefore, $T_0(\alpha) \preccurlyeq T_{\text{suff}}(\alpha)$ by Theorem 10.5. This completes the proof of the theorem.

The function $S_\alpha(t)$, $t \in T_0$, is easily computed recursively, since $S_\alpha(\text{root}(T_0)) = R(\text{root}(T_0)) - \alpha$ and $S_\alpha(t) = \min[R(t) - \alpha, S_\alpha(\text{parent}(t)) - \alpha]$ for $t \in T_0$ and $t \neq \text{root}(T_0)$.

Suppose that an initial tree $T_0 = T_{\max}$ is to be grown according to a splitting process, as described in earlier chapters; that $R(t) \geq 0$ for all $t \in T_0$; and that it is desired to find $T_0(\alpha)$ only for $\alpha \geq \alpha_{\min}$, where $\alpha_{\min}$ is a real constant. Then by Theorems 10.8, 10.9, and 10.32, it is not necessary to split any node t such that $S_{\alpha_{\min}}(t) \leq 0$.

10.3 AN EXPLICIT OPTIMAL PRUNING ALGORITHM

Consider an initial tree $T_0 = \{1, \ldots, m\}$, where m is specified, as are $\alpha_{\min}$ and the quantities $\ell(t) = \text{left}(t)$, $r(t) = \text{right}(t)$, and $R(t)$ for $1 \leq t \leq m$. Algorithm 10.1 can be used to find $T_0(\alpha)$ for $\alpha \geq \alpha_{\min}$. In this algorithm, "$k := 1$" means "set k equal to 1"; ∞ is to be interpreted as a large positive number; $N(t) = |\widetilde{T}_{kt}|$; $S(t) = R(T_{kt})$; $g(t) = g_k(t)$, and $G(t) = \min[g_k(s) : s \in T_{kt} - \widetilde{T}_{kt}]$. The statements following "repeat" are to be cycled through until the condition that $N(1) = 1$ (that is, that T_k is trivial) is checked and satisfied; at this point the algorithm is finished, $k = K$, and $g(t) = g_{K-1}(t)$ for $t \in T_0 - \widetilde{T}_0$. The write statement writes out k, $|\widetilde{T}_k|$, α_k, and $R(T_k)$ for $1 \leq k \leq K$. The small positive number ε is included to prevent computer round-off error from generating extraneous trees T_k with $\alpha_k \doteq \alpha_{k-1}$. In the intended statistical applications, ε can be chosen to be a small, positive constant times $R(1)$.

The algorithm was applied with T_0 being the tree T_7 having 10 terminal nodes that arose in the discussion of the stochastic digit recognition problem in Section 3.5.1; $R(t) = \hat{R}(t)$ is also taken from that problem; and $\alpha_{\min} = 0$. Table 10.2 shows the output of the write statement, while Table 10.3 shows the values of $g(t) = g_6(t)$ after the algorithm is finished.

ALGORITHM 10.1

```
k := 1
α := α_min
for t := m to 1 in increments of -1

  ⎧               ⎧ N(t) := 1
  ⎪  if ℓ(t) = 0  ⎨ S(t) := R(t)
  ⎪               ⎩ G(t) := ∞
  ⎪
  ⎨               ⎧ p(ℓ(t)) := t
  ⎪               ⎪ p(r(t)) := t
  ⎪  else         ⎨ N(t) := N(ℓ(t)) + N(r(t))
  ⎪               ⎪ S(t) := S(ℓ(t)) + S(r(t))
  ⎪               ⎪ g(t) := (R(t) - S(t))/(N(t) - 1)
  ⎩               ⎩ G(t) := min(g(t), G(ℓ(t)), G(r(t)))

repeat
  ⎧                        ⎧ write k, N(1), α, S(1)
  ⎪  if G(1) > α + ε       ⎨ α = G(1)
  ⎪                        ⎩ k = k + 1
  ⎪
  ⎪  if N(1) = 1 return
  ⎪  t = 1
  ⎪                        ⎧ if G(t) = G(ℓ(t))   t := 1(t)
  ⎪  while G(t) < g(t) - ε ⎨
  ⎨                        ⎩ else                t := r(t)
  ⎪  N(t) := 1
  ⎪  S(t) := R(t)
  ⎪  G(t) := ∞
  ⎪                  ⎧     t := p(t)
  ⎪                  ⎪ N(t) := N(ℓ(t)) + N(r(t))
  ⎪  while t > 1     ⎨ S(t) := S(ℓ(t)) + S(r(t))
  ⎪                  ⎪ g(t) := (R(t) - S(t))/(N(t) - 1)
  ⎩                  ⎩ G(t) := min(g(t), G(ℓ(t)), G(r(t)))
```

TABLE 10.2

k	$\lvert\widetilde{T}_k\rvert$	α_k	$R(T_k)$
1	10	.000	.285
2	9	.035	.320
3	7	.045	.410
4	6	.050	.460
5	5	.065	.525
6	2	.075	.750
7	1	.110	.860

TABLE 10.3

t	$\ell(t)$	$r(t)$	$p(t)$	$R(t)$	$g_6(t)$
1	2	3	0	.860	.110
2	4	5	1	.390	.075
3	6	7	1	.360	.075
4	8	9	2	.100	.065
5	10	11	2	.215	.045
6	0	0	3	.070	
7	12	13	3	.215	.075
8	0	0	4	.005	
9	0	0	4	.030	
10	0	0	5	.065	
11	14	15	5	.105	.045
12	0	0	7	.025	
13	16	17	7	.115	.050
14	0	0	11	.010	
15	18	19	11	.050	.035
16	0	0	13	.015	
17	0	0	13	.050	
18	0	0	15	.010	
19	0	0	15	.005	

Figure 10.2 shows the tree $T_0 = T_1$ with $g_6(t)$ written under each nonterminal node of T_0. By (10.23),

$$T_k = \{t \in T_0 : g_6(s) > \alpha_k \text{ for all ancestors } s \text{ of } t\}.$$

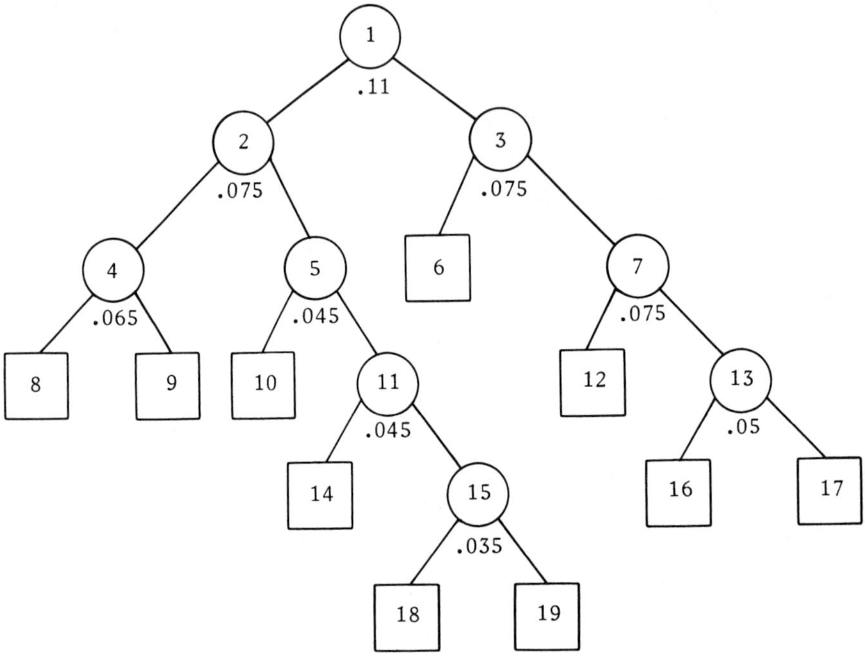

FIGURE 10.2

Thus, in going from T_1 to T_2, nodes 18 and 19 are removed; in going from T_2 to T_3, nodes 10, 11, 14, and 15 are removed; in going from T_3 to T_4, nodes 16 and 17 are removed; in going from T_4 to T_5, nodes 8 and 9 are removed; in going from T_5 to T_6, nodes 4, 5, 6, 7, 12, and 13 are removed; and in going from T_6 to T_7, nodes 2 and 3 are removed. According to (10.17), $T_0(\alpha) = T_k$ for $1 \leq k \leq 6$ and $\alpha_k \leq \alpha < \alpha_{k+1}$, while $T_0(\alpha) = T_7 = \{1\}$ for $\alpha \geq \alpha_7$. Figure 10.2 can also be used to illustrate Theorem 10.30.

In this particular example, $g_{K-1}(t) \leq g_{K-1}(s)$ whenever s, t are nonterminal nodes of T_0 and t is a descendant of s. But such a result is not true in general.

11

CONSTRUCTION OF TREES FROM A LEARNING SAMPLE

In Chapter 9, and implicitly in Chapter 10 as well, the joint distribution of $(\mathbf{X}, Y)$ was assumed to be known. In the first three sections of the present chapter, the methods described in Chapters 9 and 10 are modified and combined to yield tree construction procedures based on a learning sample. In Sections 11.4 and 11.5, methods based on test samples and cross-validation are developed for obtaining nearly unbiased estimates of the overall risk of a tree structured rule. Section 11.6 treats the selection of a particular optimally pruned subtree from among the K candidates. In Section 11.7, an alternative method based on the bootstrap is considered for eliminating the overoptimism of the resubstitution estimate of the overall risk; it is shown, however, that there exist situations in which this alternative method is defective. Section 11.8 treats the tendency of splitting rules based on a learning sample to prefer end-cut splits—that is, splits in which the estimated probability of going to the left is close to zero or one.

11.1 ESTIMATED BAYES RULE FOR A PARTITION

Recall the terminology from Sections 9.1 and 9.2. Let $(\mathbf{X}_n, Y_n)$, $n \in \eta = \{1, \ldots, N\}$, be a random (learning) sample of size N from the joint distribution of $(\mathbf{X}, Y)$. Let $\widetilde{T}$ be a partition of X, which may depend on the learning sample. Given $t \in \widetilde{T}$, set $\eta(t) = \{n \in \eta : \mathbf{X}_n \in t\}$, $N(t) = |\eta(t)|$, and estimate $P(t) = P(\mathbf{X} \in t)$ by $p(t) = N(t)/N$. Suppose that $p(t) > 0$ for all $t \in \widetilde{T}$.

Consider the estimate $d_{\widetilde{T}}$ of the Bayes rule $d_{\widetilde{T}}$ for the partition $\widetilde{T}$ that has the form $d_{\widetilde{T}}(\mathbf{x}) = \nu(\tau(\mathbf{x}))$, $\mathbf{x} \in X$, where τ is the partition function corresponding to $\widetilde{T}$ and $\nu(t)$ is defined separately for regression, classification, and class probability estimation.

In the regression problem, let

$$\bar{y}(t) = \frac{1}{N(t)} \sum_{\eta(t)} Y_n \quad \text{and} \quad s^2(t) = \frac{1}{N(t)} \sum_{\eta(t)} (Y_n - \bar{y}(t))^2$$

denote the sample mean and sample variance, respectively, of the numbers Y_n, $n \in \eta(t)$. Set $\nu(t) = \bar{y}(t)$.

In classification and class probability estimation, there are two models to consider: one model when the prior probabilities $\pi(j)$, $1 \leq j \leq J$, are unknown and estimated from the learning sample and another model when these prior probabilities take on known (or assumed) values.

Consider first the model 1 version of classification or class probability estimation in which the prior probabilities must be estimated from the learning sample. Here the random variable Y ranges over $\{1, \ldots, J\}$. For $1 \leq j \leq J$, set $\eta_j(t) = \{\eta \in \eta(t) : Y_n = j\}$, $N_j(t) = |\eta_j(t)|$, and $p(j|t) = N_j(t)/N(t)$. In the classification problem let $\nu(t)$ be the smallest value of $i \in \{1, \ldots, J\}$ that minimizes $\Sigma_j\, C(i|j)p(j|t)$. In the class probability estimation problem let $\nu(t)$ denote the vector $(p(1|t), \ldots, p(J|t))$ of estimated conditional probabilities, $p(j|t)$ being an estimate of $P(j|t) = P(Y = j|\mathbf{X} \in t)$.

Recall from Section 9.1 the definition of $L(y, a)$ in these three problems. The resubstitution estimate $R(d_{\tilde{T}})$ of the risk $R^*(d_{\tilde{T}})$ of the rule $d_{\tilde{T}}$ is defined by

$$\begin{aligned} R(d_{\tilde{T}}) &= \frac{1}{N} \sum_{\eta} L(Y_n, d_{\tilde{T}}(\mathbf{X}_n)) \\ &= \sum_{\tilde{T}} p(t) \frac{1}{N(t)} \sum_{\eta(t)} L(Y_n, \nu(t)) \\ &= \sum_{\tilde{T}} p(t) r(t) \\ &= \sum_{\tilde{T}} R(t), \end{aligned}$$

where now

$$r(t) = \frac{1}{N(t)} \sum_{\eta(t)} L(Y_n, \nu(t))$$

and $R(t) = p(t)r(t)$. In the regression problem, $r(t) = s^2(t)$; in the model 1 version of classification, $r(t) = \min_i \sum_j C(i|j)p(j|t)$; and in the model 1 version of class probability estimation, $r(t) = \sum_j p(j|t)(1 - p(j|t))$.

Consider next the model 2 version of classification or class probability estimation, in which $\pi(j)$, $1 \leq j \leq J$, are known. For $1 \leq j \leq J$, let $\mathbf{X}_n$, $n \in \eta_j$, be a random sample of size $N_j = |\eta_j|$ from the conditional distribution of $\mathbf{X}$, given that $Y = j$. (The sets η_j, $1 \leq j \leq J$, are taken to be disjoint and nonempty.) The learning sample is now given by $(\mathbf{X}_n, Y_n)$, $n \in \eta$, where $\eta = \cup_j \eta_j$ has $N = \Sigma_j N_j$ members and $Y_n = j$ for $n \in \eta_j$.

Given $t \in \tilde{T}$, set $\eta_j(t) = \{n \in \eta_j : \mathbf{X}_n \in t\}$, $N_j(t) = |\eta_j(t)|$, and

$$p(t) = \sum_j \frac{\pi(j) N_j(t)}{N_j}.$$

Suppose that $p(t) > 0$ for $t \in \tilde{T}$. Set

$$p(j|t) = \frac{\pi(j)N_j(t)/N_j}{\Sigma_j[\pi(j)N_j(t)/N_j]}.$$

In this alternative setup the estimated Bayes rule $d_{\tilde{T}}$ is again of the form $d_{\tilde{T}}(\mathbf{x}) = \nu(\tau(\mathbf{x}))$, $\mathbf{x} \in X$, where $\nu(t)$, $t \in \tilde{T}$, is now defined as follows. In the classification problem, $\nu(t)$ is the smallest value of i that minimizes

$$\sum_j C(i|j)p(j|t) = \frac{\Sigma_j C(i|j)\pi(j)N_j(t)/N_j}{\Sigma_j \pi(j)N_j(t)/N_j};$$

equivalently, $\nu(t)$ is the smallest value of i that minimizes $\Sigma_j C(i|j)\pi(j)N_j(t)/N_j$. In the class probability estimation problem, $\nu(t) = (p(1|t), \ldots, p(J|t))$, as before.

In either case the resubstitution estimate $R(d_{\tilde{T}})$ or $R^*(d_{\tilde{T}})$ is defined by

$$\begin{aligned} R(d_{\tilde{T}}) &= \sum_j \frac{\pi(j)}{N_j} \sum_{\eta_j} L(Y_n, d_{\tilde{T}}(\mathbf{X}_n)) \\ &= \sum_{\tilde{T}} p(t) \sum_j p(j|t)L(j, \nu(t)) \\ &= \sum_{\tilde{T}} p(t)r(t) \\ &= \sum_{\tilde{T}} R(t), \end{aligned}$$

where $r(t) = \Sigma_j\, p(j|t)L(j, \nu(t))$ and $R(t) = p(t)r(t)$. In the classification problem, $r(t) = \min_i \Sigma_j\, C(i|j)p(j|t)$ and in the class probability estimation problem, $r(t) = \Sigma_j\, p(j|t)(1 - p(j|t))$; these two formulas are the same as in model 1.

11.2 EMPIRICAL RISK REDUCTION SPLITTING RULE

Let a learning sample be available for either the regression problem or the model 1 or model 2 version of classification or class probability estimation. Let $\tilde{T}$ be a partition of X such that $p(t) > 0$ for $t \in \tilde{T}$.

Consider a split s of $t \in \tilde{T}$ into t_L, t_R, where $p(t_L) > 0$ and $p(t_R) > 0$. Set

$$p_L = \frac{p(t_L)}{p(t)} \quad \text{and} \quad p_R = \frac{p(t_R)}{p(t)} = 1 - p_L.$$

Let $\tilde{T}'$ be the modification to T obtained by replacing t by the pair t_L, t_R. The empirical risk reduction $\Delta R(s, t) = R(\tilde{T}) - R(\tilde{T}')$ due to the split is given by

$$\Delta R(s, t) = R(t) - R(t_L) - R(t_R) = p(t)[r(t) - p_L r(t_L) - p_R r(t)].$$

The relative empirical risk reduction

$$\Delta R(s|t) = \frac{R(\tilde{T}) - R(\tilde{T}')}{p(t)}$$

due to the split is given by

$$\Delta R(s|t) = r(t) - p_L r(t_L) - p_R r(t_R).$$

To obtain valid empirical versions of the various formulas for $\Delta R(s|t)$ in Section 9.3, replace P_L, P_R, $\sigma^2(t)$, $\mu(t)$, $P(j|t)$ by p_L, p_R, $s^2(t)$, $\bar{y}(t)$, $p(j|t)$, respectively.

The empirical risk reduction splitting rule is to choose an allowable split s of t that maximizes $\Delta R(s|t)$. This splitting rule is natural in the regression problem where it coincides with the AID splitting rule and in the class probability estimation problem where it coincides with the Gini splitting rule. It is somewhat unnatural in the classification problem, for reasons discussed in Section 4.1. Generally it seems preferable to use the Gini or twoing splitting rule in the classification problem.

11.3 OPTIMAL PRUNING

Consider a tree to be constructed from a learning sample for the purpose of regression, classification, or class probability estimation. Each node t of the tree is identified with a subset of X, and if $t_L = \text{left}(t)$ and $t_R = \text{right}(t)$, then t_L, t_R is a partition of t. Let a fixed splitting rule (for example, empirical risk reduction) be employed.

Choose $\alpha_{\min} \geq 0$. Do not consider splitting node t if $S_{\alpha_{\min}}(t) \leq 0$ (see Theorem 10.32), where $S_\alpha(t)$ is defined as in Section 10.2. Additional conditions for not splitting a node t could be used—for example, do not split t if $p(t) < \varepsilon_1$ (ε_1 being a fixed positive number); or do not split a node having more than, say, 10 ancestors. Similarly, conditions for not considering an otherwise allowable split of t into t_L, t_R might be employed—for example, do not consider the split if $p(t_L) < \varepsilon_2$ or $p(t_R) < \varepsilon_2$.

Let T_0 be the initial tree obtained by continuing the splitting process until it comes to a halt by considerations such as those in the previous paragraph. (This was denoted by $T_{\max}$ in Chapter 3.) Given a pruned subtree T of T_0, set $R(T) = \Sigma_{\tilde{T}} R(t)$. The optimal pruning algorithm applied to T_0, $\alpha_{\min}$, and $R(t)$ for $t \in T_0$ yields the following: a positive integer K; a strictly increasing sequence α_k, $1 \leq k \leq K$, of numbers such that $\alpha_1 = \alpha_{\min}$; trees T_k, $1 \leq k \leq K$, such that $T_0 \succeq T_1$, $T_k \succ T_{k+1}$ for $1 \leq k < K$, and $T_K = \{\text{root}(T_0)\} = \{X\}$. Necessarily,

$$\alpha_k = \frac{R(T_k) - R(T_{k-1})}{|\tilde{T}_{k-1}| - |\tilde{T}_k|} \quad \text{for} \quad 2 \leq k \leq K.$$

Let $1 \leq k < K$ and $\alpha_k \leq \alpha < \alpha_{k+1}$ ($\alpha_K \leq \alpha < \infty$ if $k = K$). Then $T_0(\alpha) = T_k$. That is, T_k minimizes $R(T) + \alpha|\tilde{T}|$ among all pruned subtrees of T_0; and if T' also minimizes this expression, then $T' \succeq T_k$. Let $g(t) = g_{K-1}(t)$ be determined by the optimal pruning algorithm.

For $1 \leq k \leq K$, let τ_k be the partition function corresponding to $\tilde{T}_k$ and let $d_k = d_{T_k}$ be the estimated Bayes rule corresponding to this partition. Then $d_k(\mathbf{x}) = \nu(\tau_k(\mathbf{x}))$ for $\mathbf{x} \in X$, where $\nu(t)$ is defined in Section 11.1, and $R(d_k) = R(T_k)$. Given $\mathbf{x} \in X$, the quantities $d_k(\mathbf{x})$, $1 \leq k \leq K$, can easily be determined together according to Algorithm 11.1 (see Theorem 10.30). In the description of this algorithm, $t_1 = \text{root}(T_0)$, $\ell(t) = \text{left}(t)$, $r(t) = \text{right}(t)$, and t is a terminal node if, and only if, $\ell(t) = 0$.

ALGORITHM 11.1

$t = t_1$

for $k := K$ to 1 in increments of -1

$$\begin{cases} 1 \text{ if } \ell(t) \neq 0 \text{ then} \\ \qquad \text{if } g(t) > \alpha_k \text{ then} \begin{cases} \text{if } \mathbf{x} \in \ell(t) \quad t := \ell(t) \\ \text{else } t := r(t) \\ \text{go to } 1 \end{cases} \\ d_k(\mathbf{x}) = \nu(t) \end{cases}$$

11.4 TEST SAMPLES

Let $1 \leq k \leq K$. An obvious way to cure the overoptimistic tendency of the empirical estimate $R(d_k)$ of $R^*(d_k)$ is to base the estimate of $R^*(d_k)$ on data not used to construct d_k.

In regression or the model 1 version of classification or class probability estimation, let $(\mathbf{X}_n, Y_n)$, $n \in \eta'$, be a random (test) sample of size $N' = |\eta'|$ from the joint distribution of $(\mathbf{X}, Y)$, where η' is disjoint from η. The estimate $R^{ts}(d_k)$ of $R^*(d_k)$ corresponding to the test sample is defined by

$$R^{ts}(d_k) = \frac{1}{N'} \sum_{\eta'} L(Y_n, d_k(\mathbf{X}_n)).$$

Its standard error can be estimated by

$$\mathrm{SE}(R^{ts}(d_k)) = \sqrt{s^2/N},$$

where s^2 is the sample variance

$$\frac{1}{N'}\sum_{\eta'}[L(Y_n, d_k(\mathbf{X}_n)) - R^{ts}(d_k)]^2.$$

Consider next the model 2 version of classification or class probability estimation. For $1 \leq j \leq J$, let $\mathbf{X}_n$, $n \in \eta'_j$, be a random sample of size $N'_j = |\eta'_j|$ from the conditional distribution of $\mathbf{X}$, given that $Y = J$ (the $2J$ sets η_j, $1 \leq j \leq J$, and η'_j, $1 \leq j \leq J$, are taken to be disjoint). The test sample is now given by $(\mathbf{X}_n, Y_n)$, $n \in \eta'$, where $\eta' = \cup_j \eta'_j$ has $N' = \Sigma_j N'_j$ members and $Y_n = j$ for $n \in n'_j$. The corresponding estimate $R^{ts}(d_k)$ of $R^*(d_k)$ is defined by

$$R^{ts}(d_k) = \sum_j \frac{\pi(j)}{N'_j}\sum_{\eta'_j} L(j, d_k(\mathbf{X}_n)).$$

Its standard error can be estimated by

$$\mathrm{SE}(R^{ts}(d_k)) = \left[\sum_j\left(\frac{\pi(j)s_j}{N'_j}\right)^2\right]^{1/2},$$

where s_j^2 is the sample variance of the numbers $L(j, d_k(\mathbf{X}_n))$, $n \in \eta'_j$, so that

$$s_j^2 = \frac{1}{N'_j}\sum_{\eta'_j}[L(j, d_k(\mathbf{X}_n))]^2 - \left[\frac{1}{N'_j}\sum_{\eta'_j} L(j, d_k(\mathbf{X}_n))\right]^2.$$

To evaluate the preceding formulas efficiently, for each $n \in \eta'$ it is necessary to determine $L(Y_n, d_k(\mathbf{X}_n))$, $1 \leq k \leq K$, together. This is easily done along the lines of Algorithm 11.1.

Recall that T_K is the trivial tree. In the model 2 version of the classification problem, $d_K(\mathbf{x}) = \upsilon$ for $\mathbf{x} \in X$, where υ is the smallest value of i that minimizes $\Sigma_j C(i|j)\pi(j)$, and

$$R^*(d_K) = \min_i \sum_j C(i|j)\pi(j).$$

In the model 2 version of class probability estimation, $d_K(\mathbf{x}) = \upsilon$ for all $\mathbf{x} \in X$, where ν is the vector $(\pi(1), \ldots, \pi(J))$ of probabilities, and

$$R^*(d_K) = \sum_j \pi(j)(1 - \pi(j)).$$

So it is not necessary to estimate $R^*(d_K)$ in these two situations.

In the regression context set $\mu = EY$ and temporarily set

$$\bar{Y} = \frac{1}{N'} \sum_{\eta'} Y_n \quad \text{and} \quad s^2 = \frac{1}{N'} \sum_{\eta'} (Y_n - \bar{Y})^2.$$

Recall from Chapter 8 that

$$RE^{ts}(d_k) = \frac{R^{ts}(d_k)}{s^2} = \frac{\frac{1}{N'} \sum_{\eta'} (Y_n - d_k(\mathbf{X}_n))^2}{\frac{1}{N'} \sum_{\eta'} (Y_n - \bar{Y})^2}$$

is an estimate of the relative mean squared error

$$RE^*(d_k) = \frac{R^*(d_k)}{E(Y - \mu)^2}.$$

To compute the standard error of this estimate, note that

$$\frac{1}{N'} \sum_{\eta'} (Y_n - \bar{Y})^2 = \frac{1}{N'} \sum_{\eta'} (Y_n - \mu)^2 - (\bar{Y} - \mu)^2$$

and that $(\bar{Y} - \mu)^2$ is of the order $1/N'$ by the central limit theorem. Therefore, $\bar{Y}$ can be replaced by μ in determining the standard error of $RE^{ts}(d_k)$. To determine this standard error, think of d_k as fixed and rewrite the modified expression for $RE^{ts}(d_k)$ in the form

$$H(U_1, U_2) = \frac{U_1}{U_2} = \frac{\frac{1}{N'} \sum_{\eta'} U_{1n}}{\frac{1}{N'} \sum_{\eta'} U_{2n}}.$$

Here U_1 and U_2 are consistent estimates of $\mu_1 = R^*(d_k)$ and $\mu_2 = E(Y - \mu)^2$, respectively, and $H(U_1, U_2)$ is a consistent estimate of $H(\mu_1, \mu_2) = \mu_1/\mu_2$. The asymptotic variance of $H(U_1, U_2)$ is the same as the variance of

$$\frac{\partial H}{\partial \mu_1}(U_1 - \mu_1) + \frac{\partial H}{\partial \mu_2}(U_2 - \mu_2) = \frac{\mu_1}{\mu_2}\left(\frac{U_1 - \mu_1}{\mu_1} - \frac{U_2 - \mu_2}{\mu_2}\right);$$

namely,

$$\left(\frac{\mu_1}{\mu_2}\right)^2 \left(\frac{\text{Var } U_1}{\mu_1^2} - \frac{2 \text{ Cov}(U_1, U_2)}{\mu_1 \mu_2} + \frac{\text{Var } U_2}{\mu_2^2}\right)$$

$$= \frac{1}{N'}\left(\frac{\mu_1}{\mu_2}\right)^2 \left(\frac{\sigma_1^2}{\mu_1^2} - \frac{2\sigma_{12}}{\mu_1 \mu_2} + \frac{\sigma_2^2}{\mu_2^2}\right),$$

where $\sigma_1^2 = \text{Var } U_{1n}$, $\sigma_2^2 = \text{Var } U_{2n}$, and $\sigma_{12} = \text{Cov}(U_{1n}, U_{2n})$ for $n \in \eta'$. Now σ_1^2, σ_2^2, and σ_{12} can be estimated, respectively, by

$$s_1^2 = \frac{1}{N'} \sum_{\eta'} (Y_n - d_k(\mathbf{X}_n))^4 - (R^{ts}(d_k))^2,$$

$$s_2^2 = \frac{1}{N'} \sum_{\eta'} (Y_n - \bar{Y})^4 - s^4,$$

and

$$s_{12} = \frac{1}{N'} \sum_{\eta'} (Y_n - d_k(\mathbf{X}_n))^2 (Y_n - \bar{Y})^2 - R^{ts}(d_k) s^2.$$

This leads to the formula

$$\text{SE}(R^{ts}(d_k)) = RE^{ts}(d_k) \left[\frac{1}{N'}\left(\frac{s_1^2}{(R^{ts}(d_k))^2} - \frac{2s_{12}}{R^{ts}(d_k)s^2} + \frac{s_2^2}{s^4}\right)\right]^{1/2}$$

for the standard error of $RE^{ts}(d_k)$.

11.5 CROSS-VALIDATION

The use of test samples to estimate the risk of tree structured procedures requires that one set of sample data be used to construct the procedure and a disjoint set be used to evaluate it.

When the combined set of available data contains a thousand or more cases, this is a reasonable approach. But if only a few hundred cases or less in total are available, it can be inefficient in its use of the available data; cross-validation is then preferable.

Let T_k and $d_k = d_{T_k}$, $1 \leq k \leq K$, be obtained by optimal pruning from the entire learning sample $(\mathbf{X}_n, Y_n)$, $n \in \eta$. Then $T_0(\alpha) = T_k$ for $\alpha_k \leq \alpha < \alpha_{k+1}$, where $\alpha_{K+1} = \infty$. Suppose that $\alpha_1 = \alpha_{\min} \geq 0$. Let α_k' denote the geometric mean $(\alpha_k \alpha_{k+1})^{1/2}$ of α_k and α_{k+1} with $\alpha_K' = \infty$.

Choose a positive integer $V \geq 2$. Randomly divide η into V sets η_v, $1 \leq v \leq V$, of nearly equal size. Define $v_n \in \{1, \ldots, V\}$ for $n \in \eta$ by $v_n = v$ if $n \in \eta_v$. For $1 \leq v \leq V$ and $1 \leq k \leq K$, let $T_k^{(v)}(\alpha_k')$ be the optimally pruned subtree with respect to α_k' constructed from the data $(\mathbf{X}_n, Y_n)$, $n \in \eta - \eta_v$ (with $T_K^{(v)}(\alpha_K') = T_K^{(v)}(\infty)$ being the trivial tree), and let $d_k^{(v)}$ be the rule corresponding to the tree $T_k^{(v)}(\alpha_k')$ constructed from the same data. (Observe that $d_k^{(v)}$ depends in a minor way—through α_k'—on the entire learning sample. Observe also that the data $(\mathbf{X}_n, Y_n)$, $n \in \eta_v$, were denoted by $\mathcal{L}_v$ in the first part of the book.)

Let $k \in \{1, \ldots, K\}$ be fixed. In the regression problem or the model 1 version of classification or class probability estimation, the cross-validation estimate of the risk $R(d_k)$ of d_k is defined by

$$R^{CV}(d_k) = \frac{1}{N} \sum_{\eta} L(Y_n, d_k^{(v_n)}(\mathbf{X}_n)).$$

It is not at all clear how to obtain a rigorously valid standard error estimate for $R^{CV}(d_k)$, since the random variables $L(Y_n, d_k^{(v_n)}(\mathbf{X}_n))$ are by no means independent. The heuristic device of simply ignoring this lack of independence yields a formula similar to that obtained for the corresponding test sample estimate and which appears to work reasonably well in practice; specifically,

$$\mathrm{SE}(R^{cv}(d_k)) = \sqrt{s^2/N} ,$$

where

$$s^2 = \frac{1}{N}\sum_{\eta}\left[L(Y_n, d_k^{(v_n)}(\mathbf{X}_n)) - R^{cv}(d_k)\right]^2 .$$

In the model 2 version of classification or class probability estimation, $R^{cv}(d_k)$ is defined by

$$R^{cv}(d_k) = \sum_j \frac{\pi(j)}{N_j}\sum_{\eta_j} L(j, d_k^{(v_n)}(\mathbf{X}_n)).$$

The corresponding heuristic standard error estimate is given by

$$\mathrm{SE}(R^{cv}(d_k)) = \left[\sum_j \left(\frac{\pi(j)s_j}{N_j}\right)^2\right]^{1/2} ,$$

where s_j^2 is the sample variance of the numbers $L(j, d_k^{(v_n)}(\mathbf{X}_n))$, $n \in \eta_j$, so that

$$s_j^2 = \frac{1}{N_j}\sum_{\eta_j}\left[L(j, d_k^{(v_n)}(\mathbf{X}_n))\right]^2 - \left[\frac{1}{N_j}\sum_{\eta_j} L(j, d_k^{(v_n)}(\mathbf{X}_n))\right]^2 .$$

To evaluate these formulas efficiently, for each $v \in \{1, \ldots, V\}$ and $n \in \eta^{(v)}$, the numbers $L(Y_n, d_k^{(v)}(\mathbf{X}_n))$, $1 \leq k \leq K$, should be calculated together. This is easily done along the lines of Algorithm 11.1.

Experience in the use of the preceding formulas for $\mathrm{SE}(R^{cv}(d_k))$ in the evaluation of tree structured procedures constructed from simulated data indicates that the expected value of $\mathrm{SE}^2(R^{cv}(d_k))$ is typically close to the expected value of $[R^{cv}(d_k) - R^*(d_k)]^2$. But the probability that $|R^{cv}(d_k) - R^*(d_k)|$ exceeds, say, 2 $\mathrm{SE}(R^{cv}(d_k))$ can be noticeably larger than the value .05 suggested by normal approximation. (This is especially true when there are a few relatively large values of $L(Y_n, d_k^{(v_n)}(\mathbf{X}_n))$ and when the tree has very few terminal nodes.

In the regression context, temporarily set $\bar{Y} = \frac{1}{N} \sum Y_n$ and $s^2 = \frac{1}{N} \sum (Y_n - \bar{Y})^2$. Then

$$RE^{CV}(d_k) = \frac{R^{CV}(d_k)}{s^2}$$

is an estimate of the relative mean square error $RE^*(d_k)$. By analogy with the formula derived in Section 11.4 for the standard error of $RE^{ts}(d_k)$, one can derive a heuristic formula for the standard error of $RE^{CV}(d_k)$. The result is

$$\mathrm{SE}(RE^{CV}(d_k)) = RE^{CV}(d_k)\left[\frac{1}{N}\left(\frac{s_1^2}{(R^{CV}(d_k))^2} - \frac{2s_{12}}{R^{CV}(d_k)s^2} + \frac{s_2^2}{s^4}\right)\right]^{1/2},$$

where

$$s_1^2 = \frac{1}{N} \sum_n (Y_n - d_k^{(v_n)}(\mathbf{X}_n))^4 - (R^{CV}(d_k))^2,$$

$$s_2^2 = \frac{1}{N} \sum_n (Y_n - \bar{Y})^4 - s^4,$$

and

$$s_{12} = \frac{1}{N} \sum_n (Y_n - d_k^{(v_n)}(\mathbf{X}_n))^2 (Y_n - \bar{Y})^2 - R^{CV}(d_k)s^2.$$

11.6 FINAL TREE SELECTION

Test samples or cross-validation can be used to select a particular procedure $d_k = d_{T_k}$ from among the candidates d_k, $1 \leq k \leq K$. Suppose, say, that cross-validation is used. Then k_0 can be chosen to be a value of $k \in \{1, \ldots, K\}$ that minimizes $R^{CV}(d_k)$, T_{k_0} being the final tree selected, d_{k_0} being the corresponding statistical procedure, and $R^{CV}(d_{k_0}) = R^{CV}(d_k)|_{k=k_0}$ being the final estimate of the risk of d_{k_0}.

More generally, allowance can (and should!) be made for the ultimate desire for simplicity as well as accuracy. The 1 SE rule described in Section 3.4.3 can be used. Alternatively, let k_0 now be a value of $k \in \{1, \ldots, K\}$ that minimizes $R^{CV}(d_k) + \alpha_{\text{fin}}|\widetilde{T}_k|$, where α_{fin} is a nonnegative number—for example, $\alpha_{\text{fin}} = .01 \min_k R^{CV}(d_k)$.

Strictly speaking, $R^{CV}(d_{k_0})$ is an overoptimistic estimate of $R^*(d_{k_0})$, since the cross-validation procedure does not take into account the data-adaptive choice of k_0. It is possible to correct for this overoptimism at the expense of a substantial increase in the amount of computation involved. But inspection of many simulations suggests that the overoptimism is typically small enough to be of no concern, at least when $N \geq 200$ and a reasonable choice of $\alpha_{\text{fin}} > 0$ is used.

In practice, cross-validation appears to perform much better in selecting k_0 than in estimating the risk of any particular rule. The explanation for this may well be that the difference in the cross-validation estimates of the risks of two rules tends to be much more accurate than the two estimates themselves—the difference in the estimates being thought of as an estimate of the difference in the actual risks of the two rules. (See Shibata, 1981, where this observation is made precise in a related context.)

There is a refinement to the 1 SE rule for final tree selection that is worth considering. Call T_k *allowable* if, for some $\alpha \geq 0$, k is the largest value of k' that (approximately) minimizes $R^{CV}(T_{k'}) + \alpha|\widetilde{T}_{k'}|$ or $RE^{CV}(T_{k'}) + \alpha|\widetilde{T}_{k'}|$. Algorithm 11.2 writes out the indices k of the allowable trees T_k; in this algorithm, $N_k = |\widetilde{T}_k|$, $R_k = R^{CV}(T_k)$ or $R_k = RE^{CV}(T_k)$, and ε is an appropriate small positive number. The final tree can be selected by applying the 1 SE rule to the collection of allowable T_k's instead of the collection of all T_k's.

For most purposes it should suffice to print summary statistics only for the allowable T_k's, not for all T_k's. Especially in the regression context, this can lead to a considerably more parsimonious printout with little if any loss of useful information. Indeed, consider the simulated regression example corresponding to

Table 8.2, in which there are 180 optimally pruned subtrees. Let Algorithm 11.2 with $\varepsilon = .001$ be applied directly to the rounded-off values of RE^{CV} in the table. Then only six of the trees T_k, $1 \leq k \leq 180$, emerge as being allowable, namely, T_k for $k = 168$, 174, 175, 177, 179, 180.

ALGORITHM 11.2

```
k := 0
while k < K
    g := ∞
    for k' := k + 1 to K
        if k = 0   g' := R_k'
        else g' := (R_k' - R_k)/(N_k - N_k')
        if g' ≤ g + ε   { k" := k'
                          g := g'
    k := k"
    write k
```

11.7 BOOTSTRAP ESTIMATE OF OVERALL RISK

A method based on the bootstrap (see Efron 1979, 1982) can also be used to reduce the overoptimism of the resubstitution estimate $R(d)$ of the overall risk $R^*(d)$ of a tree structured rule d based on a learning sample. For simplicity, the discussion will be confined to the regression problem or to the model 1 version of the classification or class probability estimation problem. Thus,

$$R(d) = \frac{1}{N} \sum_{n} L(Y_n, d(\mathbf{X}_n)).$$

The true risk of d is given by

$$R^*(d) = \iint L(y, d(\mathbf{x}))G^*(d\mathbf{x}, dy),$$

where G^* is the distribution of $(\mathbf{X}, Y)$. Observe that the difference $R^*(d) - R(d)$ is a random variable, since $R^*(d)$ and $R(d)$ both depend on the learning sample. Let $B^*(G^*) = E\big(R^*(d) - R(d)\big)$ denote the dependence of the indicated bias term on the true distribution of $(\mathbf{X}, Y)$.

Clearly, $R(d) + B^*(G^*)$ is a nonoveroptimistic estimate of $R^*(d)$, but this estimate cannot be evaluated in practice, since G^* is unknown. An obvious modification is the estimate $R(d) + B^*(G)$, where G is the empirical distribution of the learning sample $(\mathbf{X}_n, Y_n)$, $n \in \eta$. As Efron pointed out, an estimate $B(G)$ of $B^*(G)$ is easily obtained in practice by the Monte Carlo method. The quantity $R(d) + B(G)$ is called the bootstrap estimate of $R^*(d)$.

Although this bootstrap estimate may typically work reasonably well in practice, there exist situations in which it is clearly defective. To be specific, consider a classification problem with the usual zero-one cost matrix. Suppose that $\mathbf{X}$ and Y are independent and that the unknown prior probabilities $\pi(j)$, $1 \leq j \leq J$, all have the same value, namely, $1/J$. Then every classification rule is Bayes and has risk $(J - 1)/J$. Suppose that $\mathcal{X} = \mathbb{R}^M$, that the marginal distributions of $\mathbf{X}$ are continuous, and that only coordinate splits are to be considered. Let d be the tree structured classification rule obtained by using, say, the Gini splitting rule and continuing the splitting process down to pure nodes (that is, nodes t such that Y_n is constant as n ranges over $\eta(t)$ and hence $r(t) = 0$). Then $R(d) = 0$. Consequently, $B^*(G^*) = (J - 1)/J$. On the other hand, it is easily seen that $EB^*(G)$ equals $(J - 1)/J$ times the probability that any particular member n of η does not appear in a random sample of size N drawn with replacement from η. Thus,

$$EB^*(G) = \frac{J-1}{J}\left(1 - \frac{1}{N}\right)^N .$$

Recall that

$$\lim_{N\to\infty}\left(1 - \frac{1}{N}\right)^N = e^{-1} \doteq .37.$$

Therefore, $EB^*(G)$ can be less than 40 percent of the true bias $B^*(G^*)$. This defect has been confirmed for large trees grown from real and simulated data.

Since estimates of risk based on cross-validation have an advantage vis-à-vis bias over those based on the bootstrap, it is natural to suspect that the former estimates have a somewhat larger variance (see Glick, 1978). The presumed larger variance of cross-validation is of particular concern when the learning sample is small. In medical applications of classification (see Chapter 6), the learning sample frequently contains only a few dozen individuals in the class corresponding to adversity; if so, even when the learning sample contains many individuals in the other class, its size is effectively small.

When the learning sample is genuinely large, however, and the resubstitution estimate of risk is highly overoptimistic, the bias effect dominates the variance effect and cross-validation is superior to the bootstrap. Efron (1983) studied several proposals for modifying the bootstrap to improve its bias-correcting ability; these approaches have yet to be tried out in the context of selecting classification and regression trees.

11.8 END-CUT PREFERENCE

It has been known for some time (see Morgan and Messenger, 1973) that the empirical versions of the risk reduction splitting rule for regression and class probability estimation tend to favor end-cut splits—that is, splits in which p_L is close to zero or one. To see this in a simple setting, let X be the real line and let (X, Y) be a pair of real-valued random variables such that X has a continuous distribution function and $0 < \sigma^2 = \text{Var}(Y) < \infty$. Let (X_n, Y_n), $n \geq 1$, be a random sample from the distribution of (X, Y) and let (X_n, Y_n), $1 \leq n \leq N$, be the corresponding learning sample of size N.

Consider all splits of the root node of the form $t_L = \{x \in X : x \leq c\}$ and $t_R = \{x \in X : x > c\}$. In the regression problem, the empirical version of the risk reduction splitting rule is to choose a split that maximizes

$$p_L \bar{y}^2(t_L) + p_R \bar{y}^2(t_R) = \frac{1}{N}\left[\frac{1}{N(t_L)}\left(\sum_{\eta(t_L)} Y_n\right)^2 + \frac{1}{N(t_R)}\left(\sum_{\eta(t_R)} Y_n\right)^2\right],$$

where $\eta(t_L) = \{n : 1 \leq n \leq N \text{ and } X_n \leq c\}$, $N(t_L) = |\eta(t_L)|$, $\eta(t_R) = \{n : 1 \leq n \leq N \text{ and } X_n > c\}$, and $N(t_R) = |\eta(t_R)|$. Let $\sigma_{N1}, \ldots, \sigma_{NN}$ be that permutation of $\{X_1, \ldots, X_N\}$ such that $X_{\sigma_{N1}} < \cdots < X_{\sigma_{NN}}$. (Since X has a continuous distribution, $X_1, \ldots, X_N$ are distinct with probability one.) Set $Y_{Nn} = Y_{\sigma_{Nn}}$ for $1 \leq n \leq N$. Given the split of the root node into t_L, t_R, set $m = N(t_L)$. Then $p_L = m/N$ and

$$p_L \bar{y}^2(t_L) + p_R \bar{y}^2(t_R) = \frac{1}{N}\left[\frac{(Y_{N1} + \cdots + Y_{Nm})^2}{m} + \frac{(Y_{N,m+1} + \cdots + Y_{NN})^2}{N - m}\right].$$

Suppose now that X and Y are independent of each other. Then end-cut splits are preferred in the sense that for each $\varepsilon \in (0, 1/2)$, $P(\varepsilon \leq p_L \leq 1 - \varepsilon) \to 0$ as $N \to \infty$. This is a consequence of the following result, since $Y_{N1}, \ldots, Y_{NN}$ are independent and have the same distribution as Y.

THEOREM 11.1. *Let Y_n, $n \geq 1$, be independent and identically distributed random variables each having variance σ^2, where $0 < \sigma < \infty$. Set*

$$\Phi_{Nm} = \frac{(Y_1 + \cdots + Y_m)^2}{m} + \frac{(Y_{m+1} + \cdots + Y_N)^2}{N - m} \quad \text{for } 1 \leq m \leq N - 1.$$

Then for $0 < \varepsilon \leq 1/2$,

$$\lim_{N\to\infty} P\left(\max_{1 \leq m \leq N\varepsilon} \Phi_{Nm} > \max_{N\varepsilon < m < N(1-\varepsilon)} \Phi_{Nm}\right) = 1 \tag{11.2}$$

and

$$\lim_{N\to\infty} P\left(\max_{(1-\varepsilon)N \leq m \leq N} \Phi_{Nm} > \max_{N\varepsilon < m < N(1-\varepsilon)} \Phi_{Nm}\right) = 1. \tag{11.3}$$

PROOF. Without loss of generality, it can be assumed that Y_n's have mean zero. It suffices to prove (11.2), (11.3) then following by symmetry.

Choose $N \geq 1$ and $c > 0$. According to the Kolmogorov inequality (see Chung, 1974),

$$P\left(\max_{1 \leq m \leq N} (Y_1 + \cdots + Y_m)^2 \geq \frac{N\sigma^2}{c}\right) \leq c$$

and hence

$$P\left(\max_{N\varepsilon < m < N(1-\varepsilon)} \frac{(Y_{m+1} + \cdots + Y_N)^2}{N - m} \geq \frac{\sigma^2}{\varepsilon c}\right)$$

$$= P\left(\max_{N\varepsilon < m < N(1-\varepsilon)} \frac{(Y_1 + \cdots + Y_m)^2}{m} \geq \frac{\sigma^2}{\varepsilon c}\right) \leq c.$$

Therefore,

$$P\left(\max_{N\varepsilon < m < N(1-\varepsilon)} \Phi_{Nm} \geq \frac{2\sigma^2}{\varepsilon c}\right) \leq 2c. \tag{11.4}$$

According to the law of the iterated logarithm (see Breiman, 1968),

$$P\left(\overline{\lim_{m}} \frac{(Y_1 + \cdots + Y_m)^2}{2\sigma^2 m \log(\log m)} = 1\right) = 1.$$

Thus, for any positive integer M_0

$$\lim_{M\to\infty} P\left(\max_{M_0 \leq m \leq M} \frac{(Y_1 + \cdots + Y_m)^2}{m \log(\log m)} \geq \sigma^2\right) = 1.$$

Now M_0 can be made large enough so that $\log(\log m) > 2/(\varepsilon c)$ for $m \geq M_0$.

Consequently,

$$\lim_{N\to\infty} P\left(\max_{1 \leq m \leq N\varepsilon} \frac{(Y_1 + \cdots + Y_m)^2}{m} > \frac{2\sigma^2}{\varepsilon c}\right) = 1$$

and hence

$$\lim_{N\to\infty} P\left(\max_{1\le m\le N\varepsilon} \Phi_{Nm} > \frac{2\sigma^2}{\varepsilon c}\right) = 1. \tag{11.5}$$

Since c can be made arbitrarily small, (11.2) follows from (11.4) and (11.5). This completes the proof of the theorem.

When $J = 2$, Theorem 11.1 is also directly applicable to the Gini splitting rule for the (model 1 version of the) classification problem or, equivalently, the risk reduction splitting rule for class probability estimation. Consider splits t_L, t_R of the root node, and let $\psi(y = 1)$ equal 1 or 0 according as $y = 1$ or $y \neq 1$. The Gini splitting rule is to choose a split that maximizes

$$\begin{aligned} & p_L p^2(1|t_L) + p_R p^2(1|t_R) \\ &= \frac{1}{N}\left[\frac{1}{N(t_L)}\left(\sum_{\eta(t_L)} \psi(Y_n = 1)\right)^2 + \frac{1}{N(t_R)}\left(\sum_{\eta(t_R)} \psi(Y_n = 1)\right)^2\right]. \end{aligned}$$

If X and Y are independent, then so are X and $\psi(Y = 1)$; thus, it again follows from Theorem 11.1 that for each $\varepsilon \in (0, 1/2)$, $P(\varepsilon \le p_L \le 1 - \varepsilon) \to 0$ as $N \to \infty$.

One way to eliminate the end-cut preference of these splitting rules is to multiply the quantity to be maximized by a positive power of $p_L p_R$. Consider, for example, the twoing splitting rule for the classification problem (which coincides with the Gini splitting rule when $J = 2$). Instead of choosing the splitting rule to maximize

$$\frac{p_L p_R}{4}\left[\sum_j |p(j|t_L) - p(j|t_R)|\right]^2,$$

let it maximize

$$\Phi(s|t) = 2p_L p_R \sum_j |p(j|t_L) - p(j|t_R)|.$$

Since $p_L + p_R = 1$ and $p(j|t) = p_L p(j|t_L) + p_R p(j|t_R)$ for $1 \le j \le J$, it is easily seen that

$$\Phi(s|t) = p_L \sum_j |p(j|t_L) - p(j|t)| + p_R \sum_j |p(j|t_R) - p(j|t)|.$$

Thus, the modified twoing splitting rule—chose s to maximize $\Phi(s|t)$—coincides with the *delta splitting rule* proposed by Messenger and Mandell (1972) and incorporated into THAID (see Morgan and Messenger, 1973). That the delta splitting rule does not overly favor end-cut splits was evidently one of the main motivations for its inclusion in THAID.

12

CONSISTENCY

The tree structured classification and regression procedures discussed in this book use the learning sample to partition the measurement space. In this chapter a more general collection of such "partition-based" procedures will be considered. It is natural to desire that as the size of the learning sample tends to infinity, partition-based estimates of the regression function should converge to the true function and that the risks of partition-based predictors and classifiers should converge to the risk of the corresponding Bayes rules. If so, the procedures are said to be *consistent*. In this chapter the consistency of partition-based regression and classification procedures will be verified under surprisingly general conditions.

In Section 12.1 a required preliminary result of independent interest on the rate of convergence of a sequence of empirical distributions to its theoretical counterpart is described. In Sections 12.2 and 12.3 consistency is discussed for the regression problem and classification problem, respectively; analogous consistency results for class probability estimation is implicitly contained in these two sections. Proofs of the results in Sections 12.1 to 12.3 are given in Sections 12.4 to 12.6. A thorough understanding of these proofs requires some background in real analysis and probability theory.

For previous consistency results along the lines of those in Sections 12.2 and 12.3, see Gordon and Olshen (1978, 1980). Necessary and sufficient conditions for the consistency of a wide class of nonparametric procedures in regression and classification are contained in Stone (1977).

12.1 EMPIRICAL DISTRIBUTIONS

Throughout this chapter X is Euclidean M-dimensional space. Given a fixed positive integer M_1, let $\mathcal{B}$ denote the collection of all polyhedra in X having at most M_1 faces. Specifically, $\mathcal{B}$ is the collection of all sets $t \subset X$ that can be described as the solution set to a system of at most M_1 inequalities, each inequality being of the form $b_1x_1 + \cdots + b_Mx_M \leq c$ or $b_1x_1 + \cdots + b_Mx_M < c$, where $b_1, \ldots, b_M$ and c are real numbers. (Note that the corresponding inequalities involving $\geq$ or $>$ can be rewritten in one of these two forms.) If $M_1 \geq 2M$, then $\mathcal{B}$ includes the collection of all boxes in X; that is, the collection of all sets of the form

$$B = \{(x_1, \ldots, x_M) : x_1 \in I_1, \ldots, x_M \in I_M\},$$

where $I_1, \ldots, I_M$ are intervals of the real line, each of which may be open, closed, half-open or half-closed. The more general definition of $\mathcal{B}$ allows for linear combination splits, as described in Section 5.2.

Let $\mathbf{X}_n$, $n \geq 1$, be a random sample from a distribution P on X. For $N \geq 1$, let p_N denote the empirical distribution of $\mathbf{X}_n$, $1 \leq n \leq N$, defined by

$$p_N(t) = \frac{1}{N}|\{n : 1 \leq n \leq N \text{ and } \mathbf{X}_n \in t\}|, \quad t \subset X.$$

It follows immediately from the strong law of large numbers that for each (Borel) subset t of X, $\lim_N p_N(t) = P(t)$ with probability one. According to a general version of the Glivenko-Cantelli theorem (see Vapnik and Chervonenkis, 1971; Steele, 1978; Dudley, 1978; and Pollard, 1981),

$$\lim_N \sup_{t\in\mathcal{B}} |p_N(t) - P(t)| = 0 \text{ with probability one.} \tag{12.1}$$

The following strengthening of (12.1) plays an essential role in the consistency proofs for partition-based regression and classification that will be given later on.

THEOREM 12.2. *Given positive numbers* ε *and* c, *there is a positive number* $k = k_{\varepsilon,c}$ *such that*

$$\lim_N N^c P\left(|p_N(t) - P(t)| > \varepsilon P(t) + k\,\frac{\log N}{N} \text{ for some } t \in \mathcal{B}\right) = 0. \tag{12.3}$$

This theorem is also valid if $P(t)$ is replaced by $p_N(t)$ in the right side of the inequality in (12.3); indeed, the two forms of the theorem are easily shown to be equivalent. For an elementary application of the theorem, let k_N be a sequence of positive constants.

THEOREM 12.4. *If* $\lim_N k_N = \infty$, *then for all positive constants* ε *and* c,

$$\lim_N N^c P\left(\left|\frac{p_N(t)}{P(t)} - 1\right| > \varepsilon \text{ for some } t \in \mathcal{B} \text{ such that } p_N(t) \geq k_N\,\frac{\log N}{N}\right) = 0. \tag{12.5}$$

It is natural to conjecture that in (12.1), (12.3), and (12.5), $\mathcal{B}$ can be replaced by the collection of all polyhedra in $\mathcal{X}$. But this conjecture is not true in general. To see this, let $M = 2$ and let P be the uniform distribution on a circle. Let t be the convex hull of $\mathbf{X}_n$, $1 \leq n \leq N$, that is, the inscribed polygon having vertices $\mathbf{X}_n$, $1 \leq n \leq N$. Then $p_N(t) = 1$, but $P(t) = 0$. This example is due to Ranga Rao (1962).

Alexander (1983) has shown that Theorem 12.2 can also be derived as a consequence of his deep generalization of a theorem of Kiefer (1961). In their two consistency papers cited earlier, Gordon and Olshen made direct use of Kiefer's theorem. It will be seen shortly that Theorem 12.2 leads to improved results on consistency.

12.2 REGRESSION

Let $(\mathbf{X}, Y)$ be a pair of random variables such that $\mathbf{X} \in \mathcal{X}$, Y is real-valued, and $E|Y| < \infty$. Let P denote the distribution of $\mathbf{X}$ and let d_B denote the regression function of Y on $\mathbf{X}$, defined by $d_B(x) = E(Y|\mathbf{X} = \mathbf{x})$. Let $(\mathbf{X}_n, Y_n)$, $n \geq 1$, denote a random sample from the joint distribution of $(\mathbf{X}, Y)$ and suppose that this random sample is independent of $(\mathbf{X}, Y)$. Given $N \geq 1$ and $t \subset \mathcal{X}$, set

$$\eta_N(t) = \{n : 1 \leq n \leq N \text{ and } \mathbf{X}_n \in t\}.$$

Let p_N again denote the empirical distribution of $\mathbf{X}_n$, $1 \leq n \leq N$. Then $p_N(t) = |\eta_N(t)|/N$.

Let $\widetilde{T}_N$ denote a possibly random partition of $\mathcal{X}$ into a finite number of disjoint sets, all of which are in $\mathcal{B}$. It is assumed that $\widetilde{T}_N$ and the learning sample $(\mathbf{X}_n, Y_n)$, $1 \leq n \leq N$, are together independent of $(\mathbf{X}, Y)$. Ordinarily, $\widetilde{T}_N$ will depend only on the learning sample. But this is not required for the following theoretical results, and in some applications it is natural to let $\widetilde{T}_N$ depend on the sample values of additional random variables. Let τ_N denote the partition function corresponding to $\widetilde{T}_N$, so that $\tau_N(\mathbf{x})$ is the set $t \in \widetilde{T}_N$ containing $\mathbf{x}$. Let $\delta(t)$ denote the diameter of a set $t \subset \mathcal{X}$, defined by

$$\delta(t) = \sup_{\mathbf{x},\mathbf{x}' \in t} |\mathbf{x} - \mathbf{x}'|,$$

where $|\mathbf{x}| = (x_1^2 + \cdots + x_M^2)^{1/2}$ for $\mathbf{x} = (x_1, \ldots, x_M)$. Let $D_N(\mathbf{x}) = \delta(\tau_N(\mathbf{x}))$ denote the diameter of the set $t \in \widetilde{T}_N$ containing $\mathbf{x}$.

Finally, let d_N denote the estimator of the regression function d_B defined by $d_N(\mathbf{x}) = \bar{y}_N(\tau_N(\mathbf{x}))$, where

$$\bar{y}_N(t) = \frac{1}{|\eta_N(t)|} \sum_{\eta_N(t)} Y_n.$$

Let k_N, $N \geq 1$, be nonnegative constants such that (with probability one)

$$p_N(t) \geq k_N \frac{\log N}{N} \text{ for } N \geq 1 \text{ and } t \in \widetilde{T}_N. \tag{12.6}$$

Formula (12.9) in the next result means that $\lim_N P(D_N(\mathbf{X}) \geq \varepsilon) = 0$ for all $\varepsilon > 0$.

THEOREM 12.7. *Suppose that* $E|Y|^q < \infty$, *where* $1 \leq q < \infty$. *If*

$$\lim_N k_N = \infty \tag{12.8}$$

and

$$\lim_N D_N(\mathbf{X}) = 0 \text{ in probability,} \tag{12.9}$$

then

$$\lim_N E[|d_N(\mathbf{X}) - d_B(\mathbf{X})|^q] = 0. \tag{12.10}$$

Equation (12.10) determines the sense in which the sequence $\{d_N\}$ of estimators of the regression function d_B is consistent. The expectation in (12.10) involves the randomness in both $\mathbf{X}$ and d_N. Alternatively, this expectation can be written as

$$\int_X E[|d_N(\mathbf{x}) - d_B(\mathbf{x})|^q] P(d\mathbf{x}),$$

which involves only the randomness in d_N.

Suppose now that $EY^2 < \infty$. Given any real-valued function d on X, let $R(d) = E[(Y - d(\mathbf{X}))^2]$ denote the mean squared error of prediction of Y by $d(\mathbf{X})$. Then $R(d) \geq R(d_B)$, since the regression function d_B is the Bayes rule for the prediction problem. Observe that $R(d_N)$ is a random variable, since d_N depends on the learning sample, and that $ER(d_N) \geq R(d_B)$. The sequence $\{d_N\}$ is said to be *risk consistent* if $\lim_N ER(d_N) = R(d_B)$, that is, if for large N, $d_N(\mathbf{X})$ is nearly as good a predictor of Y as is the optimal predictor $d_B(\mathbf{X})$.

It follows easily from the properties of condition expectation that

$$E[(Y - d_B(\mathbf{X}))(d_B(\mathbf{X}) - d(\mathbf{X}))] = 0$$

and hence that

$$R(d) = R(d_B) + E[(d(\mathbf{X}) - d_B(\mathbf{X}))^2].$$

Thus, Theorem 12.7 with $q = 2$ yields the following result.

THEOREM 12.11. *Suppose that* $EY^2 < \infty$ *and that* (12.8) *and* (12.9) *hold. Then* $\{d_N\}$ *is risk consistent.*

Let $\psi(\cdot|\mathbf{x})$ denote the moment-generating function of the conditional distribution of $Y - d_B(\mathbf{x})$, given that $\mathbf{X} = \mathbf{x}$, so that

$$\psi(s|\mathbf{x}) = E\left[e^{s(Y-d_B(\mathbf{x}))} \,|\, \mathbf{X} = \mathbf{x}\right].$$

To conclude that d_N converges to d_B uniformly on compacts, the following assumption will be made.

ASSUMPTION 12.12. *For every compact set B in* X, *there is an* $s > 0$ *such that* $\psi(-s|\mathbf{x}) + \psi(s|\mathbf{x})$ *is bounded on* B.

If d_B is bounded, then Assumption 12.12 is equivalent to the same statement with $\psi(\cdot|\mathbf{x})$ the moment-generating function of the conditional distribution of Y given that $\mathbf{X} = \mathbf{x}$. Assumption 12.12 is automatically satisfied if Y is a bounded random variable. Suppose instead that Y is unbounded but that the conditional distribution of Y given $\mathbf{X} = \mathbf{x}$ is normally distributed with variance $\sigma^2(\mathbf{x})$ (and necessarily mean $d_B(\mathbf{x})$). Then Assumption 12.12 is true if, and only if, $\sigma^2(\cdot)$ is bounded on compacts.

Let g_N, $N \geq 1$, and g be real-valued functions on X. Then g_N is said to converge to g *uniformly on compacts* if

$$\lim_N \sup_{\mathbf{x}\in B} |g_N(\mathbf{x}) - g(\mathbf{x})| = 0$$

for every compact set B in X.

THEOREM 12.13. *Suppose that (12.8) and Assumption 12.12 hold; that d_B is continuous; and that, with probability one, $D_N(\cdot)$ converges to zero uniformly on compacts. Then, with probability one, d_N converges to d_B uniformly on compacts.*

12.3 CLASSIFICATION

Let $(\mathbf{X}, Y)$ be a pair of random variables such that $\mathbf{X} \in X$ and $Y \in \{1, \ldots, J\}$, where $2 \leq J < \infty$. As before, let P denote the distribution of $\mathbf{X}$. Given $1 \leq j \leq J$, set $\pi(j) = P(Y = j)$ and $P(j|\mathbf{x}) = P(Y = j|\mathbf{X} = \mathbf{x})$ for $\mathbf{x} \in X$.

Let $p_N(j|\mathbf{x})$ denote a sample-based estimate of $P(j|\mathbf{x})$ such that $0 \leq p_N(j|\mathbf{x}) \leq 1$ for $\mathbf{x} \in X$ and $1 \leq j \leq J$ and p_N is independent of $(\mathbf{X}, Y)$. Then for $1 \leq j \leq J$, the following three conditions are equivalent:

$$\lim_N E[|p_N(j|\mathbf{X}) - P(j|\mathbf{X})|^2] = 0, \tag{12.14}$$

$$\lim_N E|p_N(j|\mathbf{X}) - P(j|\mathbf{X})| = 0, \tag{12.15}$$

$$\lim_N p_N(j|\mathbf{X}) = P(j|\mathbf{X}) \text{ in probability.} \tag{12.16}$$

Let $C(i|j)$ denote the cost of classifying a class j object as class i. Given a $\{1, \ldots, J\}$-valued function d on X, let

$$R(d) = \int_X [\sum_j C(d(\mathbf{x})|j)P(j|\mathbf{x})]P(d\mathbf{x})$$

denote the risk of using d as a classifier. A Bayes rule d_B, that is, a rule that minimizes $R(d)$, is given by choosing $d_B(\mathbf{x})$ to be the smallest value of $i \in \{1, \ldots, J\}$ that minimizes $\Sigma_j\, C(i|j)$ $P(j|\mathbf{x})$. Let $d_N(\mathbf{x})$ be chosen to be the smallest value of $i \in \{1, \ldots, J\}$ that minimizes $\Sigma\, C(i|j)p_N(j|\mathbf{x})$, and note that $ER(d_N) \geq R(d_B)$. The sequence $\{d_N\}$ is again said to be risk consistent if $\lim_N ER(d_N) = R(d_B)$.

THEOREM 12.17. *If (12.15) holds, then* $\{d_N\}$ *is risk consistent.*

This theorem will now be applied to the two sampling schemes for the classification problem that have been considered in this book.

(Model 1). Let $(\mathbf{X}_n, Y_n)$, $n \geq 1$, denote a random sample from the joint distribution of $(\mathbf{X}, Y)$ and suppose that this random sample is independent of $(\mathbf{X}, Y)$. Let $\eta_N(t)$, $\widetilde{T}_N$, τ_N, p_N, and k_N be defined as in the regression problem and suppose that (12.6) holds. Also set

$$p_N(j|\mathbf{x}) = p_N(j|\tau_N(\mathbf{x})),$$

where

$$p_N(j|t) = \frac{1}{|\eta_N(t)|}|\{n \in \eta_N(t) : Y_n = j\}|,$$

and let d_N be defined in terms of $p_N(j|\mathbf{x})$ as above. Observe that

$$\sum_j C(i|j)p_N(j|\mathbf{x}) = \frac{1}{|\eta_N(t)|} \sum_{\eta_N(t)} C(i|Y_n),$$

where $t = \tau_N(\mathbf{x})$. Consequently $d_N(\mathbf{x}) = i_N(\tau_N(\mathbf{x}))$, where $i_N(t)$ is the smallest value of i that minimizes $\Sigma_{\eta_N(t)}\, C(i|Y_n)$.

(Model 2). Let the prior probabilities $\pi(j)$, $1 \leq j \leq J$, be known. For $1 \leq j \leq J$, let P_j denote the conditional distribution of $\mathbf{X}$, given that $Y = j$. Let $(\mathbf{X}_n, Y_n)$, $n \geq 1$, be a sequence of pairs of random variables satisfying the following properties: Y_n

is $\{1, \ldots, J\}$-valued for $n \geq 1$; conditioned on $Y_n = j_n$ for $n \geq 1$, the random variables $\mathbf{X}_n$, $n \geq 1$, are independent and $\mathbf{X}_n$ has conditional distribution P_{j_n}; with probability one, for each $j \in \{1, \ldots, J\}$, infinitely many of the Y_n's equal j. Set $P(j|t| = P(Y = j|\mathbf{X} \in t)$ and observe that

$$P(j|t) = \frac{\pi(j)P_j(t)}{\Sigma_j \pi(j)P_j(t)} = \frac{\pi(j)P_j(t)}{P(t)} .$$

Set

$$\eta_{jN}(t) = \{n : 1 \leq n \leq N,\ \mathbf{X}_n \in t \text{ and } Y_n = j\}$$

and

$$\eta_{jN} = \eta_{jN}(X) = \{n : 1 \leq n \leq N \text{ and } Y_n = j\}.$$

Consider the estimators

$$p_{jN}(t) = \frac{|\eta_{jN}(t)|}{|\eta_{jN}|} ,$$

$$p_N(t) = \sum_j \pi(j)p_{jN}(t),$$

$$p_N(j|t) = \frac{\pi(j)p_{jN}(t)}{\Sigma_j \pi(j)p_{jN}(t)} = \frac{\pi(j)p_{jN}(t)}{p_N(t)},$$

and

$$p_N(j|\mathbf{x}) = p_N(j|\tau_B(\mathbf{x}))$$

of the corresponding quantities with p_N replaced by P. Observe that

$$\sum_j C(i|j)p_N(j|\mathbf{x}) = \frac{\Sigma_j C(i|j)\pi(j)p_{jN}(t)}{p_N(t)} ,$$

where $t = \tau_N(\mathbf{x})$. Thus, $d_N(\mathbf{x}) = i_N(\tau_N(\mathbf{x}))$, where $i_N(t)$ is the smallest value of i that minimizes $\Sigma_j C(i|j)\pi(j)p_{jN}(t)$. Instead of (12.6), suppose that

$$p_N(t) \geq k_N \max_j \frac{\log|\eta_{jN}|}{|\eta_{jN}|}, \quad t \in \widetilde{T}_N . \tag{12.18}$$

Theorem 12.7 will be used to obtain the next result.

THEOREM 12.19. *Let model 1 and (12.6) or model 2 and (12.18) hold, and suppose that (12.8) and (12.9) are satisfied. Then* $\{d_N\}$ *is risk consistent.*

There is an alternative to model 2 for known prior probabilities that is worth considering. Let the data consist of a random sample $\mathbf{X}_n$, $n \in \eta_j$, of fixed (that is, nonrandom) size N_j from P_j for $j \in \{1, \ldots, J\}$; set $\mathbf{N} = (N_1, \ldots, N_J)$, and let $d_{\mathbf{N}}$ be defined as d_N in model 2, with p_{jN} replaced by the empirical distribution of $\mathbf{X}_n$, $n \in \eta_j$. Under appropriate conditions, $d_{\mathbf{N}}$ is Bayes risk consistent as $N_1, \ldots, N_J$ all tend to infinity. The result can be proved by assuming that the desired conclusion is false and using a subsequence argument to obtain a contradiction to Theorem 12.19 for model 2. The details are left to the interested reader.

Theorems 12.11 and 12.19 provide some theoretical justification for tree structured regression and classification—risk consistency under mild regularity conditions. But no theoretical justification has been obtained so far for any of the specific splitting rules discussed in the book, nor for optimal pruning or cross-validation.

12.4 PROOFS FOR SECTION 12.1

The inequalities for binomial probabilities in the next result are needed for the proof of Theorem 12.2.

LEMMA 12.20. *Let Z have a binomial distribution with parameters m and p. Given $\varepsilon > 0$, let $\delta_\varepsilon > 0$ be defined by $(e^{\delta_\varepsilon} - 1)/\delta_\varepsilon = 1 + \varepsilon$. Then for all $k > 0$,*

$$P(Z \geq (1 + \varepsilon)mp + k) < e^{-k\delta_\varepsilon} \tag{12.21}$$

and

$$P(Z \leq (1 - \varepsilon)mp - k) < e^{-k\delta_\varepsilon}. \tag{12.22}$$

PROOF. The moment-generating function of Z is given by $Ee^{\delta Z} = (1 - p + pe^{\delta})^m$. It is easily seen that $(e^{\delta} - 1)/\delta$ is continuous and strictly increasing on $(0, \infty)$ and has limits 1 and ∞, respectively, at 0 and ∞. Thus, given $\varepsilon > 0$, there is a unique positive number δ_ε such that $(e^{\delta_\varepsilon} - 1)/\delta_\varepsilon = 1 + \varepsilon$. Observe that

$$e^{\delta_\varepsilon p(1+\varepsilon)} - 1 > \delta_\varepsilon p(1 + \varepsilon) = p(e^{\delta_\varepsilon} - 1)$$

and hence that

$$P(Z \geq (1 + \varepsilon)mp + k) \leq e^{-\delta_\varepsilon[(1+\varepsilon)mp+k]} Ee^{\delta_\varepsilon Z}$$
$$= \left(\frac{pe^{\delta_\varepsilon} + 1 - p}{e^{\delta_\varepsilon(1+\varepsilon)p}} \right)^m e^{-k\delta_\varepsilon} < e^{-k\delta_\varepsilon};$$

so (12.21) is valid.

In proving (12.22), it can be assumed that $0 < \varepsilon < 1$ (otherwise, the result is trivially true). It is straightforward to show that $(1 - e^{-\delta_\varepsilon})/\delta_\varepsilon > 1 - \varepsilon$ and hence that

$$e^{-\delta_\varepsilon(1-\varepsilon)p} > 1 - \delta_\varepsilon(1 - \varepsilon)p > 1 - p(1 - e^{-\delta_\varepsilon}).$$

Consequently,

$$P(Z \leq (1 - \varepsilon)mp - k) \leq e^{\delta_\varepsilon[(1-\varepsilon)mp-k]} Ee^{-\delta_\varepsilon Z}$$
$$= (e^{\delta_\varepsilon(1-\varepsilon)p}(pe^{-\delta_\varepsilon} + 1 - p))^m e^{-k\delta_\varepsilon}$$
$$< e^{-k\delta_\varepsilon},$$

so (12.22) is valid.

Recall that $\mathbf{X}_1, \mathbf{X}_2, \ldots$ is a random sample from P and that P_N is the empirical distribution of $\mathbf{X}_1, \ldots, \mathbf{X}_N$. Let $\mathbf{X}_1', \mathbf{X}_2', \ldots$ be a second such random sample and let P_N' denote the empirical distribution of $\mathbf{X}_1', \ldots, \mathbf{X}_N'$. These two random samples are assumed to be independent of each other.

LEMMA 12.23. *Given positive numbers* ε *and* c, *there is a positive number* $k = k_{\varepsilon,c}$ *such that*

$$\lim_N N^c P|(p_N(t) - p'_N(t))| > \varepsilon(p_N(t) + p'_N(t)) + k\frac{\log N}{N} \quad \textit{for some } t \in \mathcal{B}) = 0.$$

PROOF*. Let $\xi_1, \xi'_1, \ldots, \xi_N, \xi'_N$ be a random sample of size $2N$ from P. Let $s_1, \ldots, s_N$ be independent and identically distributed Bernoulli (that is, zero-one) random variables each having probability .5 of equaling 1; and let the s_n's be independent of (ξ_n, ξ'_n), $n \geq 1$. Define $\mathbf{X}_n$, $\mathbf{X}'_n$ for $1 \leq n \leq N$ by

$$\mathbf{X}_n = \xi_n \text{ and } \mathbf{X}'_n = \xi'_n \text{ if } s_n = 1$$

and

$$\mathbf{X}_n = \xi'_n \text{ and } \mathbf{X}'_n = \xi_n \text{ if } s_n = 0.$$

Then $\mathbf{X}_1, \mathbf{X}'_1, \ldots, \mathbf{X}_N, \mathbf{X}'_N$ is a random sample of size $2N$ from P. Let p_N denote the empirical distribution of $\mathbf{X}_1, \ldots, \mathbf{X}_N$ (as just constructed) and let p'_N denote the empirical distribution of $\mathbf{X}'_1, \ldots, \mathbf{X}'_N$. It suffices to verify the conclusion of the lemma for this choice of p_N, p'_N.

To this end, let ψ_t denote the indicator of t defined by $\psi_t(\mathbf{x}) = 1$ if $\mathbf{x} \in t$ and $\psi_t(\mathbf{x}) = 0$ otherwise. Then

$$p_N(t) = \frac{1}{N}\sum_1^N [s_n\psi_t(\xi_n) + (1 - s_n)\psi_t(\xi'_n)]$$

and

$$p'_N(t) = \frac{1}{N}\sum_1^N [(1 - s_n)\psi_t(\xi_n) + s_n\psi_t(\xi'_n)];$$

so

$$p_N(t) + p'_N(t) = \frac{1}{N}\sum_1^N [\psi_t(\xi_n) + \psi_t(\xi'_n)]$$

---

*We wish to thank David Pollard for a clear explanation of some of the key ideas used in this proof.

and

$$p_N(t) - p'_N(t) = \frac{1}{N} \sum_1^N (2s_n - 1)[\psi_t(\xi_n) - \psi_t(\xi'_n)].$$

Now $2s_n - 1$, $1 \leq n \leq N$, are independent random variables, each taking on the values -1 and 1 with probability .5. Let m be a nonnegative integer and let Z be a binomial random variable having parameters m and $p = .5$. Observe that if $v_1, \ldots, v_N$ are constants each equal to -1, 0, or 1 and $m = \sum_1^N |v_n|$, then $\sum_1^N (2s_n - 1)v_n$ has the same distribution as $2Z - m$.

Given $\varepsilon > 0$, let δ_ε be defined as in Lemma 12.1. Choose $t \in \mathcal{B}$. It follows from the observation in the previous paragraph that on the event $\sum_1^N |\psi_t(\xi_n) - \psi_t(\xi'_n)| = m$,

$$\begin{aligned} P(|p_N(t) - p'_N(t)| &\geq \varepsilon(p_N(t) + p'_N(t)) + k \frac{\log N}{N} | (\xi_n, \xi'_n), 1 \leq n \leq N) \\ &\leq P(|\sum_1^N (2s_n - 1)(\psi_t(\xi_n) - \psi_t(\xi'_n))| \\ &\geq \varepsilon\, m + k \log N | (\xi_n, \xi'_n), 1 \leq n \leq N) \\ &= P(|2Z - m| \geq \varepsilon\, m + k \log N) \\ &= P(|Z - \frac{m}{2}| \geq \varepsilon \frac{m}{2} + \frac{k}{2} \log N) \leq 2N^{-k\delta_\varepsilon/2}. \end{aligned}$$

Consequently,

$$P(|p_N(t) - p'_N(t)| \geq \varepsilon(p_N(t) + p'_N(t)) + k \frac{\log N}{N}) \leq 2N^{-k\delta_\varepsilon/2}.$$

Think of (ξ_n, ξ'_n), $1 \leq n \leq N$, as fixed; and recall that $\mathcal{B}$ is the collection of polyhedra in R^M having at most M_1 faces. According to the pioneering combinatorial result of Vapnik and Chervonenkis (1971) (see Section 7 of Dudley, 1978), there exist subsets t_k, $1 \leq k \leq (2N)^{M_1(M+2)}$ in $\mathcal{B}$ satisfying the following property: Given each $t \in \mathcal{B}$, there is a k such that

$$\psi_t(\xi_n) = \psi_{t_k}(\xi_n) \text{ and } \psi_t(\xi'_n) = \psi_{t_k}(\xi'_n) \text{ for } 1 \leq n \leq N$$

and hence

$$p_N(t) = p_N(t_k) \text{ and } p'_N(t) = p'_N(t_k).$$

Therefore,

$$P(|p_N(t) - p'_N(t)| > \varepsilon(p_N(t) + p'_N(t)) + k \frac{\log N}{N} \text{ for some } t \in \mathcal{B})$$
$$\leq 2^{1+M_1(M+2)} N^{M_1(M+2)-k\delta_\varepsilon/2},$$

which yields the desired result. (Some measurability problems that have been ignored here are straightforward to handle for the particular collection $\mathcal{B}$ under consideration; see Section 5 of Pollard, 1981.)

PROOF OF THEOREM 12.2. Let ε and c be fixed positive numbers. According to Lemma 12.23, there is a positive constant k such that

$$\lim_N N^c P\Big(p_N(t) > \Big(1 + \frac{\varepsilon}{2}\Big)p'_N(t) + \frac{k}{2}\frac{\log N}{N} \text{ for some } t \in \mathcal{B}\Big) = 0. \tag{12.24}$$

By Lemma 12.20, there is a positive integer N_0 such that

$$P\left((1 + \varepsilon)P(t) + k \frac{\log N}{N} \geq \Big(1 + \frac{\varepsilon}{2}\Big)p'_N(t) + \frac{k}{2}\frac{\log N}{N}\right) \geq \frac{1}{2} \text{ for } t \in \mathcal{B} \text{ and } n \geq N_0. \tag{12.25}$$

Since p_N and p'_N are independent, it follows from (12.24) and (12.25) that

$$\lim_N N^c P\Big(p_N(t) > (1 + \varepsilon)P(t) + k \frac{\log N}{N} \text{ for some } t \in \mathcal{B}\Big) = 0.$$

(First replace $\mathcal{B}$ by a countable subcollection.) Similarly,

$$\lim_N N^c P\Big(p_N(t) < (1 - \varepsilon)P(t) - k \frac{\log N}{N} \text{ for some } t \in \mathcal{B}\Big) = 0,$$

so (12.3) is valid as desired.

12.5 PROOFS FOR SECTION 12.2

The proof of Theorem 12.13 begins with the following elementary result.

LEMMA 12.26. *Let $Z_1, \ldots, Z_m$ be independent real-valued random variables having mean zero and moment-generating functions $\Psi_1, \ldots, \Psi_m$, respectively, and suppose that $|\Psi_\ell(u)| \le \exp[Ku^2/2]$ for $1 \le \ell \le m$ and $|u| \le s$, where s and K are fixed positive constants. Set $\bar{Z} = (Z_1 + \cdots + Z_m)/m$. Then*

$$P(|\bar{Z}| \ge \varepsilon) \le 2e^{-\varepsilon^2 m/2K} \quad \text{for } 0 \le \varepsilon \le Ks.$$

PROOF. Observe that

$$e^{\varepsilon^2 m/K} P(\bar{Z} \ge \varepsilon) \le Ee^{\varepsilon m \bar{Z}/K} = \prod_1^m \Psi_\ell\left(\frac{\varepsilon}{K}\right) \le e^{\varepsilon^2 m/2K}$$

and hence that

$$P(\bar{Z} \ge \varepsilon) \le e^{-\varepsilon^2 m/2K}.$$

Similarly,

$$P(\bar{Z} \le -\varepsilon) \le e^{-\varepsilon^2 m/2K},$$

so the desired result is valid.

Set

$$\bar{\mu}_N(t) = \frac{1}{|\eta_N(t)|} \sum_{\eta_N(t)} d_B(\mathbf{X}_n)$$

and $\bar{d}_N(\mathbf{x}) = \bar{\mu}_N(\tau_N(\mathbf{x}))$ for $x \in X$. Suppose d_B is continuous and that, with probability one, $\lim_N D_N(\cdot) = 0$ uniformly on compacts. Then, with probability one, $\lim_N \bar{d}_N = d_B$ uniformly on compacts. Thus, Theorem 12.13 follows easily from the following result.

LEMMA 12.27. *Suppose that (12.8) and Assumption 12.12 hold. Then for every compact set B in X and every $\varepsilon > 0$ and $c > 0$,*

$$\lim_N N^c P(|\bar{y}_N(t) - \bar{\mu}_N(t)| > \varepsilon \text{ for some } t \in \mathcal{B} \text{ such that } t \subset B \text{ and } p_N(t) \ge k_N N^{-1} \log N) = 0.$$

PROOF. It can be assumed without loss of generality that $d_B = 0$. Let $\Psi(\cdot|\mathbf{x})$ denote the moment-generating function of the conditional distribution of Y, given that $\mathbf{X} = \mathbf{x}$, and let $\Psi'(u|\mathbf{x})$ and $\Psi''(u|\mathbf{x})$ refer to differentiation with respect to u. Then $\Psi(0|\mathbf{x}) = 1$, $\Psi'(0|\mathbf{x}) = 0$, and $\Psi''(u|\mathbf{x}) = E[Y^2e^{uY}|\mathbf{X} = \mathbf{x}]$. Let B, ε, and c be as in the statement of the lemma. It follows straightforwardly from Assumption 12.12, Taylor's theorem with remainder, and an argument similar to that used in proving Chebyshev's inequality that there are positive constants s and K such that $\Psi(u|\mathbf{x}) \leq \exp(Ku^2/2)$ for $\mathbf{x} \in B$ and $|u| \leq s$. Let $t \subset B$ and let $0 < \varepsilon \leq Ks$. According to Lemma 12.26,

$$
\begin{aligned}
P(|\bar{y}_N(t) - \bar{\mu}_N(t)| \geq \varepsilon|\mathbf{X}_1, \ldots, \mathbf{X}_N) &\leq 2e^{-\varepsilon^2 Np_N(t)/2K} \\
&\leq 2e^{-\varepsilon^2 k_N \log N/2K} \\
&= 2N^{-\varepsilon^2 k_N/2K}
\end{aligned}
$$

on the event that $p_N(t) \geq k_N N^{-1} \log N$. It now follows from the combinatorial result of Vapnik and Chervonenkis, as in the proof of Lemma 12.23, that

$$
\begin{aligned}
&P(|\bar{y}_N(t) - \bar{\mu}_N(t)| > \varepsilon \text{ for some } t \in \mathcal{B} \text{ such that } t \subset B \text{ and} \\
&\qquad p_N(t) \geq k_N N^{-1} \log N|\mathbf{X}_1, \ldots, \mathbf{X}_N) \\
&\qquad \leq 2^{1+M_1(M+2)} N^{M_1(M+2)-\varepsilon^2 k_N/2K}
\end{aligned}
$$

and hence that

$$
\begin{aligned}
&P(|\bar{y}_N(t) - \bar{\mu}_N(t)| > \varepsilon \text{ for some } t \in \mathcal{B} \text{ such that } t \subset B \text{ and} \\
&\qquad p_N(t) \geq k_N N^{-1} \log N) \\
&\qquad \leq 2^{1+M_1(M+2)} N^{M_1(M+2)-\varepsilon^2 k_N/2K}.
\end{aligned}
$$

Since $\lim_N k_N = \infty$, the desired conclusion holds.

The next two lemmas are preparatory to the proof of Theorem 12.7.

LEMMA 12.28. *Suppose that* $E|Y|^q < \infty$, *where* $1 \le q < \infty$, *and that (12.8) holds. Then*

$$\overline{\lim_N}\, E[|d_N(\mathbf{X})|^q] \le E[|Y|^q].$$

PROOF. By Hölder's or Jensen's inequality,

$$\frac{1}{|\eta_N(t)|}\left|\sum_{\eta_N(t)} Y_n\right|^q \le \frac{1}{|\eta_N(t)|}\sum_{\eta_N(t)}|Y_n|^q = \frac{1}{Np_N(t)}\sum_{\eta_N(t)}|Y_n|^q$$

and hence

$$\begin{aligned}
E[|d_N(\mathbf{X})|^q] &\le \frac{1}{N} E\left[\sum_{\tilde{T}_N}\frac{P(t)}{p_N(t)}\sum_{\eta_N(t)}|Y_n|^q\right] \\
&\le \frac{1}{N} E\left[\sum_1^N |Y_n|^q \sup\left\{\frac{P(t)}{p_N(t)} : t \in \mathcal{B} \text{ and } p_N(t) \ge k_N \frac{\log N}{N}\right\}\right] \\
&= E\left[|Y_N|^q \sup\left\{\frac{P(t)}{p_N(t)} : t \in \mathcal{B} \text{ and } p_N(t) \ge k_N \frac{\log N}{N}\right\}\right] \\
&\le \frac{N}{N-1} E\left[|Y_N|^q \sup\left\{\frac{P(t)}{p_{N-1}(t)} : t \in \mathcal{B} \text{ and } p_{N-1}(t) \ge k_N \frac{\log N}{N} - \frac{1}{N}\right\}\right] \\
&= \frac{N}{N-1} E[|Y|^q] E\left[\sup\left\{\frac{P(t)}{p_{N-1}(t)} : t \in \mathcal{B} \text{ and } p_{N-1}(t) \ge k_N \frac{\log N}{N} - \frac{1}{N}\right\}\right].
\end{aligned}$$

(To verify the last equality, note that Y_N is independent of p_{N-1} and has the same distribution at Y.)

Set

$$\mathcal{B}_N = \left\{ t \in \mathcal{B} : p_N(t) \geq \frac{1}{2} k_{N+1} \frac{\log N}{N} \right\}.$$

To prove the lemma, it now suffices to show that

$$\overline{\lim_N}\, E\left[\sup\left\{ \frac{P(t)}{p_N(t)} : t \in \mathcal{B}_N \right\} \right] \leq 1. \tag{12.29}$$

To this end, let $0 < \varepsilon < 1$ and let Ω_N denote the event that

$$p_N(t) \geq (1 - \varepsilon)P(t) - \frac{\varepsilon}{2} k_{N+1} \frac{\log N}{N} \text{ for all } t \in \mathcal{B}. \tag{12.30}$$

By (12.8) and Theorem 12.2,

$$\lim_N N(1 - P(\Omega_N)) = 0. \tag{12.31}$$

Observe that if $t \in \mathcal{B}_N$, then

$$\frac{P(t)}{p_N(t)} \leq \frac{2N}{K_{N+1} \log N};$$

and if $t \in \mathcal{B}_N$ and (12.30) holds, then

$$\frac{P(t)}{p_N(t)} \leq \frac{1 + \varepsilon}{1 - \varepsilon}.$$

Consequently,

$$E\left[\sup\left\{ \frac{P(t)}{p_N(t)} : t \in \mathcal{B}_N \right\} \right] \leq \frac{1 + \varepsilon}{1 - \varepsilon} + \frac{2N}{k_{N+1} \log N}(1 - P(\Omega_N)). \tag{12.32}$$

Since ε can be made arbitrarily small, (12.29) follows from (12.8), (12.31), and (12.32). This completes the proof of the lemma.

LEMMA 12.33. *Suppose that* $E[|Y|^q] < \infty$, *where* $1 \leq q < \infty$, *and let* $0 < \varepsilon < \infty$. *Then there is a bounded random variable* $Y' = H(\mathbf{X}, Y)$ *such that* $(E[|Y' - Y|^q])^{1/q} \leq \varepsilon$ *and the regression function of* Y' *on* $\mathbf{X}$ *is continuous.*

PROOF. Given a positive number K_1, define Y'' as a function of Y by

$$Y'' = \begin{cases} -K_1 & \text{if } Y < -K_1 \\ Y & \text{if } -K_1 \le Y \le K_1 \\ K_1 & \text{if } Y > K_1. \end{cases}$$

The number K_1 can be chosen so that $(E[|Y'' - Y|^q)^{1/q} \le \varepsilon/2$. Let d''_B denote the regression function of Y'' on $\mathbf{X}$. There is a bounded continuous function d'_B on $\mathcal{X}$ such that $(E[|d'_B(\mathbf{X}) - d''_B(\mathbf{X})|^q])^{1/q} \le \varepsilon/2$. (The collection of bounded continuous functions is dense in L^q.) Set $Y' = Y'' + d'_B(\mathbf{X}) - d''_B(\mathbf{X})$. Then d'_B is the regression function of Y' on $\mathbf{X}$; Y' is a bounded function of $\mathbf{X}$ and Y; and, by Minkowski's inequality, $(E[|Y' - Y|^q])^{1/q} \le \varepsilon$. This completes the proof of the lemma.

With this preparation, it is easy to complete the proof of Theorem 12.7. Suppose that $E[|Y|^q] < \infty$, where $1 \le q < \infty$. Choose $\varepsilon > 0$, let $Y' = H(\mathbf{X}, Y)$ be as in Lemma 12.33, and let d'_B denote the regression function of Y' on $\mathbf{X}$. Also set $Y'_n = H(\mathbf{X}_n, Y_n)$ for all $n \ge 1$. Set $d'_N(\mathbf{x}) = \bar{y}'_N(\tau_N(\mathbf{x}))$, where

$$\bar{y}'_N(t) = \frac{1}{|\eta_N(t)|} \sum_{\eta_N(t)} Y'_n;$$

and set $\bar{d}'_N(\mathbf{x}) = \bar{\mu}'_N(\tau_N(\mathbf{x}))$, where

$$\bar{\mu}'_N(t) = \frac{1}{|\eta_N(t)|} \sum_{\eta_N(t)} d'_B(\mathbf{X}_n).$$

Then

$$d_N - d_B = (d_N - d'_N) + (d'_N - \bar{d}'_N) + (\bar{d}'_N - d'_B) + (d'_B - d_B).$$

Thus by Minkowski's inequality, to complete the proof of Theorem 12.7 it suffices to verify the following four statements:

(i) if (12.8) holds, then $\overline{\lim_N} E[|d_N(\mathbf{X}) - d'_N(\mathbf{X})|^q] \leq \varepsilon^q$;

(ii) if (12.8) and (12.9) hold, then $\lim_N E[|d'_N(\mathbf{X}) - \bar{d}_N(\mathbf{X})|^q] = 0$,

(iii) if (12.9) holds, then $\lim_N E[|\bar{d}'_N(\mathbf{X}) - d'_B(\mathbf{X})|^q] = 0$;

(iv) $(E[|d'_B(\mathbf{X}) - d_B(\mathbf{X})|^q] \leq \varepsilon^q$.

Statement (i) follows from Lemma 12.28 with Y and Y_n, $n \geq 1$, replaced by $Y' - Y$ and $Y'_n - Y_n$, $n \geq 1$, respectively. Since Y' is a bounded random variable, it follows from (12.8), (12.9), and Lemma 12.27 that $d'_N(\mathbf{X}) - \bar{d}'_N(\mathbf{X})$ converges to zero in probability; so the conclusion to (ii) follows from the bounded convergence theorem. Since d'_B is bounded and continuous, it follows from (12.9) that $\bar{d}'_N(\mathbf{X})$ converges to $d'_B(\mathbf{X})$ in probability; so the conclusion to (iii) follows from another application of the bounded convergence theorem. It follows from the conditional form of Hölder's or Jensen's inequality that

$$E[|d'_B(\mathbf{X}) - d_B(\mathbf{X})|^q] = E[|E((Y' - Y)|\mathbf{X})|^q] \leq E[|Y' - Y|^q]$$

and hence that (iv) holds. This completes the proof of Theorem 12.7.

12.6 PROOFS FOR SECTION 12.3

PROOF OF THEOREM 12.17. Observe that

$$\sum_j [C(d_N(\mathbf{x})|j) - C(d_B(\mathbf{x})|j)]P(j|\mathbf{x})$$

$$= \sum_j [C(d_N(\mathbf{x})|j) - C(d_B(\mathbf{x})|j)]p_N(j|\mathbf{x})$$

$$+ \sum_j [C(d_N(\mathbf{x})|j) - C(d_B(\mathbf{x})|j)][P(j|\mathbf{x}) - p_N(j|\mathbf{x})]$$

$$\leq C_0 \sum_j |p_N(j|\mathbf{x}) - P(j|\mathbf{x})|,$$

where

$$C_0 = \max[C(i'|j) - C(i|j)| : i, i', j \in \{1, \ldots, J\}].$$

Thus,

$$ER(d_N) \leq R(d_B) + c_0 \sum_j E|p_N(j|\mathbf{X}) - P(j|\mathbf{X})|;$$

and hence if (12.15) holds, then $\{d_N\}$ is risk consistent as desired.

PROOF OF THEOREM 12.19 FOR MODEL 1. For $1 \leq j \leq J$, let ψ_j be the function on $\{1, \ldots, J\}$ defined by $\psi_j(i) = 1$ if $i = j$ and $\psi_j(i) = 0$ if $i \neq j$. Then $P(j|\mathbf{x}) = E(\psi_j(Y)|\mathbf{X} = \mathbf{x})$, so that $P(j|\cdot)$ is the regression function of $\psi_j(Y)$ on $\mathbf{X}$. Also,

$$p_N(j|t) = \frac{1}{|\eta_N(t)|} \sum_{\eta_N(t)} \psi_j(Y_n).$$

Suppose that (12.6), (12.8), and (12.9) hold. Then (12.14) holds by Theorem 12.7 and hence (12.15) holds. Thus, it follows from Theorem 12.17 that $\{d_N\}$ is risk consistent.

PROOF OF THEOREM 12.19 FOR MODEL 2. Recall that in model 2,

$$p_N(j|\mathbf{x}) = p_N(j|\tau_N(\mathbf{x})) = \frac{\pi(j)p_{jN}(\tau_N(\mathbf{x}))}{\sum_j \pi(j)p_{jN}(\tau_N(\mathbf{x}))}.$$

Thus, by Theorem 12.17, to verify that $\{d_N\}$ is risk consistent, it suffices to prove that for $1 \leq j \leq J$,

$$\lim_N E\left[\left|\frac{\pi(j)p_{jN}(\tau_N(\mathbf{X}))}{\sum_j \pi(j)p_{jN}(\tau_N(\mathbf{X}))} - P(j|\mathbf{X})\right|^2\right] = 0. \tag{12.34}$$

LEMMA 12.35. *If (12.9) holds, then for* $1 \leq j \leq J$,

$$\lim_N E\left[\left|\frac{\pi(j)P_j(\tau_N(\mathbf{X}))}{\sum_j \pi(j)P_j(\tau_N(\mathbf{X}))} - P(j|\mathbf{X})\right|^2\right] = 0.$$

PROOF. Let $j \in \{1, \ldots, J\}$ be fixed and set

$$\mu(t) = \frac{\pi(j)P_j(t)}{\sum_j \pi(j)P_j(t)} = E(d_B(\mathbf{X})|\mathbf{X} \in t),$$

where $d_B(\mathbf{x}) = P(j|\mathbf{x})$. Choose $\varepsilon > 0$, let d'_B be a bounded continuous function such that

$$E[|d'_B(\mathbf{X}) - d_B(\mathbf{X})|^2] \leq \varepsilon \tag{12.36}$$

(the collection of bounded continuous functions is dense in L_2), and set $\mu'(t) = E(d'_B(\mathbf{X})|\mathbf{X} \in t)$. It follows from (12.9) that $\lim_N \mu'(\tau_N(\mathbf{X})) = d'_B(\mathbf{X})$ in probability and hence that

$$\lim_N E[|\mu'(\tau_N(\mathbf{X})) - d'_B(\mathbf{X})|^2] = 0. \tag{12.37}$$

(For bounded random variables, convergence in probability and convergence in L_2 are equivalent.) Now

$$E[|\mu'(\tau_N(\mathbf{X})) - \mu(\tau_N(\mathbf{X}))|^2] \leq \varepsilon. \tag{12.38}$$

To see this, observe that

$$\begin{aligned} E[|\mu'(\tau_N(\mathbf{X})) - \mu(\tau_N(\mathbf{X}))|^2] &= E \sum_{\tilde{T}_N} |\mu'(t) - \mu(t)|^2 P(t) \\ &= E \sum_{\tilde{T}_N} |E(d'_B(\mathbf{X}) - d_B(\mathbf{X})|\mathbf{X} \in t)|^2 \; P(t) \\ &\leq E \sum_{\tilde{T}_N} E(|d'_B(\mathbf{X}) - d_B(\mathbf{X})|^2 |\mathbf{X} \in t) P(t) \\ &= E[|d'_B(\mathbf{X}) - d_B(\mathbf{X})|^2] \leq \varepsilon. \end{aligned}$$

It follows from (12.36) - (12.38) that

$$\overline{\lim_N}\, E[|\mu(\tau_N(\mathbf{X})) - d_B(\mathbf{X})|^2] \leq 4\varepsilon.$$

Since ε can be made arbitrarily small,

$$\lim_N E[|\mu(\tau_N(\mathbf{X})) - d_B(\mathbf{X})|^2] = 0,$$

which is equivalent to the conclusion of the lemma.

LEMMA 12.39. *If (12.18) and (12.8) hold, then for* $1 \leq j \leq J$.

$$\lim_N \left| \frac{\pi(j)p_{jN}(\tau_N(\mathbf{X}))}{\sum_j \pi(j)p_{jN}(\tau_N(\mathbf{X}))} - \frac{\pi(j)P_j(\tau_N(\mathbf{X}))}{\sum_j \pi(j)P_j(\tau_N(\mathbf{X}))} \right| = 0 \textit{ in probability.}$$

PROOF. Set $\mathcal{B}_N = \{t \in \mathcal{B} : \sum \pi(j)p_{jN}(t) \geq k_N c_N\}$,

where

$$c_N = \max_j \frac{\log|\eta_{jN}|}{|\eta_{jN}|}.$$

To verify the desired conclusion, it suffices to show that for $1 \leq j \leq J$,

$$\lim_N \sup_{t \in \mathcal{B}_N} \left| \frac{\pi(j)p_{jN}(t)}{\sum_j \pi(j)p_{jN}(t)} - \frac{\pi(j)P_j(t)}{\sum_j \pi(j)P_j(t)} \right| = 0 \text{ in probability.} \tag{12.40}$$

Choose $\varepsilon > 0$ and let Ω_N denote the event that

$$(1 - \varepsilon)P_j(t) - \varepsilon k_N c_N \leq p_{jN}(t) \leq (1 + \varepsilon)P_j(t) + \varepsilon k_N c_N \quad \text{for } 1 \leq j \leq J \text{ and } t \in \mathcal{B}. \tag{12.41}$$

Then $\lim_N P(\Omega_N) = 1$ by Theorem 12.2. If $t \in \mathcal{B}_N$ and (12.41) holds, then

$$(1 - \varepsilon)P_j(t) - \varepsilon \sum_j \pi(j)p_{jN}(t) \leq p_{jN}(t) \leq (1 + \varepsilon)P_j(t) + \varepsilon \sum_j \pi(j)p_{jN}(t) \tag{12.42}$$

and

$$\frac{1 - \varepsilon}{1 + \varepsilon} \sum_j \pi(j)P_j(t) \leq \sum_j \pi(j)p_{jN}(t) \leq \frac{1 + \varepsilon}{1 - \varepsilon} \sum_j \pi(j)P_j(t). \tag{12.43}$$

It follows easily from (12.42) and (12.43) that

$$\left| \frac{\pi(j)p_{jN}(t)}{\sum_j \pi(j)p_{jN}(t)} - \frac{\pi(j)P_j(t)}{\sum_j \pi(j)P_j(t)} \right| \leq \frac{4\varepsilon}{1 - \varepsilon} .$$

Since ε can be made arbitrarily small, (12.40) holds as desired.

It follows from Lemmas 12.35 and 12.39 that if (12.18), (12.8), and (12.9) hold, then (12.34) is satisfied and hence $\{d_N\}$ is risk consistent. This completes the proof of Theorem 12.19 for model 2.

BIBLIOGRAPHY

Alexander, K. S. 1983. Rates of growth for weighted empirical processes. *Proceedings of the Neyman-Kiefer Conference.* In press.

Anderson, J. A., and Philips, P. R. 1981. Regression, discrimination and measurement models for ordered categorical variables. *Appl. Statist.*, *30*: 22-31.

Anderson, T. W. 1966. Some nonparametric multivariate procedures based on statistically equivalent blocks. In *Multivariate analysis*, ed. P. R. Krishnaiah. New York: Academic Press, 5-27.

Bellman, R. E. 1961. *Adaptive control processes.* Princeton, N.J.: Princeton University Press.

Belsley, D. A.; Kuh, E.; and Welsch, R. E. 1980. *Regression diagnostics.* New York: Wiley.

Belson, W. A. 1959. Matching and prediction on the principle of biological classification. *Appl. Statist.*, *8*: 65-75.

Beta-Blocker Heart Attack Trial Study Group. 1981. Beta-blocker heart attack trial. *J. Amer. Med. Assoc.*, *246*: 2073-2074.

Breiman, L. 1968. *Probability.* Reading, Mass.: Addison-Wesley.

Breiman, L. 1978. Description of chlorine tree development and use. Technical report. Santa Monica, Calif.: Technology Service Corporation.

Breiman, L. 1981. Automatic identification of chemical spectra. Technical report. Santa Monica, Calif.: Technology Service Corporation.

Breiman, L., and Stone, C. J. 1978. Parsimonious binary classification trees. Technical report. Santa Monica, Calif.: Technology Service Corporation.

Bridges, C. R. 1980. *Binary decision trees and the diagnosis of acute myocardial infarction.* Masters thesis, Massachusetts Institute of Technology, Cambridge.

Chung, K. L. 1974. *A course in probability theory.* 2d ed. New York: Academic Press.

Collomb, G. 1981. Estimation non paramétrique de la regression: review bibliographique. *Internat. Statist. Rev.*, *49*: 75-93.

Cover, T. M., and Hart, P. E. 1967. Nearest neighbor pattern classification. *IEEE Trans. Information Theory*, *IT-13*: 21-27.

Darlington, R. B. 1968. Multiple regression in psychological research and practice. *Psychological Bull.*, *69*: 161-182.

Dillman, R. O., and Koziol, J. A. 1983. Statistical approach to immunosuppression classification using lymphocyte surface markers and functional assays. *Cancer Res.*, *43*: 417-421.

Dillman, R. O.; Koziol, J. A.; Zavanelli, M. I.; Beauregard, J. C.; Halliburton, B. L.; and Royston, I. 1983. Immunocompetence in cancer patients—assessment by *in vitro* stimulation tests and quantification of lymphocyte subpopulations. *Cancer.* In press.

Doyle, P. 1973. The use of automatic interaction detector and similar search procedures. *Operational Res. Quart.*, *24*: 465-467.
Duda, R. O., and Shortliffe, E. H. 1983. Expert systems research. *Science*, *220*: 261-268.
Dudley, R. M. 1978. Central limit theorems for empirical measures. *Ann. Probability*, *6*: 899-929.
DuMouchel, W. H. 1981. Documentation for DREG. Technical report. Cambridge: Massachusetts Institute of Technology.
Efron, B. 1979. Bootstrap methods: Another look at the jackknife. *Ann. Statist.*, *7*: 1-26.
Efron, B. 1982. *The jackknife, the bootstrap and other resampling plans*. Philadelphia: Society for Industrial and Applied Mathematics.
Efron, B. 1983. Estimating the error rate of a prediction rule: improvements on cross-validation. *J. Amer. Statist. Assoc.*, *78*: 316-331.
Einhorn, H. 1972. Alchemy in the behavioral sciences. *Pub. Op. Quart.*, *36*: 367-378.
Feinstein, A. 1967. *Clinical judgment*. Baltimore: Williams and Wilkins.
Fielding, A. 1977. Binary segmentation: the automatic interaction detector and related techniques for exploring data structure. In *The analysis of survey data*, Vol. I, ed. C. A. O'Muircheartaigh and C. Payne. Chichester: Wiley.
Fisher, W. D. 1958. On grouping for maximum homogeneity. *J. Amer. Statist. Assoc.*, *53*: 789-798.
Fix, E., and Hodges, J. 1951. Discriminatory analysis, nonparametric discrimination: consistency properties. Technical report. Randolph Field, Texas: USAF School of Aviation Medicine.
Friedman, J. H. 1977. A recursive partitioning decision rule for nonparametric classification. *IEEE Trans. Computers*, *C-26*: 404-408.
Friedman, J. H. 1979. A tree-structured approach to nonparametric multiple regression. In *Smoothing techniques for curve estimation*, eds. T. Gasser and M. Rosenblatt. Berlin: Springer-Verlag.
Gilpin, E.; Olshen, R.; Henning, H.; and Ross, J., Jr. 1983. Risk prediction after myocardial infarction: comparison of three multivariate methodologies. *Cardiology*, *70*: 73-84.
Glick, N. 1978. Additive estimators for probabilities of correct classification. *Pattern Recognition*, *10*: 211-222.
Gnanadesikan, R. 1977. *Methods for statistical data analysis of multivariate observations*. New York: Wiley.
Goldman, L.; Weinberg, M.; Weisberg, M.; Olshen, R.; Cook, F.; Sargent, R. K.; Lamas, G. A.; Dennis, C.; Deckelbaum, L.; Fineberg, H.; Stiratelli, R.; and the Medical Housestaffs at Yale-New Haven Hospital and Brigham and Women's Hospital. 1982. A computer-derived protocol to aid in the diagnosis of emergency room patients with acute chest pain. *New England J. Med.*, *307*: 588-596.

Gordon, L., and Olshen, R. A. 1978. Asymptotically efficient solutions to the classification problem. *Ann. Statist.*, *6*: 515-533.

Gordon, L., and Olshen, R. A. 1980. Consistent nonparametric regression from recursive partitioning schemes. *J. Multivariate Anal.*, *10*: 611-627.

Hand, D. J. 1981. *Discrimination and classification*. Chichester: Wiley.

Hand, D. J. 1982. *Kernel discriminant analysis*. Chichester: Research Studies Press, Wiley.

Harrison, D., and Rubinfeld, D. L. 1978. Hedonic prices and the demand for clean air. *J. Envir. Econ. and Management*, *5*: 81-102.

Henning, H.; Gilpin, E. A.; Covell, J. W.; Swan, E. A.; O'Rourke, R. A.; and Ross, J., Jr. 1976. Prognosis after acute myocardial infarction: multivariate analysis of mortality and survival. *Circulation*, *59*: 1124-1136.

Henrichon, E. G., and Fu, K. -S. 1969. A nonparametric partitioning procedure for pattern classification. *IEEE Trans. Computers*, *C-18*: 614-624.

Hills, M. 1967. Discrimination and allocation with discrete data. *Appl. Statist.*, *16*: 237-250.

Hooper, R., and Lucero, A. 1976. Radar profile classification: a feasibility study. Technical report. Santa Monica, Calif.: Technology Service Corporation.

Jennrich, R., and Sampson, P. 1981. Stepwise discriminant analysis. In *BMDP statistical software 1981*, ed. W. J. Dixon. Berkeley: University of California Press.

Kanal, L. 1974. Patterns in pattern recognition: 1968-1974. *IEEE Trans. Information Theory*, *IT-20*: 697-722.

Kiefer, J. 1961. On large deviations of the empiric d.f. of vector chance variables and a law of iterated logarithm. *Pacific J. Math.*, *11*: 649-660.

Light, R. J., and Margolin, B. H. 1971. An analysis of variance for categorical data. *J. Amer. Statist. Assoc.*, *66*: 534-544.

Mabbett, A.; Stone, M.; and Washbrook, J. 1980. Cross-validatory selection of binary variables in differential diagnosis. *Appl. Statist.*, *29*: 198-204.

McCullagh, P. 1980. Regression models for ordinal data. *J. Roy. Statist. Soc. Ser. B*, *42*: 109-142.

McLafferty, F. W. 1973. *Interpretation of mass spectra*. Reading, Mass.: Benjamin.

Meisel, W. S. 1972. *Computer-oriented approaches to pattern recognition*. New York: Academic Press.

Meisel, W. S., and Michalpoulos, D. A. 1973. A partitioning algorithm with application in pattern classification and the optimization of decision trees. *IEEE Trans. Computers*, *C-22*: 93-103.

Messenger, R. C., and Mandell, M. L. 1972. A model search technique for predictive nominal scale multivariate analysis. *J. Amer. Statist. Assoc.*, *67*: 768-772.

Morgan, J. N., and Messenger, R. C. 1973. *THAID: a sequential search program for the analysis of nominal scale dependent variables*. Ann Arbor: Institute for Social Research, University of Michigan.

Morgan, J. N., and Sonquist, J. A. 1963. Problems in the analysis of survey data, and a proposal. *J. Amer. Statist. Assoc.*, *58*: 415-434.

Narula, S. C., and Wellington, J. F. 1982. The minimum sum of absolute errors regression: a state of the art survey. *Internat. Statist. Rev.*, *50*: 317-326.

Norwegian Multicenter Study Group. 1981. Timolol-induced reduction in mortality and reinfarction in patients surviving acute myocardial infarction. *New England J. Med.*, *304*: 801-807.

Pollard, D. 1981. Limit theorems for empirical processes. *Z. Wahrscheinlichkeitstheorie verw. Gebiete*, *57*: 181-195.

Pozen, M. W.; D'Agostino, R. B.; and Mitchell, J. B. 1980. The usefulness of a predictive instrument to reduce inappropriate admissions to the coronary care unit. *Ann. Internal Med.*, *92*: 238-242.

Ranga Rao, R. 1962. Relations between weak and uniform convergence of measures with applications. *Ann. Math. Statist.*, *33*: 659-680.

Rounds, E. M. 1980. A combined nonparametric approach to feature selection and binary decision tree design. *Pattern Recognition*, *12*: 313-317.

Shibata, R. 1981. An optimal selection of regression variables. *Biometrika*, *68*: 45-54.

Sklansky, J. 1980. Locally trained piecewise linear classifiers. *IEEE Trans. Pattern Analysis Machine Intelligence*, *PAMI-2*: 101-111.

Sonquist, J. A. 1970. *Multivariate model building: the validation of a search strategy*. Ann Arbor: Institute for Social Research, University of Michigan.

Sonquist, J. A.; Baker, E. L.; and Morgan, J. N. 1973. *Searching for structure*. Rev. ed. Ann Arbor: Institute for Social Research, University of Michigan.

Sonquist, J. A., and Morgan, J. N. 1964. *The detection of interaction effects*. Ann Arbor: Institute for Social Research, University of Michigan.

Steele, J. M. 1978. Empirical discrepancies and subadditive processes. *Ann. Probability*, *6*: 118-127.

Stone, C. J. 1977. Consistent nonparametric regression (with discussion). *Ann. Statist.*, *5*: 595-645.

Stone, C. J. 1981. Admissible selection of an accurate and parsimonious normal linear regression model. *Ann. Statist.*, *9*: 475-485.

Stone, M. 1977. Cross-validation: a review. *Math. Operationforsch. Statist. Ser. Statist.*, *9*: 127-139.

Sutherland, D. H., and Olshen, R. A. 1984. *The development of walking*. London: Spastics International Medical Publications. To appear.

Sutherland, D. H.; Olshen, R.; Cooper, L.; and Woo, S. L.-Y. 1980. The development of mature gait. *J. Bone Joint Surgery, 62A*: 336-353.

Sutherland, D. H.; Olshen, R.; Cooper, L.; Wyatt, M.; Leach, J.; Mubarak, S.; and Schultz, P. 1981. The pathomechanics of gait in Duchenne muscular dystrophy. *Developmental Med. and Child Neurology, 39*: 598-605.

Szolovits, P. 1982. Artificial intelligence in medicine. In *Artificial intelligence in medicine*, ed. P. Szolovits. Boulder, Colo.: Westview Press.

Toussaint, G. T. 1974. Bibliography on estimation of misclassification. *IEEE Trans. Information Theory, IT-20*: 472-479.

Van Eck, N. A. 1980. Statistical analysis and data management highlights of OSIRIS IV. *Amer. Statist., 34*: 119-121.

Vapnik, V. N.; and Chervonenkis, A. Ya. 1971. On the uniform convergence of relative frequencies of events to their probabilities. *Theor. Probability Appl., 16*: 264-280.

Zeldin, M., and Cassmassi, J. 1978. Development of improved methods for predicting air quality levels in the South Coast Air Basin. Technical report. Santa Monica, Calif.: Technology Service Corporation.

NOTATION INDEX

$\mathbf{x}$, $\mathbf{X}$	3, 221, 266, 267	measurement vector
$\mathcal{X}$	3, 221, 266	measurement space, the set of all measurement vectors
$C = \{1,\ldots,J\}$	3	set of classes
$d(\mathbf{x})$	4, 221, 267	classifier, classification rule, decision rule, prediction rule, predictor
$\mathcal{L}$	5, 221	learning sample
$R^*(d)$	8, 9, 73, 122, 256, 304	expected risk of the rule d
$(\mathbf{X}, Y)$	10	a random pair with $\mathbf{X} \in \mathcal{X}$ and $Y \in C$
$\chi(\cdot)$	11	indicator function of the event in parentheses
$R^{ts}(d)$	11, 223, 226, 257, 303, 304	test sample estimate of $R^*(d)$
$R^{cv}(d)$	12, 223, 257, 307	cross-validation estimate of $R^*(d)$
$d^{(v)}$	12	a classifier based on $\mathcal{L} - \mathcal{L}_v$, where $\mathcal{L}$ is the disjoint union $\mathcal{L} = \mathcal{L}_1 \cup \ldots \cup \mathcal{L}_V$
$d_B(\cdot)$	13, 14, 121, 222, 256, 268	a Bayes rule or predictor, a rule which minimizes $R^*(d)$
$\pi(j)$	13	$P(Y = j)$, the prior probability of the class j
R_B	13, 14	Bayes misclassification rate
$f_j(\mathbf{x})$	14	conditional density of X given that Y is in class j
$\hat{f}_j(\mathbf{x})$	15	estimate of $f_j(\mathbf{x})$
$\mathcal{s}$	21, 25, 32	a split

α	66	complexity parameter
$R_\alpha(T)$	66, 233, 284	cost-complexity measure: $R(T) + \alpha\lvert\tilde{T}\rvert$
$T(\alpha)$	66, 285	smallest subtree of T_{max} which minimizes $R_\alpha(T)$
$R(T_t)$	68	$\sum_{t' \in \tilde{T}_t} R(t')$
α_k, $k = 2,3,\ldots$	70	value of α at which the $(k-1)st$ change of preference occurs in the pruning process
$T_k = T(\alpha_k)$	70-72	kth optimally pruned subtree
$\hat{R}(T_K)$	72, 78	estimate of risk of a rule based on T_k
$Q^*(i\|j)$	73	probability that a case in class j is classified into class i
$R^*(j)$	73	expected cost of misclassification for class j items
$\mathcal{L}_v$	73-75	vth subset of the learning sample, $\mathcal{L}$; in the test sample case $\mathcal{L}_1$ is the new learning sample, $\mathcal{L}_2$ the test sample
$N_j^{(2)}$	73	number of class j cases in $\mathcal{L}_2$
$N_{ij}^{(2)}$	73, 74	number of class j cases in $\mathcal{L}_2$ whose predicted class is class i
$Q^{ts}(i\|j)$	74	$N_{ij}^{(2)} / N_j^{(2)}$
$\mathcal{L}^{(v)}$	75	$\mathcal{L} - \mathcal{L}_v$
$T_{max}^{(v)}$	75, 76	T_{max} grown from $\mathcal{L}^{(v)}$
$T^{(v)}(\alpha)$, $v = 1, \ldots, V$	75, 76	minimal cost-complexity subtrees of $T_{max}^{(v)}$
$N_{ij}^{(v)}$	76	number of class j cases in $\mathcal{L}_v$ classified as i by $T^{(v)}(\alpha)$
N_{ij}	76	$\sum_v N_{ij}^{(v)}$
$Q^{CV}(i\|j)$	76	N_{ij} / N_j

$R^{CV}(j)$	76	cross-validation estimate of the cost of misclassifying class j cases
$R^{CV}(T(\alpha))$	76	cross-validation estimate of the risk of $T(\alpha)$
T_{k_0}	79	the tree from among $\{T_k\}$ which minimizes $\hat{R}(T_k)$
T_{k_1}	80	k_1 is the smallest k such that $\hat{R}(T_{k_1}) \leq \hat{R}(T_{k_0}) + \mathrm{SE}(\hat{R}(T_{k_0}))$
R^*	88	Bayes optimal misclassification risk
$\pi'(j)$	114	altered prior probability of class j
$C'(i\|j)$	114	altered cost of misclassifying a class j object as a class i object
MSE	122, 222	mean squared error
$\mathbf{d}^{(v)}$	125	decision rule associated with $\mathcal{L}^{(v)}$
S_m	140	set of all splits on the variable x_m
$\tilde{s}_m$	141	surrogate split on x_m
$\lambda(s^*\|\tilde{s}_m)$	141	predictive measure of association between s^* and $\tilde{s}_m$
s^*_m	142	best split on variable x_m
$M(x_m)$	147	measure of importance of the variable x_m
$r^{ts}(d)$	151	test sample estimate of $r^*(t)$
$\hat{r}(t)$	152, 153	estimate of $r^*(t)$
$T_{\text{suff}}(\alpha)$	160, 292	a tree sufficiently large so as to contain $T(\alpha)$
$S(t)$	161	criterion by which it is determined whether $t \in T_{\text{suff}}(\alpha)$
$RE^*(d)$	224, 257, 305	relative mean squared error or relative absolute error

$RE^{ts}(d)$	225, 257, 305	test sample estimate of $RE^*(d)$
$SE(R^{ts})$	226, 304	estimate of the standard error of R^{ts}
$\bar{y}(t)$	230, 298	empirical average of the y's whose corresponding **x**'s belong to the node t
$R(d)$	230, 257, 267	resubstitution estimate of $R^*(d)$, risk (in Chaps. 9 and 12)
$\Delta R(s,t)$	231, 272	$R(t) - R(t_L) - R(t_R)$
$s^2(t)$	232, 298	empirical second moment of the y's whose corresponding **x**'s belong to t about $\bar{y}(t)$
$RE(T)$	232	resubstitution estimate of $RE^*(T)$, where $RE^*(T) = RE^*(d)$ if the decision rule d corresponds to the tree T
$RE^{CV}(T)$	234	cross-validation estimate of $RE^*(T)$
$\nu(\cdot)$	256	median of the distribution of data specified in parentheses
$R(\hat{\nu})$	257	$1/N \sum_n \lvert y_n - \hat{\nu}\rvert$, where $\hat{\nu}$ is a sample median of $y_1, \ldots, y_N$
$\bar{d}(t)$	259	average absolute deviation of those y's in node t from $\nu(t)$
A	267	set of possible actions
$L(y,a)$	267	"loss" if y is the actual value of the response and a is the action taken
$\mu(\mathbf{x})$	268	$E(Y \mid \mathbf{X} = \mathbf{x})$
P	266	distribution of a generic measurement vector
τ	269	partition function associated with $\tilde{T}$
$d_{\tilde{T}}$	270	Bayes rule corresponding to the partition $\tilde{T}$
$\mu(t)$	270	$E(Y \mid \mathbf{X} \in t)$
$\nu(t)$	270	a value of a that minimizes $E[L(Y,a) \mid \mathbf{X} \in t]$

SUBJECT INDEX